Wideband CDMA for Third Generation Mobile Communications

The Artech HouseUniversal Personal Communications Series

Ramjee Prasad, Series Editor

CDMA for Wireless Personal Communications, Ramjee Prasad

Universal Wireless Personal Communications, Ramjee Prasad

Wideband for Third Generation Mobile Communications, Tero Ojanperä and Ramjee Prasad

For further information on these and other Artech House titles, including previously considered out-of-print books now available through our In-Print-Forever® (IPF®) program, contact:

Artech House
685 Canton Street
Norwood, MA 02062
Phone: 781-769-9750
Fax: 781-769-6334
e-mail: artech@artechhouse.com

Artech House
46 Gillingham Street
London SW1V 1AH UK
Phone: +44 (0)20 7596-8750
Fax: +44 (0)20 7630-0166
e-mail: artech-uk@artechhouse.com

Find us on the World Wide Web at:
www.artechhouse.com

For a recent listing of titles in the *Artech House Mobile Communications Library*, turn to the back of the book.

Wideband CDMA for Third Generation Mobile Communications

Tero Ojanperä
Ramjee Prasad
editors

Artech House
Boston • London

Library of Congress Cataloging-in-Publication Data
Ojanperä, Tero
  Wideband CDMA for third generation mobile communications / Tero Ojanperä,
  Ramjee Prasad, editors.
    p.   cm.
  Includes bibliographical references and index.
  ISBN 0-89006-735-X (alk. paper)
  1. Code division multiple access.   2. Mobile communication systems.
  3. Broadband communication systems.   I. Prasad, Ramjee.   II. Title.
TK5103.45.O34   1998
621.3845—dc21   98-33857
                                                                      CIP

British Library Cataloguing in Publication Data
Wideband CDMA for third generation mobile communications
    1. Code division multiple access   2. Broadband communication systems
    3. Global system for mobile communications
    I. Ojanperä, Tero   II. Prasad, Ramjee
6213'84'56

ISBN 0-89006-735-X

Cover design by Lynda Fishbourne

International Standard Book Number: 0-89006-735-X
Library of Congress Catalog Card Number: 98-33857

10 9 8 7 6 5

Books are the joy of life, and learning is a never-ending experience.

To my wife Tiina, and to our son Eerik
To my brother Aki, to my mother Maiju and to the memories of my brother Juha and father Iisakki
—Tero Ojanperä

To my wife Jyoti, to our daughter Neeli, and to our sons Anand and Rajeev
—Ramjee Prasad

CONTENTS

CHAPTER 5   CDMA AIR INTERFACE DESIGN   101

Tero Ojanperä

Tero Ojanperä, Antti Toskala, and Harri Holma

CHAPTER 8   HIERARCHICAL CELL STRUCTURES   249
Seppo Hämäläinen, Harri Lilja, and Tero Ojanperä

CHAPTER 9   TIME DIVISION DUPLEX DS-CDMA   261
Antti Toskala, Harri Holma, and Tero Ojanperä

CHAPTER 12 NETWORK ASPECTS   351

Tero Ojanperä

PREFACE

Third generation mobile radio networks, often dubbed as 3G, have been under intense research and discussion recently and will emerge around the year 2000. In the International Telecommunication Union (ITU), third generation networks are called International Mobile Telecommunications – 2000 (IMT-2000), and in Europe, Universal Mobile Telecommunication System (UMTS). IMT-2000 will provide a multitude of services, especially multimedia and high bit rate packet data. Wideband code division multiple access (CDMA) has emerged as the mainstream air interface solution for the third generation networks. In Europe, Japan, Korea, and the United States, wideband CDMA systems are currently being standardized.

This book provides a comprehensive introduction to wideband CDMA and third generation networks. It provides the technical background necessary to understand how wideband CDMA air interfaces are designed, starting from system requirements, applications, and radio environments, combined with a detailed treatment of technical solutions for spreading codes, coding, modulation, RAKE receiver, and soft handover. It also provides a review of the wideband CDMA air interface proposals including WCDMA in Europe and Japan, cdma2000 in the United States, and wideband CDMA in Korea. Radio network planning is a key competence for network operators. This book gives an introduction to the art of network planning with problems and solutions specific to CDMA. The standardization and regulation environment for development of third generation networks are very complex. This book provides insight to the structure and operations of different standardization bodies, industry interest groups, and regulatory bodies.

Figure 1 illustrates the coverage of the book. Chapter 1 introduces basic definitions and the background of third generation systems development. Chapter 2 explains the basic principles of CDMA. Chapter 3 introduces IMT-2000 service targets and applications. Chapter 4 presents radio environments and their characteristics. Chapter 5 covers all the

xvii

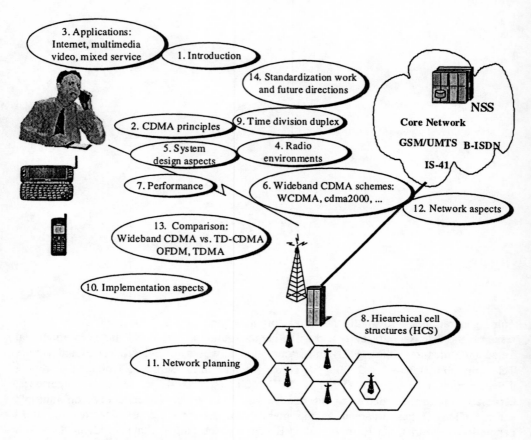

Figure 1   Illustration of the coverage of the book.

main aspects of wideband CDMA air interface design starting from frame structures and physical channel design, including spreading codes and their properties, into radio resource management aspects such as handover and power control. Chapter 6 reviews the main wideband CDMA air interfaces: WCDMA in Europe and Japan, cdma2000 in the United States, Korean wideband CDMA schemes, CODIT, and IS-655. In Chapter 7, radio performance of wideband CDMA, including spectrum efficiency and range, are discussed. Hierarchical cell structures (HCS) are discussed in Chapter 8. Chapter 9 presents the time division duplex concept. Chapter 10 describes the implementation of wideband CDMA mobiles and base stations. CDMA network planning is discussed in Chapter 11. Chapter 12 presents network aspects, different core network solutions, the generic radio access network concept, and the evolution of second generation networks such as GSM and IS-41. Chapter 13 compares wideband CDMA to time division multiple access (TDMA) and other third generation air interfaces. Finally, standardization organizations and future directions of wideband CDMA are discussed in Chapter 14.

This book is intended for everybody involved in the field of mobile radio systems. It provides different levels of material suitable for managers, researchers, system designers, and graduate students. The structure of the book facilitates two reading approaches. A general overview of wideband CDMA and third generation systems, suitable for managers, can be obtained by reading Chapter 1, Chapter 2, Sections 3.1 to 3.6, Sections 4.1 to 4.3, Chapter 6 (overview parts), Chapter 12 selectively, and Chapter 14. A more thorough reader interested in technical details should read all sections. Section 3.7 provides a more mathematical view to application covering traffic models. Sections 4.4 to 4.7 provide detailed propagation and mobility models. Chapter 5 contains an in-depth treatment of wideband CDMA system design aspects.

The views expressed in this book are the view of the authors and do not necessarily represent the views of their employers.

ACKNOWLEDGMENTS

The material in this book originates from several projects at Nokia and Delft University of Technology (DUT) with a common goal of defining third generation of mobile communications. The research program to produce wideband CDMA specifications within Nokia dates back to the early 1990s when an experimental wideband CDMA testbed project was started. Tero was intensively involved with the start of CDMA system studies in 1993. Based on these studies, a project called CSS2000 was started in 1994. Tero had the pleasure of being a project leader with an ambitious, at that time almost overwhelming, goal of producing a wideband CDMA air interface for UMTS. Later the research was expanded to a research program investigating several aspects such as 2 Mbps transmission, radio resource management, and packet aspects for UMTS. The book *CDMA for Wireless Personal Communications* (Artech, 1996) by Ramjee proved to be a good starting point for this book. Ramjee started the CDMA research activities in DUT in 1989 and currently the Centre for Wireless Personal Communications of DUT is very active in the research and development of wideband CDMA.

In 1995, the FRAMES project brought together a number of talented individuals from several companies, laboratories, and universities, namely CNET, CSEM, Chalmers University of Technology, Ericsson, ETHZ, Instituto Superior Técnico, Nokia, Siemens, Roke Manor Research, The Royal Institute of Technology, DUT, University of Kaiserslautern, and University of Oulu. The goal of FRAMES was to study and define a proposal for UMTS air interface. FRAMES also created the seeds for this book by getting us acquainted with each other. We decided to put the results of various CDMA and third generation related projects into a book format.

During the research and writing of this book, several individuals contributed to the success. Especially, the efforts of the other contributors of this book – Timo Eriksson, Harri Holma, Seppo Hämäläinen, Harri Lilja, and Antti Toskala – are highly appreciated.

António Trindade from DUT helped to prepare the complete manuscript. He also provided very constructive suggestions to improve the book and corrected numerous mistakes. Without his hard dedication it would not have been possible to complete this book.

Within Nokia we would like to thank the following individuals: Ari Hottinen, Chi-Zhun Honkasalo, Hannu Häkkinen, Kari Kalliojärvi, Ilkka Keskitalo, Jorma Lilleberg, Peter Muszynki, Kari Pehkonen, and Kari Rikkinen, who have been the technical forces involved in several CDMA projects. We are deeply indebted to a number of individuals — Kwang-Cheng Chen, Stefano Faccin, George Fry, Steven Gray, Pertti "Bertil" Huuskonen, Markku Juntti, Matti Latva-aho, Tom Leskinen, Khiem Le, Pertti Lukander, Janne Parantainen, Riku Pirhonen, Harri Posti, Gordon Povey, Rauno Ruismäki, Tom Sexton, Alan Shu, and Markku Verkama — who helped to make this book possible by providing helpful comments and improvements. We also wish to thank Nokia management, especially Dr. Heikki Huomo, for opportunity to write this book. Furthermore, we appreciate the support of DUT management. The FRAMES project was one of the key projects that made UMTS happen. We would like to thank colleagues from companies, laboratories, and universities participating in FRAMES.

Tero wishes to thank for the support and constructive comments of his dear friends Pekka Koponen and Kari Pulli. In addition, he wants to thank his mother-in-law Aila and father-in-law Markku Kesti for encouragement and support when writing this book.

After the first CDMA air interface project, CSS2000, we studied several multiple access techniques. A circle was completed by the selection of wideband CDMA for UMTS in January 1998. We hope this book will help to explain rationale behind the UMTS wideband CDMA, as well as other wideband CDMA air interfaces and third generation systems in general.

Tero Ojanperä
Ramjee Prasad
September 1998

Chapter 1

INTRODUCTION

Recently, extensive investigations have been carried out into the application of a code division multiple access (CDMA) system as an air interface multiple access scheme for IMT-2000 / UMTS (International Mobile Telecommunications System 2000 / Universal Mobile Telecommunications System). It appears that CDMA is the strongest candidate for the third generation wireless personal communication systems. Many research and development (R&D) projects in the field of wideband CDMA have been going on in Europe, Japan, the United States, and Korea [1–4]. It seems that wideband CDMA will be an appropriate answer to the question: "What will be the multiple access scheme for IMT-2000 / UMTS?"

1.1   MULTIPLE ACCESS

The basis for any air interface design is how the common transmission medium is shared between users, that is, the multiple access scheme. The basic multiple access schemes are illustrated in Figure 1.1. In frequency division multiple access (FDMA), the total system bandwidth is divided into frequency channels that are allocated to the users. In time division multiple access (TDMA), each frequency channel is divided into time slots and each user is allocated a time slot. In CDMA, multiple access is achieved by assigning each user a pseudo-random code (also called pseudo-noise codes due to noise-like autocorrelation properties) with good auto- and cross-correlation properties. This code is used to transform a user's signal into a wideband spread spectrum signal. A receiver then transforms this wideband signal into the original signal bandwidth using the same pseudo-random code. The wideband signals of other users remain wideband signals. Possible narrowband interference is also suppressed in this process. TDMA and CDMA usually use FDMA to divide the frequency band into smaller frequency channels, which are then divided in a time or code division fashion.

In addition to FDMA, TDMA, and CDMA, orthogonal frequency division multiplexing (OFDM), a special form of multicarrier modulation, and space division multiple access (SDMA) are used as multiple access schemes in wireless

1

communications. In OFDM, densely spaced subcarriers with overlapping spectra are generated using fast Fourier transform (FFT) [4]. Signal waveforms are selected in such a way that the subcarriers maintain their orthogonality despite the spectral overlap. Usually, CDMA or TDMA is used with OFDM to achieve multiple access capability. SDMA separates the users spatially, typically using beam-forming techniques. It can be applied to all other multiple access schemes.

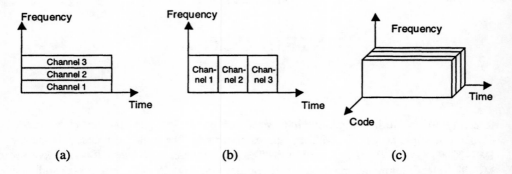

(a)   (b)   (c)

Figure 1.1   Multiple access schemes: (a) FDMA, (b) TDMA, and (c) CDMA.

1.1.1   Classification of CDMA

There are several ways to classify CDMA schemes. The most common is the division based on the modulation method used to obtain the wideband signal. This division leads to three types of CDMA: direct sequence (DS), frequency hopping (FH), and time hopping (TH) as illustrated in Figure 1.2 [1]. In DS-CDMA, spectrum is spread by multiplying the information signal with a pseudo-noise sequence, resulting in a wideband signal. In the frequency hopping spread spectrum, a pseudo-noise sequence defines the instantaneous transmission frequency. The bandwidth at each moment is small, but the total bandwidth over, for example, a symbol period is large. Frequency hopping can either be fast (several hops over one symbol) or slow (several symbols transmitted during one hop). In the time hopping spread spectrum, a pseudo-noise sequence defines the transmission moment. Furthermore, combinations of these techniques are possible. In this book we focus on DS-CDMA, since it is the technique used for third generation wideband CDMA proposals. Wideband CDMA is defined as a direct sequence spread spectrum multiple access scheme where the information is spread over a bandwidth of approximately 5 MHz or more. A detailed classification of CDMA is discussed in Chapter 2.

Frequency

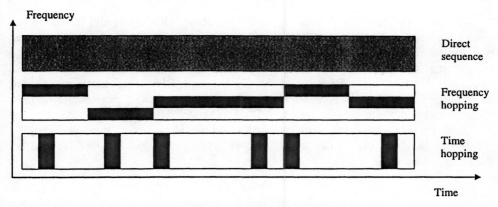

Direct
sequence

Frequency
hopping

Time
hopping

Time

Figure 1.2   Direct sequence, frequency hopping, and time hopping CDMA.

1.2   CDMA: PAST, PRESENT, AND FUTURE

The origins of spread spectrum are in the military field and navigation systems. Techniques developed to counteract intentional jamming have also proved suitable for communication through dispersive channels in cellular applications. In this section we highlight the milestones for CDMA development starting from the 1950s after the invention of the Shannon theorem [5]. An extensive overview of spread spectrum history is given in [6].

In 1949, John Pierce wrote a technical memorandum where he described a multiplexing system in which a common medium carries coded signals that need not be synchronized. This system can be classified as a time hopping spread spectrum multiple access system [6]. Claude Shannon and Robert Pierce introduced the basic ideas of CDMA in 1949 by describing the interference averaging effect and the graceful degradation of CDMA [7]. In 1950, De Rosa-Rogoff proposed a direct sequence spread spectrum system and introduced the processing gain equation and noise multiplexing idea [6]. In 1956, Price and Green filed for the antimultipath "RAKE" patent [6]. Signals arriving over different propagation paths can be resolved by a wideband spread spectrum signal and combined by the RAKE receiver. The near-far problem (i.e., a high interference overwhelming a weaker spread spectrum signal) was first mentioned in 1961 by Magnuski [6].

The cellular application spread spectrum was suggested by Cooper and Nettleton in 1978 [8]. During the 1980s, Qualcomm investigated DS-CDMA techniques, which finally led to the commercialization of cellular spread spectrum communications in the form of the narrowband CDMA IS-95 standard in July 1993. Commercial operation of IS-95 systems started in 1996. Multiuser detection (MUD) has been subject to extensive research since 1986 when Verdu formulated an optimum multiuser detection for the additive white Gaussian noise (AWGN) channel, maximum likelihood sequence estimator (MLSE) [9].

During the 1990s, wideband CDMA techniques with a bandwidth of 5 MHz or more have been studied intensively throughout the world, and several trial systems have been built and tested [2]. These include FRAMES FMA2 (FRAMES Multiple Access) in Europe, Core-A in Japan, the European/Japanese harmonized WCDMA scheme, cdma2000 in the United States, and the TTA I and TTA II (Telecommunication Technology Association) schemes in Korea. Introduction of third generation wireless communication systems using wideband CDMA is expected around the year 2000.

Based on the above description, the CDMA era is divided in three periods: (1) pioneer CDMA era, (2) narrowband CDMA era, and (3) wideband CDMA era, as shown in Table 1.1.

Table 1.1
CDMA Era

Pioneer Era

1949	John Pierce: time hopping spread spectrum
1949	Claude Shannon and Robert Pierce: basic ideas of CDMA
1950	De Rosa-Rogoff: direct sequence spread spectrum
1956	Price and Green: antimultipath "RAKE" patent
1961	Magnuski: near-far problem
1970s	Several developments for military field and navigation systems

Narrowband CDMA Era

1978	Cooper and Nettleton: cellular application of spread spectrum
1980s	Investigation of narrowband CDMA techniques for cellular applications
1986	Formulation of optimum multiuser detection by Verdu
1993	IS-95 standard

Wideband CDMA Era

1995 -	Europe : FRAMES FMA2 Japan : Core-A USA : cdma2000 Korea : TTA I, TTA II	} WCDMA
2000s	Commercialization of wideband CDMA systems	

1.3 MOBILE CELLULAR ERA

In 1980 the mobile cellular era started. Mobile communications have undergone significant changes and experienced enormous growth since then. First generation mobile systems using analog transmission for speech services were introduced in the 1980s. Several standards were developed: AMPS (Advanced Mobile Phone Service) in the United States, TACS (Total Access Communication System) in the United Kingdom, NMT (Nordic Mobile Telephones) in Scandinavia, NTT (Nippon Telephone and Telegraph) in Japan, and so on.

Second generation systems using digital transmission were introduced in the late 1980s. They offer higher spectrum efficiency, better data services, and more advanced roaming than the first generation systems. GSM (Global System for Mobile Communications), PDC (Personal Digital Cellular), IS-136 (D-AMPS), and IS-95 (US CDMA system) belong to the second generation systems. The services offered by these systems cover speech and low bit rate data.

The second generation systems will further evolve towards third generation systems and offer more advanced services such as bit rates of 100 to 200 Kbps for circuit and packet switched data. These evolved systems are commonly referred to as generation 2.5. Third generation systems will offer even higher bit rates, more flexibility, simultaneous multiple services for one user, and services with different quality of service classes.

Figure 1.3 shows the number of subscribers for the first and second generation digital systems from 1991 to 2002. Today, the largest system is GSM with over 100 million subscribers and with a predicted growth of over 350 million by the year 2002 (i.e., over 50% of the total subscribers).

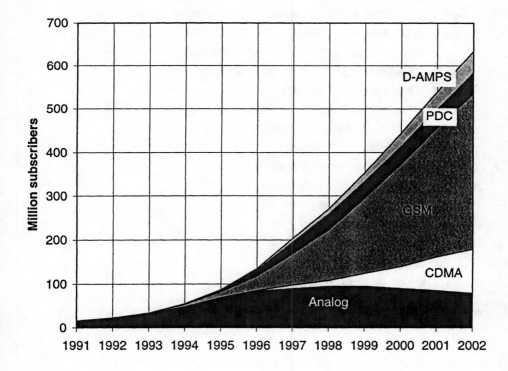

Figure 1.3   The number of subscribers for the first and second generation systems. Note: data from 1998 to 2002 are estimates (*Source*: EMC Research, December 1997).

Figure 1.4 depicts the penetration rates of mobile cellular radio in different countries. Since the introduction of digital cellular systems in 1992, the growth has been enormously fast. The highest penetration, over 40%, is in Finland. High penetration rates require high-capacity networks and large amounts of frequency spectrum. This enormous growth of the second generation systems paves the way for the introduction of third generation networks. More and more people will become familiar with wireless access available anywhere, anytime.

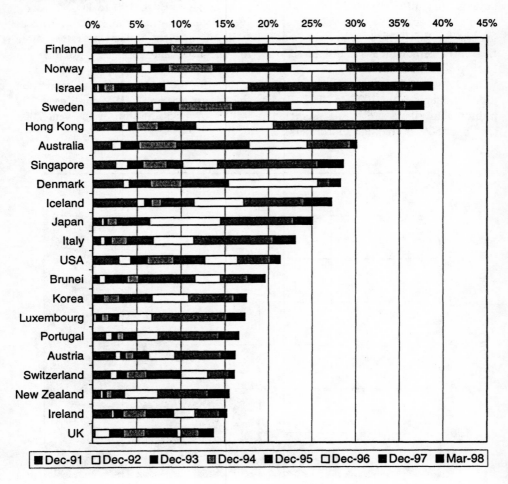

Figure 1.4 Penetration of mobile radio systems (*Source*: World Bank EMC, March 1998).

Table 1.2 shows the technical parameters of second generation systems. The details of the IS-95 standard are discussed in more detail in Chapter 2. All these systems are frequency division duplex (FDD) systems: they transmit and receive in different frequency bands. Time division duplex (TDD) systems are discussed in Chapter 9. In the table, the maximum data rate is the data rate the standard currently supports. The actual data rate available in commercial systems is usually much smaller. In 1998, GSM supports 14.4 Kbps, IS-95 9.6 Kbps, IS-136 9.6 Kbps, and PDC 9.6 Kbps. For a more detailed discussion of these standards, refer to [10].

Table 1.2
Second Generation Digital Systems

	GSM	IS-136	IS-95	PDC
Multiple access	TDMA	TDMA	CDMA	TDMA
Modulation	GMSK[a]	$\pi/4$-DQPSK[b] Coherent $\pi/4$-DQPSK Coherent 8-PSK	QPSK/ O-QPSK[c]	$\pi/4$-DQPSK
Carrier spacing	200 kHz	30 kHz	1.25 MHz	25 kHz
Carrier bit rate	270.833 Kbps	48.6 Kbps ($\pi/4$-PSK and $\pi/4$-DQPSK) 72.9 Kbps (8-PSK)	1.2288 Mchip/s[d]	42 Kbps
Frame length	4.615 ms	40 ms	20 ms	20 ms
Slots per frame	8/16	6	1	3/6
Frequency band (uplink/ downlink) (MHz)	880-915 /935-960 1720-1785 / 1805-1880 1930-1990 / 1850-1910	824-849 / 869-894 1930-1990 / 1850-1910	824-849 / 869-894 1930-1990 / 1850-1910	810-826 / 940-956 1429-1453 / 1477-1501
Speech codec	RPE-LTP[e] 13 Kbps Half rate 6.5 Kbps Enhanced full rate (EFR) 12.2 Kbps	VSELP[f] 8 Kbps IS-641-A: 7.4 Kbps ACELP)[g] US1: 12.2 Kbps (ACELP)	QCELP 8 Kbps CELP 8 Kbps CELP 13 Kbps	VCELP 6.7 Kbps
Maximum possible data rate	HSCSD:115.2 Kbps GPRS: 115.2 – 182.4 Kbps (depending on the coding)	IS-136+: 43.2 Kbps	IS95A: 14.4 Kbps IS95B: 115.2 Kbps	28.8 Kbps
Frequency hopping	Yes	No	N/A	No
Handover	Hard	Hard	Soft	Hard

[a] Gaussian minimum shift keying.
[b] Differential quadrature phase shift keying.
[c] Offset QPSK.
[d] A "chip" is used to denote a spread symbol in DS-CDMA systems.
[e] Regular pulse excited long term prediction.
[f] Vector sum excited linear predictive.
[g] Algebraic code excited linear predictive.

1.3.1   GSM

Table 1.3 depicts the time schedule of GSM development. The first step towards GSM was the allocation of a common frequency band in 1978, twice 25 MHz, at around 900

MHz for mobile communications in Europe [11]. The actual development of the GSM standard started in 1982 when *Conference Europénne des Postes et Télécommunications* (CEPT) established Groupe Special Mobile (the origin of the term GSM, which now means Global System for Mobile Communications) to develop a future Pan-European cellular mobile network. In 1984, three working parties (WP) were established to define services (WP1), radio interface (WP2), and other issues such as transmission, signaling protocols, interfaces, and network architecture (WP3). In 1986, a so-called permanent nucleus (PN), consisting of a team of full-time members, was established to continuously coordinate the GSM specification work. The European Telecommunications Standards Institute (ETSI) was established 1988, and GSM became an ETSI technical committee Special Mobile Group (TC SMG), the working parties became subtechnical committees (STCs), and the permanent nucleus became the GSM project team number 12 (PT12). For a description of the SMG structure, refer to Chapter 14.

In 1990, the GSM specifications for 900 MHz were frozen. In 1990 it was decided that GSM1800,[1] the high frequency version of GSM, was to be standardized and the specifications were completed by 1991. The official commercial launch of GSM was in 1992. Since the GSM standardization process could not be completed for all aspects before the service launch, it was decided to divide the specifications into phases:

Table 1.3
GSM Development Time Schedule

1982	Groupe Special Mobile established within CEPT
1984	Several proposals for GSM multiple access: wideband TDMA, narrowband TDMA, DS-CDMA, hybrid CDMA/FDMA, narrowband FDMA
1986	Eight prototype systems tested in CNET laboratories in France Permanent nucleus is set up
1987	Basic transmission principles selected: 8-slot TDMA, 200-kHz carrier spacing, frequency hopping
1987	MoU signed
1988	GSM becomes an ETSI technical committee
1990	GSM phase 1 specifications frozen (drafted 1987-1990) GSM1800 standardization begins
1991	GSM1800 specifications are frozen
1992	GSM900 commercial operation starts
1992	GSM phase 2+ development starts
1995	GSM submitted as a PCS technology candidate to the United States
1995	PCS1900 standard adopted in the United States
1996	Enhanced full rate (EFR) speech codec standard ready
1996	14.4-Kbps standard ready GSM1900 commercial operation starts
1997	HSCSD standard ready GSM cordless system (home base station) standardization started EDGE standardization started
1998	GPRS standard ready WCDMA selected as the third generation air interface

---

[1] GSM1800 was originally termed DCS1800 (Digital Cellular System 1800).

phase 1 and phase 2 [11]. The evolution needs went beyond even these two phases so a workshop held in 1992 in Finland launched the idea for the phase 2+ standardization. The phase 2+ features are updated on a regular basis according to market needs and availability of specification. So, the original standard was future proof and capable for evolution. The standardization process for the U.S. equivalent, PCS1900 (Personal Communication System, now also called GSM1900) began in 1992. The GSM1900 service was launched 1996.

A Memorandum of Understanding (MoU) was signed by telecommunication network operator organizations in 1987. The GSM MoU covers areas such as time scales for the procurement and deployment of the system, compatibility of numbering and routing plans, and concerted service introduction [11]. Today, there are 293 MoU signatories in 120 countries. From its European origin, GSM has grown into a truly global system, and in 1997 the second largest GSM market, after Italy, was actually China.

1.3.1.1   GSM Air Interface Development

In the beginning of GSM air interface development, several multiple access schemes were proposed. The main alternatives were single channel per carrier FDMA, narrowband TDMA, wideband TDMA, and slow frequency-hopped (SFH) CDMA [12]. Comparative testing of eight prototypes (four from France and West Germany and four from Scandinavia) was carried out in Centre National d'Études des Télecommunication (CNET) laboratories in France during 1986. In 1987, based on the comparison results, the following basic principles were decided to be adopted for the GSM radio interface:

- 8 channels per carrier;
- 200-kHz carrier bandwidth;
- Slow frequency hopping.

Ten years later in 1997, a similar evaluation process was repeated in the UMTS air interface definition process. As a result, the conclusions expressed in 1985 in [12] that "it seems that DS/CDMA systems have no chance at all to be adopted for the European system" were reversed, as the WCDMA was selected for the UMTS air interface.

1.3.1.2   GSM Phase 2+

The services offered by the evolved second generation systems will be the reference for third generation systems. The third generation systems have to exceed the capabilities of the second generation systems to be commercially viable. In this section we discuss the main features of GSM phase 2+ in order to better understand what services will be in the market by the year 2000. The most important GSM phase 2+ work items from the radio access system point of view are:

- Enhanced full rate speech codec (EFR) [13];

- Adaptive multirate codec (AMR);
- 14.4-Kbps data service [14];
- High speed circuit switched data (HSCSD);
- General packet radio service (GPRS) [15];
- Enhanced data rates using optimized modulation (EDGE);
- GSM cordless system (home base station).

The EFR speech codec was first developed for the U.S. market and subsequently adopted also by ETSI. It offers landline quality speech using 12.2 Kbps for speech coding and 10.6 Kbps for error protection, resulting in 22.8Kbps [13]. AMR further improves the GSM speech service and takes advantage of changing channel conditions. Thus, while maintaining good speech quality, radio resources can be used more efficiently.

The 14.4-Kbps data service provides higher throughput per timeslot by puncturing (i.e., by deleting certain bits) the original rate half channel coding. HSCSD offers circuit switched data services from 9.6 Kbps with one slot up to 115.2 Kbps using eight slots. The HSCSD standard was completed in 1997, and the first infrastructure and terminal products are expected in 1998.

General Packet Radio Service provides the mobile user with a connectionless packet access to data networks. The most important application of GPRS could be access to applications like the World Wide Web, e-mail, and Telnet. The short call setup time will facilitate transaction type and telemetric applications, like toll road systems and credit card validation. The GPRS air interface is based on the same principles as the current GSM system. In addition, new radio link control (RLC) and medium access control (MAC) layer functions allow flexible retransmission capabilities needed for packet access. GPRS offers four different bit rates per timeslot (9.06, 13.4, 15.6, and 21.4 Kbps) depending on the applied channel coding. Link adaptation selects among the four possible rates the bit rate that would result in the highest throughput according to channel and interference conditions. The GPRS standard was finalized in 1998, and commercial deployment should start by 1999.

The GSM evolution towards higher bit rates is further boosted by the enhanced data rates for GSM/global evolution (EDGE) concept, which has also been selected for IS-136 based third generation systems as discussed in Section 1.6.2. Also, more spectrum-efficient detection methods are currently under study. Joint detection detects the desired signal and dominant co-channel interferer jointly, thus improving the performance [16].

The GSM home base station concept is under standardization as well. It will offer cordless access using a standard GSM phone, thus removing the need for dual mode GSM/cordless terminals.

For more details on GSM phase 2+ and on GSM's evolution towards third generation, refer to [17].

1.3.2   US-TDMA (IS-54/IS-136)

In September 1988, Cellular Telecom Industry Association (CTIA) published requirements for the next generation of cellular technology in the United States. The first standard that was adopted was time division multiple access, called US-TDMA.[2] In January 1989, the TIA/CTIA (Telecommunications Industry Association / Cellular Telecom Industry Association) selected TDMA over FDMA technology. In 1995, an upbanded version of the US-TDMA was adopted for the US PCS frequency band. The US-TDMA standard with an analog control channel is termed IS-54, and the US-TDMA with digital control channel is termed IS-136, which is now the commonly used term when referring to the US-TDMA. The TIA technical committee TR45.3 is responsible for IS-136 standardization. Today, IS-136 is deployed in North and South America, and some parts of Asia.

The IS-136 is a dual-mode standard, since it specifies both analog (AMPS) and digital (US-TDMA) modes. Due to backward compatibility to AMPS, the US-TDMA carrier spacing is 30 kHz

The Universal Wireless Communications Consortium (UWCC) was established in 1996 to support IS-136 standards. In 1997, within the UWCC, the Global TDMA Forum set up the IS-136+ program aiming to implement higher bit rates within the 30-kHz channel. The TIA TR45.3 standardization group adopted IS-136+ in 1998. IS-136+ has two modulation schemes: coherent π/4-QPSK and 8-PSK. The control channel uses π/4-DQPSK modulation. The user information bit rates for the π/4-QPSK modulation are 9.6 Kbps (single rate), 19.2 Kbps (double rate), and 28.8 Kbps (triple rate); and for the coherent 8-PSK modulation 14.4 Kbps (single rate), 28.8 Kbps (double rate), and 43.2 Kbps (triple rate).

1.3.3   IS-95

During the 1980s, Qualcomm carried out an in-house development and validation of a direct sequence CDMA air interface concept. In November 1989, the first IS-95 narrowband CDMA field trials were carried out using two cell sites and one mobile station in San Diego [18]. In September 1990, Qualcomm released the first version of its CDMA "Common Air Interface" Specification. On December 5, 1991, at CTIA's "Presentation of the Results of the Next Generation Cellular Field Trials," CDMA field trial results were presented. On January 6, 1992, CTIA requested that TIA should start preparing for CDMA standardization. Following that, in March 1992 a new subcommittee, TR45.5, was formed to develop the IS-95 standard. IS-95 was completed by 1993 and revised in 1995 (IS-95A). An up-banded version of IS-95 for the 1.9-GHz PCS frequency band was standardized in 1995.

IS-95 is also a dual-mode standard with an AMPS analog mode. In designing the IS-95 system, the AMPS backward compatibility was taken into account. For example, a handover from the digital to the analog mode is possible.

Further development of IS-95 towards higher bit rate services was started in

---

[2] US-TDMA is also referred as D-AMPS (Digital AMPS).

1996. This led to the completion of the IS-95B standard in 1998 [19]. While the IS-95A standard uses only one spreading code per traffic channel, IS-95B can concatenate up to eight codes for the transmission of higher bit rates. This provides a maximum data rate of 115.2 Kbps. For more details on IS-95, see Chapter 2.

CDMA Development Group (CDG) was established to promote IS-95 CDMA systems in 1994. At the beginning of 1998, IS-95 was commercially operated in the United States, Hong Kong, Singapore, and Korea, the latter being the largest market with over 5 million subscribers. The success of IS-95 in Korea is based on the adoption of IS-95 as a national standard in the beginning of the 1990s. A joint effort by a number of companies under the Electronics and Telecommunications Research Institute (ETRI) led to the fast development of equipment and deployment of commercial networks.

1.3.4   Personal Digital Cellular[3]

Based on the work of a study committee organized by the Ministry of Posts and Telecommunications (MPT), the requirements for PDC were defined in 1990. In May 1990, the Japanese Research and Development Center for Radio Systems (RCR) standards committee organized subcommittees for the standardization of PDC [20]. The commercial service was started by NTT in 1993 in the 800-MHz band and in 1994 in the 1.5-GHz band. Recently, packet data has been developed for the PDC standard [21].

The PDC system, originally known as JDC (Japanese Digital Cellular) is only operated in Japan. RCR TSD27 is a common air interface specification for PDC, and it has some similarities with the American IS-54 standard. PDC operates in two frequency bands, 800 MHz and 1.5 GHz.

1.4   THIRD GENERATION SYSTEMS

1.4.1   Objectives and Requirements

Emerging requirements for higher rate data services and better spectrum efficiency are the main drivers identified for the third generation mobile radio systems. In the ITU, third generation networks are called IMT-2000, and in Europe, UMTS. Since 1985, the ITU has been developing IMT-2000, previously termed Future Public Land Mobile Telephone System (FPLMTS). In ETSI, UMTS standardization started 1990 when subtechnical committee SMG5 was established. The main objectives for the IMT-2000 air interface can be summarized as:

- Full coverage and mobility for 144 Kbps, preferably 384 Kbps;
- Limited coverage and mobility for 2 Mbps;
- High spectrum efficiency compared to existing systems;
- High flexibility to introduce new services.

---

[3] Sometimes PDC is also referred as Pacific Digital Cellular.

The bit rate targets have been specified according to the Integrated Services Digital Network (ISDN) rates. The 144-Kbps data rate provides the ISDN 2B+D channel, 384 Kbps provides the ISDN H0 channel, and 1920 Kbps provides the ISDN H12 channel.[4] However, it may be that the main IMT-2000 services are not ISDN-based services. It has to be noted that these figures have been subject to considerable debate. Ultimately, market demand will determine what data rates will be offered in commercial systems. Figure 1.5 describes the relation between bit rates and mobility for the second and third generation systems.

The targets of third generation systems are wide and, depending on the main driver, system solutions will be different. The maturity of second generation mobile radio systems varies, ranging from over 40% penetration in Scandinavia to a very low penetration in developing countries, where the cellular systems are in the beginning of their lifecycle. Therefore, it is clear that the need to develop a new system varies, and the different views and needs may result in several different variants of IMT-2000. In addition, different backward compatibility requirements influence the technology applied to third generation systems.

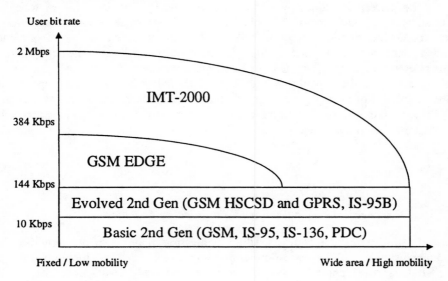

Figure 1.5   IMT-2000 user bit rate versus coverage and mobility.

1.4.2   IMT-2000 Air Interface Selections

Main regional standards bodies have already decided the preferred technology for IMT-2000. The fast development during recent years has been due to the Japanese

---

[4] Even though 2 Mbps is used generally as the upper limit for IMT-2000 services, the exact service is specified to be 1.92 or 2.048 Mbps.

initiative. In the beginning of 1997, the Association for Radio Industry and Business (ARIB), a standardization body responsible for Japan's radio standardization, decided to proceed with detailed standardization of wideband CDMA. The technology push from Japan accelerated standardization in Europe and the United States. During 1997 joint parameters for Japanese and European wideband CDMA proposals were agreed upon. The air interface is now commonly referred as WCDMA. In January 1998, strong support behind wideband CDMA led to the selection of WCDMA as the UMTS terrestrial air interface scheme for FDD frequency bands in ETSI. The selection of wideband CDMA was also backed by Asian and American GSM operators. For TDD bands, a time division CDMA (TD-CDMA) concept was selected. In the United States in March 1998, the TIA (Telecommunications Industry Association) TR45.5 committee, responsible for IS-95 standardization, adopted a framework for wideband CDMA backward compatible to IS-95, called cdma2000. TR45.3, responsible for IS-136 standardization, adopted a TDMA-based third generation proposal, UWC-136 (Universal Wireless Communications), based on the recommendation from the UWCC in February 1998. Korea is still considering two wideband CDMA technologies, one similar to WCDMA and the other similar to cdma2000.

The preferred technology for third generation systems depends on technical, political, and business factors. Technical factors include issues such as provision of required data rates, and performance. Political factors involve reaching an agreement between standards bodies and taking into account the different starting points of different countries and regions. On one hand, the investments into the existing systems motivate a backward compatibility approach. On the other, new business opportunities or the possibility of changing the current situation might motivate a new approach. In the following sections we discuss the research and standardization activities that led to the selections described above.

1.4.3   Europe

1.4.3.1   European Third Generation Radio Research Activities

Figure 1.6 illustrates the European third generation radio research activities. The development towards third generation standards can be divided into three main phases: basic studies, development of system concepts, and comparison/consolidation. The driving force behind the European development has been the European Commission funded research programs RACE (Research of Advanced Communication Technologies in Europe) and ACTS (Advanced Communication Technologies and Services) [22–24]. Their value has been in producing technical results for standardization, but most of all in bringing together different parties from the industry.

In Europe, the RACE I program, launched in 1988 and lasting until June 1992, started the third generation research activities. Main studies of RACE I concentrated on individual technologies such as cellular propagation studies, handover, dynamic resource allocation, modulation, equalization, coding, channel management, and fixed network mobile functions. Many of these individual technologies were later used as a basis for the system concept development in RACE II.

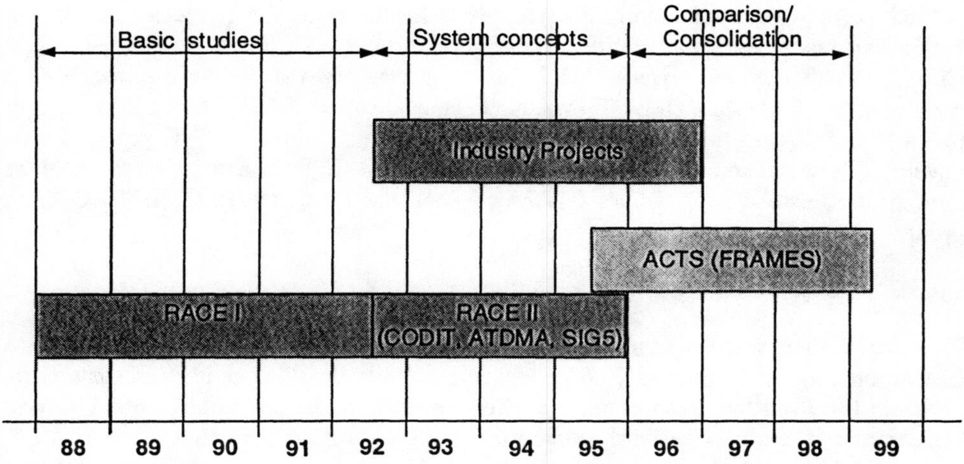

Figure 1.6   European third generation radio reseach activities.

Between 1992 and 1995, in the RACE II program, the CODIT (Code Division Multiple Testbed) [25] and ATDMA (advanced TDMA) [26] projects developed air interface proposals and testbeds for UMTS radio access. The ATDMA testbed provided up to a 64-Kbps user data rate with rate 1/4 coding, and the CODIT testbed provided a user data rate up to 128 Kbps with rate 1/2 coding [25]. Laboratory tests were performed for both systems. After the RACE II program, field tests for the CODIT testbed were performed [27]. A special interest group in RACE (SIG5) compared the CODIT and ATDMA air interfaces using both quantitative and qualitative criteria [28]. Depending on the selected radio environment and service scenario, each of the schemes showed advantage, but no decision to favor either one as a main candidate for UMTS was made.

ACTS was launched at the end of 1995 to support collaborative mobile research and development. Within ACTS, the FRAMES (Future Radio Wideband Multiple Access System) project investigated hybrid multiple access technologies in order to select the best combination for a basis for further detailed development of the UMTS radio access system. Based on this evaluation, a harmonized multiple access platform, FRAMES multiple access (FMA), was designed consisting of two modes: FMA1, a wideband TDMA scheme with and without spreading, and FMA2, a wideband CDMA scheme [29–34]. The FMA1 without spreading has roots in the ATDMA scheme and its subsequent developments [35]. The comparison results of FRAMES are discussed in Chapter 13.

In addition to the RACE and ACTS programs, several industrial projects have developed technologies for UMTS and IMT-2000. From 1992 to 1995, a wideband

CDMA concept reported in [36,37] was developed by Nokia. Furthermore, a wideband CDMA testbed was developed from 1992 to 1995 with transmission capabilities up to 128 Kbps for video application [38]. Field tests for the system were carried out during 1996. The test system bandwidth is configurable up to 30 MHz. Nokia's wideband CDMA was submitted to FRAMES where it was used as the basis for FMA2 development. A hybrid CDMA/TDMA concept, also called TD-CDMA, was studied in [39–42] and formed the basis of FMA1 with spreading. The UK LINK program also developed a wideband CDMA concept and testbed [43]. Telia developed the OFDM concept and testbed [44]. Telia's OFDM concept was submitted to the ETSI Gamma group.

1.4.3.2   European Standardization Activities

The tremendous success of GSM will strongly impact the standardization of third generation systems in Europe. Europe is trying to build the third generation systems as an evolution from the GSM system to offer a smooth transition path from the second generation systems into the third generation.

Within ETSI, the technical committee SMG has the standardization mandate for UMTS. Within SMG, there are several subcommittees that are responsible for the detailed technical standardization of UMTS. The subcommittee SMG2, responsible for UMTS radio access system standardization, started the definition process of UMTS terrestrial radio access (UTRA) in a UMTS workshop held in December 1996. The milestones of the process are presented below:

- M1 6/97: Definition of a limited number of UTRA concepts based on the proposed access technologies or combinations of them;
- M2 12/97: Selection of one UTRA concept;
- M3 6/98: Definition of key technical aspects of the UTRA (including carrier bandwidth, modulation, channel coding, channel types, frame structure, access protocols, channel allocation and handover mechanisms, and cell selection mechanisms).

The submitted air interface concepts were grouped into five different concept groups. FMA2 and three wideband CDMA schemes from Japan were submitted to the Alpha group. The Beta group evaluated FMA1 without spreading together with some other TDMA ideas. Two OFDM concepts were submitted to the Gamma group. The Delta group evaluated FMA1 with spreading, which has been also called the TD-CDMA scheme. In the beginning of the evaluation, another hybrid scheme, CTDMA (code time division multiple access), was also considered in the Delta group. The Epsilon group considered ODMA (opportunity driven multiple access), which is a relay technology applicable in principle to all multiple access schemes.

In the concept group evaluation process, the FMA2 proposal was considered together with the wideband CDMA schemes from Japan. At the same time, the Co-ordination Group (CG) in ARIB had discussions on the harmonization of different CDMA proposals. These two efforts led to the harmonization of the parameters for

ETSI and ARIB wideband CDMA schemes [45]. The main parameters of the current scheme are based in the uplink on the FMA2 scheme and in the downlink on the ARIB wideband CDMA. Also, contributions from other proposals and parties were incorporated to further enhance the concept. In January 1998, ETSI SMG reached a consensus agreement on UTRA concept [46]:

- In the paired band (FDD - frequency division duplex) of UMTS the system adopts the radio access technique formerly proposed by the W-CDMA group.
- In the unpaired band (TDD - time division duplex) the UMTS system adopts the radio access technique proposed formerly by the TD-CDMA group.
- In implementing this solution, ETSI SMG members pursue, together, the specification of UMTS with the objective of providing low cost terminals, ensuring harmonization with GSM and providing FDD/TDD dual mode operation terminals.

Consequently, technical work was initiated to achieve fully integrated UTRA concept. In April 1998, the FDD and TDD modes were harmonized to have the same chip rate, frame length, and number of slots per frame.

1.4.4 Japan

The IMT-2000 Study Committee in ARIB was established in April 1993 to coordinate the Japanese research and development activities for IMT-2000. In October 1994, the Radio Transmission Special Group for radio transmission was formed to perform technical studies and to develop draft specifications for IMT-2000. The Special Group consisted of two ad hoc groups: CDMA and TDMA [47].

For TDMA, there were eight proposals. From these the group compiled a single carrier TDMA system, MTDMA (multimode and multimedia TDMA). For the MTDMA system, a prototype equalizer with a carrier bit rate of 1.536 Mbps and a user bit rate of 512 Kbps has been built and tested [48]. One organization proposed an OFDM scheme called BDMA (band division multiple access), and it was decided to conduct a study of this scheme. The BDMA scheme was also proposed in ETSI. However, during 1997 it became clear that the MTDMA and BDMA schemes would not be standardized for IMT-2000 within ARIB.

In Japan, several companies have developed wideband CDMA air interface proposals. Originally, 13 different wideband CDMA radio interfaces for FDD were presented to the IMT-2000 Study Committee. In early 1995, they were condensed into three FDD proposals (Core A, B, and C) and into one TDD proposal. The Core-B system is based on the wideband CDMA standardized for US PCS known as IS-665. At the end of 1996, the four schemes were further combined into a single proposal where the main parameters were from Core-A, which is presented in more detail in [49–53]. Three of the original wideband CDMA proposals from Japan, two FDD and one TDD concept, were also contributed to ETSI by their respective companies. In addition, the TDD proposal was submitted to TIA TR45.5 with modified parameters.

Both laboratory tests and field trials have been conducted for Japanese wideband

CDMA proposals during 1995 and 1996. Core-A has been tested with speech and 384-Kbps video transmission in field trials [51]. Transmission of 2 Mbps has also been tested in laboratories using a bandwidth of 20 MHz. In the second phase of the Core-A trials, interference cancellation has been tested. The Core-B proposal has only been tested for speech service in the framework of PCS standardization in the United States [47]. Core-C has been tested for data rates up to 32 Kbps. Furthermore, the CDMA/TDD proposal was tested in laboratory conditions [48].

The main conclusion of the IMT-2000 Radio Transmission Special Group was that detailed studies on CDMA should be started. This meant that CDMA was the main technology choice for IMT-2000. Actually, a clear thrust for wideband CDMA came when the world's largest cellular operator, NTT DoCoMo, decided to proceed with wideband CDMA development and issued a trial system tender in 1996. All major manufacturers submitted proposals for terminal and infrastructure development. For terminal development, Matsushita, Motorola, NEC, and Nokia were selected, and for infrastructure development, Ericsson, Lucent, Matsushita, and NEC were selected. This was the first real commitment from an operator to go forward with third generation, and it created global momentum for wideband CDMA.

1.4.5   The United States

The U.S. standardization situation is more diverse compared to Europe and Japan. Main standardization activities for wireless systems are carried out in TIA Engineering Committees TR45 and TR46, and in the T1 committee T1P1. TR45.5 is responsible for the IS-95 standardization, TR45.3 for the IS-136 standardization, and T1P1 with TR46 are jointly responsible for GSM1900 and some other technologies. In addition to standardization bodies, there are industry forums that elaborate on policy issues related to standards: UWCC for IS-136, CDMA Development Group (CDG) for IS-95, and the GSM Alliance and GSM North America for GSM.

In April 1997, CDG issued the so-called Advanced Systems initiative to develop the IS-95-based third generation air interface proposal. During 1997, several wideband CDMA proposals from different companies, including Hughes, Lucent, Motorola, Nokia, Nortel, Qualcomm, and Samsung, were submitted to TR45.5 for the development of cdma2000[5] [54]. A common characteristic of all of these schemes was backward compatibility to IS-95. In March 1998, TR45.5 agreed upon the basic framework for cdma2000.

In the beginning of 1997, within UWCC, GTF (Global TDMA Forum) established the High Speed Data (HSD) group to evaluate air interface candidates for the IS-136 evolution towards third generation. Several schemes including TDMA, wideband CDMA and two OFDM schemes, were submitted to the HSD group. Based on these proposals, UWCC developed the UWC-136 concept which was accepted by TR45.3 in February 1998. UWC-136 consists of an enhanced IS-136 30 kHz carrier,

---

[5] Recently, cdma2000 was adopted as the new name for wideband CDMA based on IS-95. This way of writing the name, cdma2000, is used throughout the book. cdma2000 is also referred to as Wideband cdmaOne, Wideband IS-95, or IS95 3G.

200 kHz high speed data carrier and 1.6 MHz wideband TDMA carrier for an indoor radio environment. The 200 kHz HSD carrier has the same parameters as the enhanced GSM carrier (EDGE - enhanced data rates for GSM/global evolution) currently a work item in the GSM phase 2+ standardization [55]. The wideband TDMA is based on the FRAMES FMA1 without spreading [34].

TR46.1 is developing a wideband CDMA air interface (WIMS, Wireless Multimedia and Messaging Services Wideband CDMA) for wireless local loop applications. The basis for the wideband CDMA scheme in TR46.1 is the US PCS standard IS-665, wideband CDMA with bandwidth of 5 MHz, which has also been used as a basis for the Core-B proposal in Japan [48]. Recently, WIMS has been harmonized with the WCDMA scheme.

1.4.6   Korea

ETRI has established an R&D consortium to define the Korean proposal for IMT-2000 during 1997 and 1999. A wideband CDMA proposal has been developed within ETRI [56]. This formed the basis for the TTA I scheme.

SK Telecom (previously KMT, Korea Mobile Telecom) decided in 1994 to develop a wideband CDMA air interface for IMT-2000 with a focus on a low-tier system for a pedestrian radio environment [57]. A common air interface specification was written and a test system was built. The test system provides a data rate of 36 Kbps using 5-MHz bandwidth. The final air interface specification was completed at the end of 1996. Currently, a precommercial system development is being developed [57]. SK Telecom's system was merged with other industrial proposals at the beginning of 1998 and formed the TTA II scheme [58].

For the ITU radio transmission technology submission in June 1998, TTA I and TTA II have been renamed Global CDMA I and II, respectively.

1.5   FREQUENCY ALLOCATION FOR THIRD GENERATION SYSTEMS

For IMT-2000, a total of 230 MHz was identified at World Administrative Radio Conference (WARC'92). However, this will be allocated in different ways in different regions and countries. Figure 1.7 depicts the IMT-2000 frequency allocation in the United States, Europe, and Japan. In the United States, a part of the IMT-2000 frequency allocation is already used for PCS systems. Also, it is unlikely that the upper band of WARC'92 allocation would be taken into use. Thus, third generation systems have to fit into the current PCS frequency allocations. In Japan and Europe, the frequency allocation for third generation is almost identical except that the PHS spectrum allocation partially overlaps the identified UMTS TDD spectrum.

Frequency allocations have an impact on technology choices and air interface design. Licensing policy will define the use of the IMT-2000 spectrum. If the licenses are given to the existing operators, the use of spectrum might be different than if they are given to new operators. Also, the way licenses are issued will have an impact on the market. In the United States the PCS spectrum was auctioned, which led to large investments before the actual operation started. Spectrum allocation for UMTS and spectrum licensing are further discussed in Sections 9.3.1 and 14.5, respectively.

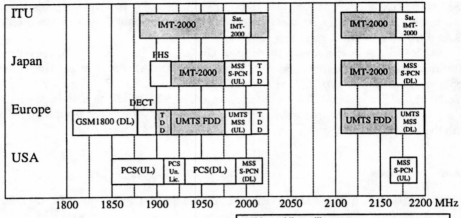

Figure 1.7   IMT-2000 frequency allocations.

1.6   AIR INTERFACE TECHNOLOGIES FOR THIRD GENERATION

In the search for the most appropriate multiple access technology for third generation wireless systems, a number of new multiple access schemes have been proposed. In this section, an overview of the air interface technologies selected for the third generation systems (wideband CDMA schemes, UWC-136 TDMA-based scheme, and TD-CDMA) is given. In addition, we briefly review OFDM technology.

1.6.1   Wideband CDMA

Wideband CDMA has a bandwidth of 5 MHz or more. The nominal bandwidth for all third generation proposals is 5 MHz. There are several reasons for choosing this bandwidth. First, data rates of 144 and 384 Kbps, the main targets of third generation systems, are achievable within 5 MHz bandwidth with a reasonable capacity. Even a 2-Mbps peak rate can be provided under limited conditions. Second, lack of spectrum calls for reasonably small minimum spectrum allocation, especially if the system has to be deployed within the existing frequency bands occupied already by second generation systems. Third, the 5-MHz bandwidth can resolve (separate) more multipaths than narrower bandwidths, increasing diversity and thus improving performance. Larger bandwidths of 10, 15, and 20 MHz have been proposed to support higher data rates more effectively.

Several wideband CDMA proposals have been made for third generation wireless systems. They can be characterized by the following new advanced properties:

- Provision of multirate services;
- Packet data;
- Complex spreading;
- A coherent uplink using a user dedicated pilot;
- Additional pilot channel in the downlink for beamforming;
- Seamless interfrequency handover;
- Fast power control in the downlink;
- Optional multiuser detection.

The third generation air interface standardization for the schemes based on CDMA seems to focus on two main types of wideband CDMA: network asynchronous and synchronous. In network asynchronous schemes the base stations are not synchronized, while in network synchronous schemes the base stations are synchronized to each other within a few microseconds. As discussed, there are two network asynchronous CDMA proposals: WCDMA in ETSI and in ARIB, and TTA II in Korea have similar parameters. A network synchronous wideband CDMA scheme has been adopted by TR45.5 (cdma2000) and is being considered by Korea (TTA I). Table 1.4 presents the features of wideband CDMA schemes. Chapter 6 describes the different wideband CDMA schemes in more detail.

Table 1.4
Features of Wideband CDMA Schemes

	WCDMA	Korea TTA II	Korea TTA I	cdma2000
Chip rate	(1.024) / 4.096 / 8.192 / 16.384 Mcps	1.024 / 4.096 / (8.192) / 16.384Mcps	0.9216 / 3.6864 / 14.7456 Mcps	1.2288 / 3.6864 / 7.3728/ 11.0593 / 14.7456 Mcps for direct spread n×1.2288 Mcps (n=1,3,6,9,12) for multicarrier
Carrier spacing	(1.25), 5, 10, 20 MHz	1.25, 5, (10), 20 MHz	1.25, 5, 20 MHz	1.25, 5, 10, 15, 20 MHz
Frame length	10 ms	10 ms	20 ms	20 ms
Inter base station synchronization	Asynchronous	Asynchronous	Synchronous	Synchronous
Coherent detection	User dedicated time multiplexed pilot (downlink and uplink), and common pilot in downlink	UL: Pilot symbols time multiplexed with power control (PC) bits DL: Common pilot channel	UL: Pilot symbols time multiplexed with PC bits DL: Common pilot channel	UL: Pilot symbols time multiplexed with PC bits DL: Common continuous pilot channel and auxiliary pilot

Note: UL=uplink, DL=downlink.

1.6.2 TDMA

As already discussed, several TDMA schemes have been studied for the third generation air interface: GSM evolution using a higher level modulation [59,60], FMA1 without spreading (wideband TDMA) [32], ATDMA [26], GSM compatible ATDMA [61], and MTDMA [48]. Based on these studies, TDMA will emerge into the third generation era as an evolution of the IS-136 standard. Furthermore, TDMA will be part of the continuing GSM evolution in the form of the EDGE concept.

The UWCC targets for the IS-136 evolution are to meet IMT-2000 requirements and an initial deployment within 1-MHz spectrum allocation. UWC-136 meets these targets via modulation enhancement to the existing 30 kHz 136 channel (136+) and by defining complementary wider band TDMA carriers with bandwidths of 200 kHz and 1.6 MHz (136 HS). The 200-kHz carrier, 136 HS (Vehicular/Outdoor), with the same parameters as EDGE, provides medium bit rates up to 384 Kbps; and the 1.6 MHz carrier, 136 HS (Indoor), provides bit rates up to 2 Mbps. The parameters of the 136 HS proposal are listed in Table 1.5, and the different carrier types of UWC-136 are shown in Figure 1.8.

Table 1.5
Parameters of 136 HS and GSM EDGE

	136 HS (Vehicular/Outdoor) / GSM EDGE	136 HS (Indoor)
Duplex method	FDD	FDD and TDD
Carrier Spacing 200 kHz	1.6 MHz	Modulation
	B-O-QAM Q-O-QAM GMSK	B-O-QAM Q-O-QAM
Gross bit rate	722.2 Kbps (Q-O-QAM) 361.1 Kbps (B-O-QAM) 270.8 Kbps (GMSK)	5200 Kbps (Q-O-QAM) 2600 Kbps (B-O-QAM)
Payload	521.6 Kbps (Q-O-QAM) 259.2 Kbps(B-O-QAM) 182.4 Kbps (GMSK)	4750 Kbps (Q-O-QAM) 2375 Kbps (B-O-QAM)
Frame length	4.615 ms	4.615 ms
Number of slots	8	64 (72 μs) 16 (288 μ s)
Coding	Convolutional 1/2, 1/4, 1/3, 1/1 ARQ	Convolutional 1/2, 1/4, 1/3, 1/1 Hybrid Type II ARQ
Frequency Hopping	Optional	Optional
Dynamic Channel Allocation	Optional	Optional

Note: the modulation parameters are for the original EDGE and UWC-136 proposals (B-O-QAM and Q-O-QAM [62]). In 1998, a new modulation scheme (8-PSK) replacing the original schemes has been adopted.

The motivation for the 200-kHz carrier is twofold. First, the adoption of the same physical layer for 136 HS (Vehicular/Outdoor) and GSM data carriers provides economics of scale and therefore cheaper equipment and faster time to market. Second, the 200-kHz carrier with higher order modulation can provide bit rates of 144 and 384

Kbps with reasonable range and capacity, fulfilling IMT-2000 requirements for pedestrian and vehicular environments. The 136 HS (Indoor) carrier can provide 2-Mbps user data rate with a reasonably strong channel coding.

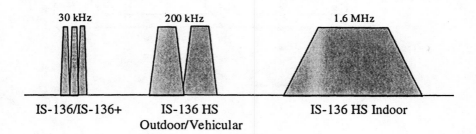

Figure 1.8   UWC-136 carrier types.

The 136 HS (Vehicular/Outdoor) data frame length is 4.615 ms, and one frame consists of eight slots. The burst structure is suitable for transmission in a high delay spread environment.

The frame and slot structures of the 136 HS (Indoor) carrier were optimized for cell coverage for high bit rates. The HS-136 (Indoor) supports both FDD and TDD duplex methods. Figure 1.9 illustrates the frame and slot structure. The frame length is 4.615 ms and it can consist of:

- 64 1/64 time slots of length 72 μs;
- 16 1/16 time slots of length 288 μs.

In the TDD mode, the same burst types as defined for the FDD mode are used. The 1/64 slot can be used for low to medium rate services. The 1/16 slot is used for medium to high rate data services. Figure 1.9 illustrates also the dynamic allocation of resources between the uplink and the downlink in the TDD mode.

The UWC-136 multirate scheme is based on a variable slot, code, and modulation structure. Data rates up to 43.2 Kbps can be offered using the 136+ 30-kHz carrier and multislot transmission. 136 HS (Outdoor/Vehicular) data services are shown in Table 1.6. Depending on the user requirements and channel conditions, suitable combination of modulation, coding, and number of data slots is selected. Asymmetrical data rates are provided by allocating a different number of time slots in the uplink and downlink. For packet swiched services, the RLC/MAC protocol provides fast medium access via a reservation based medium access scheme, supplemented by selective ARQ for efficient retransmission.

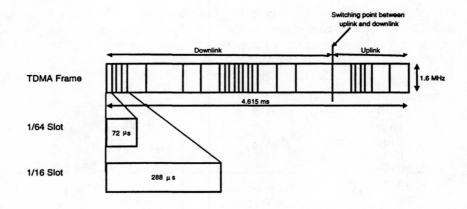

Figure 1.9 Wideband TDMA frame and slot structure.

Similar to 136 HS (Outdoor/Vehicular), the 136 HS (Indoor) uses two modulation schemes and different coding schemes to provide variable data rates. In addition, two different slot sizes can be used. Error control for packet data services is based on the Type II hybrid automatic repeat request (ARQ) scheme [34]. The basic idea is to first transmit a part of the coded data block, which is decodable separately. If decoding fails, a part of the redundant information is transmitted and decoding is tried in the receiver. This goes on until all bursts containing data of the coded data block are sent once. After this, a burst (or bursts) having the worst quality estimate (e.g., signal-to-interference ratio) is retransmitted, and combined with the previously transmitted information in the receiver. This kind of ARQ procedure can be used because of the RLC/MAC protocol's ability to allocate resources quickly and to send transmission requests in the feedback channel [34].

Table 1.6
Overview of Data Services for 136 HS

Service name	Code rate	Modulation	Gross rate	Radio interface rate*
ECS-1	0.51	Q-O-QAM	65.2 Kbps	33.0 Kbps
ECS-2	0.63	Q-O-QAM	65.2 Kbps	41.0 Kbps
ECS-3	0.74	Q-O-QAM	65.2 Kbps	48.0 Kbps
ECS-4	1	Q-O-QAM	65.2 Kbps	65.2 Kbps
ECS-5	0.35	B-O-QAM	32.4 Kbps	11.2 Kbps
ECS-6	0.45	B-O-QAM	32.4 Kbps	14.5 Kbps
ECS-7	0.52	B-O-QAM	32.4 Kbps	16.7 Kbps
ECS-8	0.70	B-O-QAM	32.4 Kbps	22.8 Kbps

* The radio interface rate includes the signaling overhead for the RLC/MAC layer.
Note: only single time slot rates shown.

1.6.3   Hybrid CDMA/TDMA

In the ETSI air interface selection, the TDD solution was decided to be based on the TD-CDMA principles. In this section the original TD-CDMA scheme is presented. The role of CDMA in TD-CDMA is to multiplex the different channels within a timeslot. The parameters of TD-CDMA are listed in Table 1.7.

The spreading ratio of TD-CDMA is small, and, thus, if more than a few users are desired per frame, joint detection is needed to remove the intracell interference. Joint detection is also required due to slow power control, which results in large variations in the received signal levels. Since the joint detection is a mandatory feature, it is more critical compared to wideband CDMA. However, if the number of users is small, the complexity of joint detection may not be excessive.

Table 1.7
TD-CDMA Parameters

Multiple access method	TDMA/CDMA
Basic channel spacing	1.6 MHz
Carrier chip/bit rate	2.167 Mchip/s
Time slot structure	8 slots / TDMA frame
Spreading	Orthogonal, 16 chips/symbol
Frame length	4.615 ms
Multirate concept	Multislot and multicode
Channel coding	Convolutional codes R=1/4 to 1 puncturing/repetition
Data modulation	QPSK / 16QAM
Spreading modulation	Linearized GMSK
Pulse shaping	(Linearized GMSK)
Detection	Coherent, based on training sequence
Other diversity means	Frequency/time hopping per frame or slot

Source: [34].

1.6.4   Orthogonal Frequency Division Multiplexing

OFDM applies the principle of multicarrier modulation (MCM) by dividing the data stream into several bit streams, each of which has much lower bit rate, and using these substreams to modulate several carriers. First applications of MCM were in military radios in the late 1950s and the early 1960s [4]. The OFDM principle was patented in 1970 [63]. The research of OFDM for Digital Audio Broadcasting (DAB) was started at the end of the 1980s, and commercial DAB systems are currently being introduced. Recently, OFDM has been applied to HIPERLAN type II and wireless ATM systems [64].

Research on OFDM for cellular applications has intensified during the 1990s. Both CDMA- and TDMA-based OFDM have been studied. A multicarrier CDMA (MC-CDMA) transmitter spreads the original signal by using a given spreading code in the frequency domain. It is crucial for multicarrier transmissions to have frequency nonselective fading over each subcarrier [1]. Introduction of OFDM into the cellular world has been driven by two main benefits:

- Flexibility: each transceiver has access to all subcarriers within a cell layer;
- Easy equalization: OFDM symbols are longer than the maximum delay spread, resulting in a flat fading channel that can be easily equalized.

The main drawback of OFDM is the high peak to average power. This is especially severe for the mobile station and for long-range applications. Different encoding techniques have been investigated to overcome this problem. Furthermore, the possibility of accessing all the resources within the system bandwidth results in an equally complex receiver for all services, regardless of the bit rate. Of course, a partial FFT for only one OFDM block is possible for low bit rate services, but this would require an RF synthesizer for frequency hopping.

Table 1.8 shows the parameters of the OFDM air interface proposed by the Gamma group of ETSI. The Gamma concept has been developed based on contributions from different OFDM schemes described in [44,48,65].

Table 1.8
Main Features of OFDM Proposal

Bandslot width	100 kHz
Number of subcarriers	24
Subcarrier bandwidth	4.17 kHz
Modulation period	288.8
Block size	24 carriers / 1 symbol
Frame length	4.615 ms

Source: [65].

1.7 PREVIEW OF THE BOOK

This book consists of 14 chapters and concentrates on a study of wideband CDMA from the radio access design point of view. Propagation and application requirements for the design of wireless access systems are covered. A detailed study of wideband CDMA air interface solutions is presented. Network aspects are addressed to give an overview of a complete wireless network architecture.

Basic principles of CDMA are explained in Chapter 2. First, the CDMA concept is discussed. Basic CDMA functions such as the RAKE receiver, power control, soft handover, interfrequency handover, and multi-user detection are explained. The narrowband IS-95 CDMA standard is introduced.

In Chapter 3, UMTS applications and user scenarios are introduced. The characteristics of UMTS teleservices such as video telephony and Internet services are discussed, and bearer service requirements are introduced. Traffic models for simulation purposes are developed. Detailed know-how on radio propagation is a prerequisite for any radio system design.

Chapter 4 presents radio environments and their characteristics. Wideband channel models are discussed in detail, as are pathloss and mobility models.

Chapter 5 covers all aspects of the wideband CDMA air interface design starting from frame structures and physical channel design (including spreading codes and their properties) to radio resource management aspects (handover and power control).

Several different wideband CDMA proposals have been proposed in different standardization bodies and research projects. These are reviewed in Chapter 6. The main proposals for wideband CDMA, WCDMA, cdma2000, TTA I, and TTA II are described. In addition, CODIT and the IS-665 standard are discussed.

Chapter 7 covers performance aspects of wideband CDMA. Simulation assumptions and the simulator used in the performance evaluation are introduced. Modeling of different CDMA functions in both the link level and the system level are discussed. The impact of power control into system performance is analyzed. Spectrum efficiency and range are two major performance measures that are considered.

Hierarchical cell structures (HCS) are discussed in Chapter 8. Since the spectrum spreading caused by nonlinear power amplifiers impacts the deployment of hierarchical cell structures, nonlinear power amplifier characteristics and power amplifier efficiency are discussed first. Two options for HCS deployment are covered: micro and macro cells at the same frequency, and micro and macro cells at different frequencies. Since the latter is more probable, its system level performance is analyzed.

In Chapter 9, TDD aspects are covered. Reasons to use TDD, such as asymmetric services and dedicated spectrum, are reviewed. Interference scenarios caused by the use of TDD are discussed. Different TDD designs for wideband CDMA are discussed and their main parameters explained.

Implementation issues are dealt with in Chapter 10. Optimization criteria for transceiver design are introduced. Modularity issues are discussed. Receiver and transmitter designs both in baseband and RF are explained in detail. Based on this, a typical wideband CDMA mobile terminal architecture is presented. Multimode terminals are discussed in final section.

CDMA network planning issues are discussed in Chapter 11. First, basic radio network planning procedure is explained. Next, micro cell planning in CDMA is covered. Co-existence of wideband CDMA with other air interfaces is discussed, including intermodulation aspects, deployment scenarios, guard zones and bands, and handover between systems. Finally, the transition from second generation systems into wideband CDMA is considered from the operator's perspective.

Chapter 12 presents network aspects including design methodology, different core network solutions, the ITU family of systems concept, the generic radio access network concept for UMTS, and network technologies. Furthermore, the evolution of second generation networks, in particular GSM and IS-95, is discussed.

Chapter 13 presents a comparison of wideband CDMA with TDMA and other third generation air interfaces with respect to spectrum efficiency, coverage, and other evaluation criteria. Main air interface comparisons are reviewed.

Finally in Chapter 14, the standardization environment, different standards bodies, and their role in third generation standardization are described. In addition, future directions of wideband CDMA and third generation wireless networks are discussed.

REFERENCES

[1] Prasad, R., *CDMA for Wireless Personal Communications*, Boston-London: Artech House, 1996.

[2] Ojanperä, T., "Overview of Research Activities for Third Generation Mobile Communication," in *Wireless Communications TDMA vs. CDMA*, S. G. Glisic and P. A. Leppanen (eds.), Dordrecht: Kluwer Academic Publishers, 1997, pp. 415–446.

[3] Ojanperä, T., and R. Prasad, "Overview of Air Interface Multiple Access for IMT-2000/UMTS," *IEEE Communications Magazine*, September 1998.

[4] Prasad, R., *Universal Wireless Personal Communications*, Boston: Artech House, 1998.

[5] Shannon, C. E., "A Mathematical Theory of Communication," *Bell System Technical Journal*, Vol. 27, 1948, pp. 379–423 and 623–656.

[6] Scholtz, R. A., "The Evolution of Spread-Spectrum Multiple-Access Communications," in *Code Division Multiple Access Communications*, S. G. Glisic and P. A. Leppänen (eds.), Kluwer Academic Publishers, 1995.

[7] "A conversation with Claude Shannon," *IEEE Communications Magazine*, Vol. 22, No. 5, May 1984, pp.123–126.

[8] Cooper, G. R., and R. W. Nettleton, "A spread-spectrum technique for high-capacity mobile communications," *IEEE Trans. Veh. Tech.*, Vol. 27, No. 4, November 1978, pp. 264–275.

[9] Verdu, S., "Minimum Probability of Error for Asynchronous Gaussian Multiple Access," *IEEE Trans. on IT.*, Vol. IT-32, No. 1, January 1986, pp. 85–96.

[10] Goodman, D. J., *Wireless Personal Communications Systems*, Reading, MA: Addison-Wesley, 1997.

[11] Mouly, M., and M.-B. Pautet, *The GSM System for Mobile Communications*, published by the authors, 1992

[12] Failli, R., and P. P. Giusto, "Alternative Architectures for a European Digital Mobile Radio System in the 900 MHz band," *Proceedings of Nordic Seminar on Digital Land Mobile Radio Communications*, Espoo, Finland, February 1985.

[13] Jarvinen, K., J. Vainio, P. Kapanen, T. Honkanen, P. Haavisto, R. Salami, C. Laflamme, and J.-P. Adoul, "GSM Enhanced Full Rate Speech Codec," *Proceedings of ICASSP'97*, April 1997, pp. 771–774.

[14] Pirhonen, R., and P.Ranta, "Single Slot 14.4 Kbps Service for GSM," *Proceedings of Seventh IEEE International Symposium on Personal, Indoor and Mobile Radio Communications (PIMRC'96)*, Taipei, Taiwan, October 1–8, 1996, pp. 943–947.

[15] Hämäläinen, J., *Design of GSM High Speed Data Services*, PhD Thesis, Tampere University of Technology, Finland, 1996.

[16] Ranta, P.-A., A. Lappeteläinen, and Z.-C. Honkasalo, "Interference cancellation by Joint Detection in Random Frequency Hopping TDMA Networks," *Proceedings of ICUPC'96 conference*, Vol. 1, Cambridge, Massachusetts, USA, September/October 1996, pp. 428–432.

[17] Jung, P., Z. Zvonar, and K. Kammerlander (eds.), *GSM: Evolution Towards 3rd Generation*, Kluwer Academic Publishers, 1998.

[18] Ross, A. H. M., and K. L. Gilhousen, "CDMA Technology and the IS-95 North American Standard," in *The Mobile Communications Handbook*, J. D. Gipson (ed.), Boca Raton, Florida: CRC Press, 1996, pp. 430–448.

[19] TIA/EIA IS-95B, "Mobile Station-Base Station Compatibility Standard for Dual-Mode Wideband Spread Spectrum Cellular Systems," 1998.

[20] Kinoshita, K., and M. Nagakawa "Japanese Cellular Standard," in *The Mobile Communication Handbook*, J. D. Gibson (ed.), Boca Raton, Florida: CRC Press, 1996, pp. 449–461.

[21] Murase, A., A. Maebara, I. Okajima, and S. Hirata, "Mobile Radio Packet Data Communications in a TDMA Digital Cellular System," *Proceedings of VTC'97*, Phoenix, Arizona, USA, May 1997, pp. 1034–1038.

[22] "The European Path Towards UMTS," *IEEE Pers. Commun.*, Special Issue, Vol. 2, No. 1, February 1995.

[23] DaSilva, J. S., B. Arroyo, B. Barani, and D. Ikonomou, "European Third-Generation Mobile Systems," *IEEE Communications Magazine*, Vol. 34, No. 10, October 1996, pp. 68–83.

[24] Prasad, R., J. S. DaSilva, and B. Arroyo, "ACTS Mobile Programme in Europe," *guest editorial in IEEE Communications Magazine*, Vol. 36, No. 2, Feb. 1998.

[25] Andermo, P.-G., (ed.), "UMTS Code Division Testbed (CODIT)," *CODIT Final Review Report*, September 1995.

[26] Urie, A., M. Streeton, and C. Mourot, "An Advanced TDMA Mobile Access System for UMTS," *IEEE Personal Communications*, Vol. 2, No. 1, February 1995, pp. 38–47.

[27] Ewerbring, M., J. Ferjh, and W. Granzow, "Performance Evaluation of a Wideband Testbed Based on CDMA," *Proceedings of VTC'97*, Phoenix, Arizona, USA, May 1997, pp. 1009–1013.

[28] M. Pizarroso and J. Jiménez (eds.), "Preliminary Evaluation of ATDMA and CODIT System Concepts," *SIG5 deliverable MPLA/TDE/SIG5/DS/P/002/b1*, September 1995.

[29] Ojanperä, T. M. Gudmundson, P. Jung, J. Sköld, R. Pirhonen, G. Kramer, and A. Toskala, "FRAMES-Hybrid Multiple Access Technology," *Proceedings of ISSSTA'96*, Vol. 1, Mainz, Germany, September 1996, pp. 320–324.

[30] Ojanperä, T., P. O. Anderson, J. Castro, L. Girard, A. Klein, and R. Prasad, "A Comparative Study of Hybrid Multiple Access Schemes for UMTS," *Proceedings of ACTS Mobile Summit Conference*, Vol. 1, Granada, Spain, October 1996, pp. 124–130.

[31] Ojanperä, T., J. Sköld, J. Castro, L. Girard, and A. Klein, "Comparison of Multiple Access Schemes for UMTS," *Proceedings of VTC97*, Vol. 2, Phoenix, Arizona, USA, May 1997, pp. 490–494.

[32] Klein, A., R. Pirhonen, J. Sköld, and R. Suoranta, "FRAMES Multiple Access Mode 1–Wideband TDMA with and without Spreading," *Proceedings of PIMRC97*, Helsinki, September 1997, pp. 37–41.

[33] Ovesjö, F., E. Dahlman, T. Ojanperä, A.Toskala, and A. Klein, "FRAMES Multiple Access Mode 2–Wideband CDMA," *Proceedings of PIMRC97*, Helsinki, September 1997, pp. 42–46.

[34] Nikula, E., A. Toskala, E. Dahlman, L. Girard, and A. Klein, "FRAMES Multiple Access for UMTS and IMT-2000," *IEEE Pers. Commun.*, April 1998, pp. 16–24.

[35] Nikula, E., and E. Malkamäki, "High Bit Rate Services for UMTS Using Wideband TDMA Carriers," *Proceedings of ICUPC'96*, Vol. 2, Cambridge, Massachusetts, USA, September/October 1996, pp. 562–566.

[36] Ojanperä, T., K. Rikkinen, H. Häkkinen, K. Pehkonen, A. Hottinen, and J. Lilleberg, "Design of a 3rd Generation Multirate CDMA System with Multiuser Detection, MUD-CDMA," *Proceedings of ISSSTA'96*, Vol. 1, Mainz, Germany, 1996, pp. 334–338.

[37] Westman, T., and H. Holma, "CDMA System for UMTS High Bit Rate Services," *Proceedings of VTC'97*, Phoenix, Arizona, USA, May 1997, pp. 825–829.

[38] Pajukoski, K., and J. Savusalo, "Wideband CDMA Test System," *Proceedings of PIMRC'97*, Helsinki, Finland, September 1997, pp. 825–829.

[39] Klein, A., and P. W. Baier, "Simultaneous cancellation of cross interference and ISI in CDMA mobile radio communications," *Proceedings of IEEE International Symposium on Personal, Indoor and Mobile Radio Communications (PIMRC'92)*, Boston, Massachusetts,1992, pp. 118–122.

[40] Klein, A., and P. W. Baier, "Linear unbiased data estimation in mobile radio systems applying CDMA," *IEEE Journal on Selected Areas in Communications*, Vol. SAC-11, 1993, pp. 1058–1066.

[41] Naßhan, M. M., P. Jung, A. Steil, and P. W. Baier, "On the effects of quantization, nonlinear amplification and band limitation in CDMA mobile radio systems using joint detection," *Proceedings of the Fifth Annual International Conference on Wireless Communications WIRELESS'93*, Calgary, Canada, 1993, pp. 173–186.

[42] Jung, P., J. J. Blanz, M. M. Naßhan, and P. W. Baier, "Simulation of the uplink of JD-CDMA mobile radio systems with coherent receiver antenna diversity," *Wireless Personal Communications*, Vol. 1, 1994, pp. 61–89.

[43] Wales, S. C., "The U.K. LINK Personal Communications Programme: A DS-CDMA Air Interface for UMTS," *Proceedings of RACE Mobile Telecommunications Summit*, Cascais, Portugal, November 1995.

[44] ETSI SMG2, "Description of Telia's OFDM based proposal," *TD 180/97 ETSI SMG2*, May 1997.

[45] Ishida, Y., "Recent Study on Candidate Radio Transmission Technology for IMT-2000," *First Annual CDMA European Congress*, London, UK, October 1997.

[46] ETSI Press Release, "Agreement Reached on Radio Interface for Third Generation Mobile System, UMTS," January 29, 1998.

[47] Sasaki, A., "A Perspective of Third Generation Mobile Systems in Japan," *IIR Conference Third Generation Mobile Systems: The Route Towards UMTS*, London, UK, February 1997.

[48] ARIB FPLMTS Study Committee, "Report on FPLMTS Radio Transmission Technology SPECIAL GROUP, (Round 2 Activity Report)," Draft v.E1.1, January 1997.

[49] Adachi, F., K. Ohno, M. Sawahashi, and A. Higashi, "Multimedia mobile radio access based on coherent DS-CDMA," *Proceedings of 2nd International Workshop on Mobile Multimedia Commun.*, Vol. A2.3, Bristol University, UK,

April 1995.

[50] Ohno, K., M. Sawahashi, and F. Adachi, "Wideband coherent DS-CDMA," *Proceedings of IEEE VTC'95*, Chicago, Illinois, USA, July 1995, pp. 779–783.

[51] Dohi, T., Y. Okumura, A. Higashi, K. Ohno, and F. Adachi, "Experiments on Coherent Multicode DS-CDMA," *Proceedings of IEEE VTC'96*, Atlanta, Georgia, USA, pp. 889–893.

[52] Adachi, F., M. Sawahashi, and K. Ohno, "Coherent DS-CDMA: Promising Multiple Access for Wireless Multimedia Mobile Communications," *Proceedings of IEEE ISSSTA'96*, Mainz, Germany, September 1996, pp. 351–358.

[53] Onoe, S., K. Ohno, K. Yamagata, and T. Nakamura, "Wideband-CDMA Radio Control Techniques for Third Generation Mobile Communication Systems," *Proceedings of VTC'97*, Vol. 2, Phoenix, Arizona, USA, May 1997, pp. 835–839.

[54] Chia, S., "Will cdmaOne be the third choice," *CDMA Spectrum*, September 1997, pp. 30–34.

[55] ETSI SMG2, "Improved Data Rates through Optimized Modulation, EDGE Feasibility Study," *Work Item 184*, Version 1.0, 1997.

[56] Hang, Y., S. C. Bang, H.-R. Park, and B.-J. Kang, "Performance of Wideband CDMA System for IMT-2000," *Proceedings of The 2nd CDMA International Conference*, Seoul, Korea, October 1997, pp. 583–587.

[57] Koo, J. M., E. K. Hong, and J.I. Lee, "Wideband CDMA technology for FPLMTS," *Proceedings of the 1st CDMA International Conference*, Seoul Korea, November 1996.

[58] Hong, E. K., "Intercell Asynchronous W-CDMA System for IMT-2000," *ITU-R TG 8/1 workshop*, Toronto, Canada, September, 1997.

[59] Sköld, J., P. Schramm, P.-O. Anderson, and M. Gudmundson, "Cellular Evolution into Wideband Services," *Proceedings of VTC'97*, Vol. 2, Phoenix, Arizona, USA, May 1997, pp. 485–489.

[60] Honkasalo, H., "The technical evolution of GSM," *Proceedings of Telecom'95*, Geneva, October 1995, pp. 15–19.

[61] Urie, A., "Advanced GSM: A Long Term Future Scenario for GSM," *Proceedings of Telecom'95*, Vol. 2, Geneva, Switzerland, October 1995, pp. 33–37.

[62] Malkamäki, E., "Binary and Multilevel Offset QAM, Spectrum Efficient Modulation Schemes for Personal Communications," *Proceedings of VTC'92*, Denver, Colorado, USA, May 1–3, 1992, pp. 325–328.

[63] Chang, R.W., "Orthogonal Frequency Division Multiplexing," *U.S. Patent 3,488,455*, filed 1966, issued Jan 6, 1970.

[64] Mikkonen, J. K., J. P. Aldis, G. A. Awater, A. S. Lunn, and D. Hutchinson, "The Magic WAND-Functional Overview," *IEEE Journal on Selected Areas in Communications*, to appear in 1998.

[65] Alikhani, H., R. Böhnke, and M. Suzuki, "BDMA Band Division Multiple Access — A New Air-Interface for 3rd Generation Mobile System in Europe," *Proceedings of ACTS Summit*, Aalborg, Denmark, October 1997, pp. 482–488.

Chapter 2

BASIC PRINCIPLES OF CDMA

2.1   INTRODUCTION

This chapter illustrates the basic principles of CDMA. The scope of the chapter is to give generic understanding of CDMA without overwhelming mathematical details. Readers should refer to Chapters 5 and 6 for a more extensive discussion on direct sequence CDMA air interface design aspects and wideband CDMA air interface proposals, respectively. Furthermore, [1–11] provide details of spread-spectrum (SS) and CDMA technologies.

This chapter is divided into three sections. Section 2.1 introduces the CDMA concept in general. Which criteria the transmitted signal has to fulfill in order to constitute a spread-spectrum modulation are explained. Processing gain is defined. The fundamental properties of CDMA signals, namely multiple access capability, protection against multipath interference, privacy, interference rejection, anti-jamming capability, and low probability of interception, are introduced. Different modulation methods for CDMA are treated in detail. These are direct sequence spread-spectrum, frequency hopping spread-spectrum, time hopping spread-spectrum, and hybrid modulation. Each modulation scheme is described with the help of block diagrams for the transmitter and the receiver. In addition, how each spread-spectrum modulation scheme achieves the above listed six properties of CDMA signals is discussed.

In Section 2.2 we review the fundamental elements of direct sequence CDMA and its application into third generation systems, namely RAKE receiver, power control, soft handover, interfrequency handover, and multiuser detection.

In Section 2.3 the IS-95 standard is used to illustrate the downlink and uplink channel structures, as well as the power control principles of direct sequence CDMA.

2.2   CDMA CONCEPTS

In CDMA each user is assigned a unique code sequence (spreading code) it uses to encode its information-bearing signal. The receiver, knowing the code sequences of the user, decodes a received signal after reception and recovers the original data. This is possible since the crosscorrelations between the code of the desired user and the codes of the other users are small. Since the bandwidth of the code signal is chosen to be much larger than the bandwidth of the information-bearing signal, the encoding process enlarges (spreads) the spectrum of the signal and is therefore also known as spread-spectrum modulation. The resulting signal is also called a spread-spectrum signal, and CDMA is often denoted as spread-spectrum multiple access (SSMA).

The spectral spreading of the transmitted signal gives to CDMA its multiple access capability. It is therefore important to know the techniques necessary to generate spread-spectrum signals and the properties of these signals. A spread-spectrum modulation technique must fulfill two criteria:

1. The transmission bandwidth must be much larger than the information bandwidth.
2. The resulting radio-frequency bandwidth is determined by a function other than the information being sent (so the bandwidth is statistically independent of the information signal). This excludes modulation techniques like frequency modulation (FM) and phase modulation (PM).

The ratio of transmitted bandwidth to information bandwidth is called the *processing gain* G_p of the spread-spectrum system,

$$G_p = \frac{B_t}{B_i} \qquad (2.1)$$

where B_t is the transmission bandwidth and B_i is the bandwidth of the information-bearing signal.

The receiver correlates the received signal with a synchronously generated replica of the spreading code to recover the original information-bearing signal. This implies that the receiver must know the code used to modulate the data.

Because of the coding and the resulting enlarged bandwidth, SS signals have a number of properties that differ from the properties of narrowband signals. The most interesting from the communication systems point of view are discussed below. To have a clear understanding, each property has been briefly explained with the help of illustrations, if necessary, by applying direct sequence spread-spectrum techniques.

1. *Multiple access capability.* If multiple users transmit a spread-spectrum signal at the same time, the receiver will still be able to distinguish between the users provided each user has a unique code that has a sufficiently low cross-correlation with the other codes. Correlating the received signal with a code signal from a certain user will then only despread the signal of this user, while the other spread-spectrum signals will remain spread over a large bandwidth. Thus, within the information bandwidth the power of the desired user will be larger than the interfering power provided there are not too many interferers, and the desired signal can be extracted. The multiple access capability is illustrated in Figure 2.1. In Figure 2.1(a), two users

generate a spread-spectrum signal from their narrowband data signals. In Figure 2.1(b) both users transmit their spread-spectrum signals at the same time. At the receiver 1 only the signal of user 1 is coherently summed by the user 1 despreader and the user 1 data recovered.

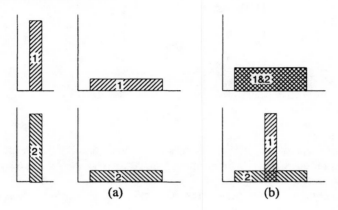

Figure 2.1   Principle of spread-spectrum multiple access.

2. *Protection against multipath interference.* In a radio channel there is not just one path between a transmitter and receiver. Due to reflections (and refractions), a signal will be received from a number of different paths. The signals of the different paths are all copies of the same transmitted signal but with different amplitudes, phases, delays, and arrival angles. Adding these signals at the receiver will be constructive at some of the frequencies and destructive at others. In the time domain, this results in a dispersed signal. Spread-spectrum modulation can combat this multipath interference; however, the way in which this is achieved depends very much on the type of modulation used. In the next section, where CDMA schemes based on different modulation methods are discussed, we show for each scheme how multipath interference rejection is obtained.
3. *Privacy.* The transmitted signal can only be despread and the data recovered if the code is known to the receiver.
4. *Interference rejection.* Cross-correlating the code signal with a narrowband signal will spread the power of the narrowband signal thereby reducing the interfering power in the information bandwidth. This is illustrated in Figure 2.2. The receiver observes spread-spectrum signal (s) summed with a narrowband interference (i). At the receiver, the SS signal is "despread" while the interference signal is spread, making it appear as background noise compared to the despread signal. Demodulation will be successful if the resulting background is of sufficiently weak energy in the despread information bandwidth.
5. *Anti-jamming capability, especially narrowband jamming.* This is more or less the same as interference rejection except the interference is now willfully inflicted on the system. It is this property, together with the next one, that makes spread-

spectrum modulation attractive for military applications.

6. *Low probability of interception (LPI).* Because of its low power density, the spread-spectrum signal is difficult to detect and intercept by a hostile listener.

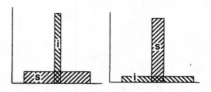

Figure 2.2   Interference rejection.

A general classification of CDMA is given in Figure 2.3. There are a number of modulation techniques that generate spread-spectrum signals. We briefly discuss the most important ones.

Direct sequence spread-spectrum. The information-bearing signal is multiplied directly by a high chip rate spreading code.

Frequency hopping spread-spectrum. The carrier frequency at which the information-bearing signal is transmitted is rapidly changed according to the spreading code.

Time hopping spread-spectrum. The information-bearing signal is not transmitted continuously. Instead the signal is transmitted in short bursts where the times of the bursts are decided by the spreading code.

Hybrid modulation. Two or more of the above-mentioned SS modulation techniques can be used together to combine the advantages and, it is hoped, to combat their disadvantages. Furthermore, it is possible to combine CDMA with other multiple access methods: TDMA, multicarrier (MC) or multitone (MT) modulation. In the case of MC-CDMA, spreading is done along the frequency axis, while for MT-CDMA, spreading is done along the time axis. Note that MC-CDMA and MT-CDMA are based on orthogonal frequency division multiplexing (OFDM).

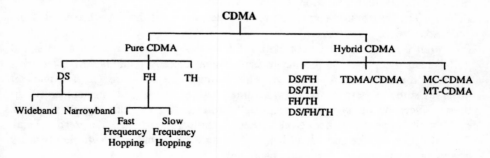

Figure 2.3   Classification of CDMA.

In the next section the above-mentioned pure CDMA modulation techniques are used to show the multiple access capability of CDMA. However, the remainder of the chapters will mainly concentrate on direct sequence (DS)-CDMA and its related subjects.

2.2.1   Spread-Spectrum Multiple Access

2.2.1.1 Direct Sequence

In DS-CDMA the modulated information-bearing signal (the data signal) is directly modulated by a digital, discrete time, discrete valued code signal. The data signal can be either an analog signal or a digital one. In most cases it is a digital signal. In the case of a digital signal, the data modulation is often omitted and the data signal is directly multiplied by the code signal and the resulting signal modulates the wideband carrier. It is from this direct multiplication that the direct sequence CDMA gets its name.

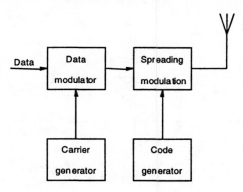

Figure 2.4   Block diagram of a DS-SS transmitter.

In Figure 2.4 a block diagram of a DS-CDMA transmitter is given. The binary data signal modulates a RF carrier. The modulated carrier is then modulated by the code signal. This code signal consists of a number of code bits called "chips" that can be either +1 or −1. To obtain the desired spreading of the signal, the chip rate of the code signal must be much higher than the chip rate of the information signal. For the spreading modulation, various modulation techniques can be used, but usually some form of phase shift keying (PSK) like binary phase shift keying (BPSK), differential binary phase shift keying (D-BPSK), quadrature phase shift keying (QPSK), or minimum shift keying (MSK) is employed.

If we omit the data modulation and use BPSK for the code modulation, we get the block diagram given in Figure 2.5. The DS-SS signal resulting from this transmitter is shown in Figure 2.6. The rate of the code signal is called the *chip rate*; one chip denotes one symbol when referring to spreading code signals. In this figure, 10 code chips per information symbol are transmitted (the code chip rate is 10 times the data rate) so the processing gain is equal to 10.

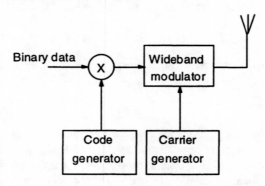

Figure 2.5 Modified block diagram of a DS-SS transmitter.

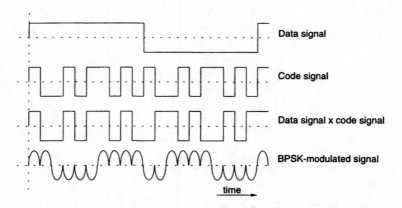

Figure 2.6 Generation of a BPSK-modulated SS signal.

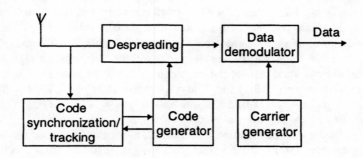

Figure 2.7 Receiver of a DS-SS signal.

After transmission of the signal, the receiver (shown in Figure 2.7) despreads the SS signal using a locally generated code sequence. To be able to perform the despreading operation, the receiver must not only know the code sequence used to spread the signal, but the codes of the received signal and the locally generated code must also be synchronized. This synchronization must be accomplished at the beginning of the reception and maintained until the whole signal has been received. The code synchronization/tracking block performs this operation. After despreading, a data modulated signal results, and after demodulation, the original data can be recovered.

In the previous section a number of advantageous properties of spread-spectrum signals were mentioned. The most important of those properties from the viewpoint of CDMA is the multiple access capability, the multipath interference rejection, the narrowband interference rejection, and with respect to secure/private communication, the LPI. We explain these four properties for the case of DS-CDMA.

- *Multiple access*: If multiple users use the channel at the same time, there will be multiple DS signals overlapping in time and frequency. At the receiver, despreading is used to remove the spreading code. This operation concentrates the power of the desired user in the information bandwidth. If the crosscorrelations between the code of the desired user and the codes of the interfering users are small, coherent detection will only put a small part of the power of the interfering signals into the information bandwidth.
- *Multipath interference*: If the code sequence has an ideal autocorrelation function, then the correlation function is zero outside the interval $[-T_c, T_c]$, where T_c is the chip duration. This means that if the desired signal and a version that is delayed for more than $2T_c$ are received, despreading will treat the delayed version as an interfering signal, putting only a small part of the power in the information bandwidth.
- *Narrowband interference*: The coherent detection at the receiver involves a multiplication of the received signal by a locally generated code sequence. However, as we saw at the transmitter, multiplying a narrowband signal with a wideband code sequence spreads the spectrum of the narrowband signal so that its power in the information bandwidth decreases by a factor equal to the processing gain.
- *LPI*: Because the direct sequence signal uses the whole signal spectrum all the time, it will have a very low transmitted power per hertz. This makes it very difficult to detect a DS signal.

Apart from the above-mentioned properties, DS-CDMA has a number of other specific properties that we can divide into advantageous (+) and disadvantageous (−) behavior:

+ The generation of the coded signal is easy. It can be performed by a simple multiplication.
+ Since only one carrier frequency has to be generated, the frequency synthesizer (carrier generator) is simple.
+ Coherent demodulation of the DS signal is possible.

+ No synchronization among the users is necessary.
- It is difficult to acquire and maintain the synchronization of the locally generated code signal and the received signal. Synchronization has to be kept within a fraction of the chip time.
- For correct reception, the synchronization error of locally generated code sequence and the received code sequence must be very small, a fraction of the chip time. This combined with the nonavailability of large contiguous frequency bands practically limits the bandwidth to 10 to 20 MHz.
- The power received from users close to the base station is much higher than that received from users further away. Since a user continuously transmits over the whole bandwidth, a user close to the base will constantly create a lot of interference for users far from the base station, making their reception impossible. This near-far effect can be solved by applying a power control algorithm so that all users are received by the base station with the same average power. However this control proves to be quite difficult due to feedback delays, imperfect power estimates, errors in the feedback channel, and traffic conditions.

2.2.1.2   Frequency Hopping

In frequency hopping CDMA, the carrier frequency of the modulated information signal is not constant but changes periodically. During time intervals T, the carrier frequency remains the same, but after each time interval the carrier hops to another (or possibly the same) frequency. The hopping pattern is decided by the spreading code. The set of available frequencies the carrier can attain is called the *hop-set*.

The frequency occupation of an FH-SS system differs considerably from a DS-SS system. A DS system occupies the whole frequency band when it transmits, whereas an FH system uses only a small part of the bandwidth when it transmits, but the location of this part differs in time.

The difference between the FH-SS and the DS-SS frequency usage is illustrated in Figure 2.8. Suppose an FH system is transmitting in frequency band 2 during the first time period. A DS system transmitting in the same time period spreads its signal power over the whole frequency band so the power transmitted in frequency band 2 will be much less than that of the FH system. However, the DS system transmits in frequency band 2 during all time periods while the FH system only uses this band part of the time. On average, both systems will transmit the same power in the frequency band.

The block diagram for an FH-CDMA system is given in Figure 2.9. The data signal is baseband modulated. Using a fast frequency synthesizer that is controlled by the code signal, the carrier frequency is converted up to the transmission frequency.

The inverse process takes place at the receiver. Using a locally generated code sequence, the received signal is converted down to the baseband. The data is recovered after (baseband) demodulation. The synchronization/tracking circuit ensures that the hopping of the locally generated carrier synchronizes to the hopping pattern of the received carrier so that correct despreading of the signal is possible.

Within frequency hopping CDMA a distinction is made that is based on the hopping rate of the carrier. If the hopping rate is (much) greater than the symbol rate,

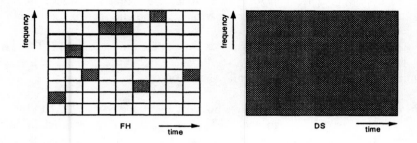

Figure 2.8   Time/frequency occupancy of FH and DS signals.

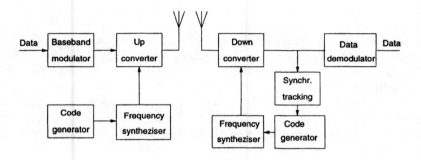

Figure 2.9   Block diagram of an FH-CDMA transmitter and receiver.

the modulation is considered to be fast frequency hopping (F-FH). In this case the carrier frequency changes a number of times during the transmission of one symbol, so that one bit is transmitted in different frequencies. If the hopping rate is (much) smaller than the symbol rate, one speaks of slow frequency hopping (S-FH). In this case multiple symbols are transmitted at the same frequency.

The occupied bandwidth of the signal on one of the hopping frequencies depends not only on the bandwidth of the information signal but also on the shape of the hopping signal and the hopping frequency. If the hopping frequency is much smaller than the information bandwidth (which is the case in slow frequency hopping), then the information bandwidth is the main factor that decides the occupied bandwidth. If, however, the hopping frequency is much greater than the information bandwidth, the pulse shape of the hopping signal will decide the occupied bandwidth at one hopping frequency. If this pulse shape is very abrupt (resulting in very abrupt frequency changes), the frequency band will be very broad, limiting the number of hop frequencies. If we make sure that the frequency changes are smooth, the frequency band at each hopping frequency will be about $1/T_h$ times the frequency bandwidth, where T_h

is equal to the hopping frequency. We can make the frequency changes smooth by decreasing the transmitted power before a frequency hop and increasing it again when the hopping frequency has changed.

As has been done for DS-CDMA, we discuss the properties of FH-CDMA with respect to multiple access capability, multipath interference rejection, narrowband interference rejection, and probability of interception.

- *Multiple access*: It is easy to visualize how the F-FH and S-FH CDMA obtain their multiple access capability. In the F-FH, one symbol is transmitted in different frequency bands. If the desired user is the only one to transmit in most of the frequency bands, the received power of the desired signal will be much higher than the interfering power and the signal will be received correctly.

  In the S-FH, multiple symbols are transmitted at one frequency. If the probability of other users transmitting in the same frequency band is low enough, the desired user will be received correctly most of the time. For those times that interfering users transmit in the same frequency band, error-correcting codes are used to recover the data transmitted during that period.
- *Multipath interference*: In the F-FH CDMA the carrier frequency changes a number of times during the transmission of one symbol. Thus, a particular signal frequency will be modulated and transmitted on a number of carrier frequencies. The multipath effect is different at the different carrier frequencies. As a result, signal frequencies that are amplified at one carrier frequency will be attenuated at another carrier frequency and vice versa. At the receiver the responses at the different hopping frequencies are averaged, thus reducing the multipath interference. Since usually noncoherent combining is used, this is not as effective as the multipath interference rejection in a DS-CDMA system, but it still gives quite an improvement.
- *Narrowband interference*: Suppose a narrowband signal is interfering on one of the hopping frequencies. If there are G_p hopping frequencies (where G_p is the processing gain), the desired user will (on the average) use the hopping frequency where the interferer is located $1/G_p$ percent of the time. The interference is therefore reduced by a factor G_p.
- *LPI*: The difficulty in intercepting an FH signal lies not in its low transmission power. During a transmission, it uses as much power per hertz as a continuous transmission. But the frequency at which the signal is going to be transmitted is unknown, and the duration of the transmission at a particular frequency is quite small. Therefore, although the signal is more readily intercepted than a DS signal, it is still a difficult task to perform.

Apart from the above-mentioned properties, the FH-CDMA has a number of other specific properties that we can divide into advantageous (+) and disadvantageous (−) behavior:

+ Synchronization is much easier with FH-CDMA than with DS-CDMA. FH-CDMA synchronization has to be within a fraction of the hop time. Since spectral spreading is not obtained by using a very high hopping frequency but by using a large hop-set, the hop time will be much longer than the chip time of a DS-CDMA system. Thus, an FH-CDMA system allows a larger synchronization error.

+ The different frequency bands that an FH signal can occupy do not have to be contiguous because we can make the frequency synthesizer easily skip over certain parts of the spectrum. Combined with the easier synchronization, this allows much higher spread-spectrum bandwidths.

+ The probability of multiple users transmitting in the same frequency band at the same time is small. A user transmitting far from the base station will be received by it even if users close to the base station are transmitting, since those users will probably be transmitting at different frequencies. Thus, the near-far performance is much better than that of DS.

+ Because of the larger possible bandwidth a FH system can employ, it offers a higher possible reduction of narrowband interference than a DS system.

− A highly sophisticated frequency synthesizer is necessary.

− An abrupt change of the signal when changing frequency bands will lead to an increase in the frequency band occupied. To avoid this, the signal has to be turned off and on when changing frequency.

− Coherent demodulation is difficult because of the problems in maintaining phase relationships during hopping.

2.2.1.3   Time Hopping

In time hopping CDMA the data signal is transmitted in rapid bursts at time intervals determined by the code assigned to the user. The time axis is divided into frames, and each frame is divided into M time slots. During each frame the user will transmit in one of the M time slots. Which of the M time slots is transmitted depends on the code signal assigned to the user. Since a user transmits all of its data in one, instead of M time slots, the frequency it needs for its transmission has increased by a factor M. A block diagram of a TH-CDMA system is given in Figure 2.10.

Figure 2.11 shows the time-frequency plot of the TH-CDMA systems. Comparing Figure 2.11 with Figure 2.8, we see that the TH-CDMA uses the whole wideband spectrum for short periods instead of parts of the spectrum all of the time.

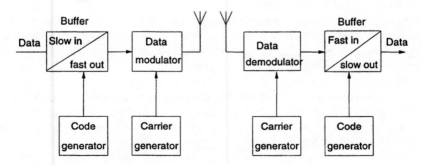

Figure 2.10 Block diagram of a TH-CDMA transmitter and receiver.

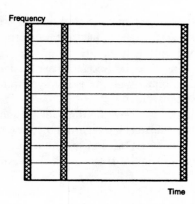

Figure 2.11 Time-frequency plot of the TH-CDMA.

Following the same procedure as for the previous CDMA schemes, we discuss the properties of TH-CDMA with respect to multiple access capability, multipath interference rejection, narrowband interference rejection, and probability of interception.

- *Multiple access*: The multiple access capability of TH-SS signals is acquired in the same manner as that of the FH-SS signals; namely, by making the probability of users' transmissions in the same frequency band at the same time small. In the case of time hopping, all transmissions are in the same frequency band, so the probability of more than one transmission at the same time must be small. This is again achieved by assigning different codes to different users. If multiple transmissions do occur, error-correcting codes ensure that the desired signal can still be recovered.

    If there is synchronization among the users, and the assigned codes are such that no more than one user transmits at a particular slot, then the TH-CDMA reduces to a TDMA scheme where the slot in which a user transmits is not fixed but changes from frame to frame.
- *Multipath interference*: In the time hopping CDMA, a signal is transmitted in reduced time. The signaling rate, therefore, increases, and dispersion of the signal will now lead to overlap of adjacent bits. Therefore, no advantage is to be gained with respect to multipath interference rejection.
- *Narrowband interference*: A TH-CDMA signal is transmitted in reduced time. This reduction is equal to $1/G_p$, where G_p is the processing gain. At the receiver we will only receive an interfering signal during the reception of the desired signal. Thus, we only receive the interfering signal $1/G_p$ percent of the time, reducing the interfering power by a factor G_p.
- *LPI*: With TH-CDMA, the frequency at which a user transmits is constant but the times at which a user transmits are unknown, and the durations of the transmissions are very short. Particularly when multiple users are transmitting, this makes it

difficult for an intercepting receiver to distinguish the beginning and end of a transmission and to decide which transmissions belong to which user.

Apart from the above-mentioned properties, the TH-CDMA has a number of other specific properties that we can divide into advantageous (+) and disadvantageous (−) behavior:

+ Implementation is simpler than that of FH-CDMA.
+ It is a very useful method when the transmitter is average-power limited but not peak-power limited since the data are transmitted is short bursts at high power.
+ As with the FH-CDMA, the near-far problem is much less of a problem since most of the time a terminal far from the base station transmits alone, and is not hindered by transmissions from stations close by.
− It takes a long time before the code is synchronized, and the time in which the receiver has to perform the synchronization is short.
− If multiple transmissions occur, a large number of data bits are lost, so a good error-correcting code and data interleaving are necessary.

2.2.1.4   Hybrid Systems

The hybrid CDMA systems include all CDMA systems that employ a combination of two or more of the above-mentioned spread-spectrum modulation techniques or a combination of CDMA with some other multiple access technique. By combining the basic spread-spectrum modulation techniques, we have four possible hybrid systems: DS/FH, DS/TH, FH/TH, and DS/FH/TH; and by combining CDMA with TDMA or multicarrier modulation we get two more: CDMA/TDMA and MC-CDMA.

The idea of the hybrid system is to combine the specific advantages of each of the modulation techniques. If we take, for example, the combined DS/FH system, we have the advantage of the anti-multipath property of the DS system combined with the favorable near-far operation of the FH system. Of course, the disadvantage lies in the increased complexity of the transmitter and receiver. For illustration purposes, we give a block diagram of a combined DS/FH CDMA transmitter in Figure 2.12.

The data signal is first spread using a DS code signal. The spread signal is then modulated on a carrier whose frequency hops according to another code sequence. A code clock ensures a fixed relation between the two codes.

2.3   BASIC DS-CDMA ELEMENTS

In this section, we review the fundamental elements for understanding direct sequence CDMA and its application into third generation systems, namely, RAKE receiver, power control, soft handover, interfrequency handover, and multiuser detection.

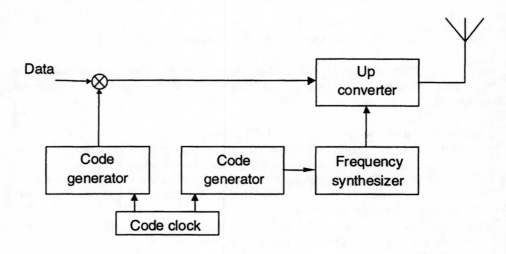

Figure 2.12 Hybrid DS-FH transmitter.

2.3.1 RAKE Receiver

A DS spread-spectrum signal waveform is well matched to the multipath channel. In a multipath channel, the original transmitted signal reflects from obstacles such as buildings, and mountains, and the receiver receives several copies of the signal with different delays. If the signals arrive more than one chip apart from each other, the receiver can resolve them. Actually, from each multipath signal's point of view, other multipath signals can be regarded as interference and they are suppressed by the processing gain. However, a further benefit is obtained if the resolved multipath signals are combined using *RAKE receiver*. Thus, the signal waveform of CDMA signals facilitates utilization of multipath diversity. Expressing the same phenomenon in the frequency domain means that the bandwidth of the transmitted signal is larger than the coherence bandwidth of the channel and the channel is frequency selective (i.e., only part of the signal is affected by the fading).

RAKE receiver consists of correlators, each receiving a multipath signal. After despreading by correlators, the signals are combined using, for example, maximal ratio combining. Since the received multipath signals are fading independently, diversity order and thus performance are improved. Figure 2.13 illustrates the principle of RAKE receiver. After spreading and modulation, the signal is transmitted and it passes through a multipath channel, which can be modeled by a tapped delay line (i.e., the reflected signals are delayed and attenuated in the channel). In Figure 2.13 we have three multipath components with different delays (τ_1, τ_2, and τ_3) and attenuation factors (a_1, a_2, and a_3), each corresponding to a different propagation path. The RAKE receiver has a receiver finger for each multipath component. In each finger, the received signal is correlated by a spreading code, which is time-aligned with the delay of the multipath signal. After despreading, the signals are weighted and combined. In Figure 2.13, maximal ratio combining is used, that is, each signal is weighted by the path gain (attenuation factor). Due to the mobile movement, the scattering environment will change, and thus, the delays and attenuation factors will change as well. Therefore, it is necessary to measure the tapped delay line profile and to reallocate RAKE fingers whenever the delays have changed by a significant amount. Small-scale changes, less than one chip, are taken care of by a code tracking loop, which tracks the time delay of each multipath signal.

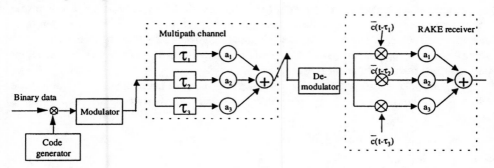

Figure 2.13 Principle of RAKE receiver.

2.3.2   Power Control

In the uplink of a DS-CDMA system, the requirement for power control is the most serious negative point. The power control problem arises because of the multiple access interference. All users in a DS-CDMA system transmit the messages by using the same bandwidth at the same time and therefore users interfere with one another. Due to the propagation mechanism, the signal received by the base station from a user terminal close to the base station will be stronger than the signal received from another terminal located at the cell boundary. Hence, the distant users will be dominated by the close user. This is called the *near-far effect*. To achieve a considerable capacity, all signals, irrespective of distance, should arrive at the base station with the same mean power. A solution to this problem is power control, which attempts to achieve a constant received mean power for each user. Therefore, the performance of the transmitter power control (TPC) is one of the several dependent factors when deciding on the capacity of a DS-CDMA system.

In contrast to the uplink, in the downlink all signals propagate through the same channel and thus are received by a mobile station with equal power. Therefore, no power control is required to eliminate near-far problem. The power control is, however, required to minimize the interference to other cells and to compensate against the interference from other cells. The worst-case situation for a mobile station occurs when the mobile station is at the cell edge, equidistant from three base stations. However, the interference from other cells does not vary very abruptly.

In addition to being useful against interfering users, power control improves the performance of DS-CDMA against fading channel by compensating the fading dips. If it followed the channel fading perfectly, power control would turn a fading channel into AWGN channel by eliminating the fading dips completely.

There exist two types of power control principles: open loop and closed loop. The open loop power control measures the interference conditions from the channel and adjusts the transmission power accordingly to meet the desired frame error rate (FER) target. However, since the fast fading does not correlate between uplink and downlink, open loop power control will achieve the right power target only on average. Therefore, closed loop power control is required. The closed loop power control measures the signal-to-interference ratio (SIR) and sends commands to the transmitter on the other end to adjust the transmission power.

Section 2.4.3 illustrates the different power control principles by explaining the IS-95 power control scheme. Power control is discussed in more detail in Chapters 5 and 7.

2.3.3 Soft Handover

In soft handover a mobile station is connected to more than one base station simultaneously. Soft handover is used in CDMA to reduce the interference into other cells and to improve performance through macro diversity. Softer handover is a soft handover between two sectors of a cell.

Neighboring cells of a cellular system using either FDMA or TDMA do not use the frequencies used by the given cell (i.e., there is spatial separation between cells using the same frequencies). This is called the frequency reuse concept. Because of the processing gain, such spatial separation is not needed in CDMA, and a frequency reuse factor of one can be used. Usually, a mobile station performs a handover when the signal strength of a neighboring cell exceeds the signal strength of the current cell with a given threshold. This is called *hard handover*. Since in a CDMA system the neighboring cell frequencies are the same as in the given cell, this type of approach would cause excessive interference into the neighboring cells and thus a capacity degradation. In order to avoid this interference, an instantaneous handover from the current cell to the new cell would be required when the signal strength of the new cell exceeds the signal strength of the current cell. This is not, however, feasible in practice. The handover mechanism should always allow the mobile station to connect into a cell, which it receives with the highest power (i.e., with the lowest pathloss). Since in soft handover the mobile station is connected to either two or more base stations, its transmission power can be controlled according to the cell, which the mobile station receives with the highest signal strength. A mobile station enters the soft handover state when the signal strength of a neighboring cell exceeds a certain threshold but is still below the current base station's signal strength.

Fortunately, the signal structure of CDMA is well suited for the implementation of soft handover. This is because in the uplink, two or more base stations can receive the same signal because of the reuse factor of one; and in the downlink the mobile station can coherently combine the signals from different base stations since it sees them as just additional multipath components. This provides an additional benefit called macro diversity (i.e., the diversity gain provided by the reception of one or more additional signals). A separate pilot channel is usually used for the signal strength measurements for handover purposes (see Section 2.4.1).

In the downlink, however, soft handover creates more interference to the system since the new base station now transmits an additional signal for the mobile station. It is possible that the mobile station cannot collect all of the energy that the base station transmits due to a limited number of RAKE fingers. Thus, the gain of soft handover in the downlink depends on the gain of macro diversity and the loss of performance due to increased interference.

Figure 2.14 illustrates the soft handover principle with two base stations involved. In the uplink the mobile station signal is received by the two base stations, which, after demodulation and combining, pass the signal forward to the combining point, typically to the base station controller (BSC). In the downlink the same information is transmitted via both base stations, and the mobile station receives the information from two base stations as separate multipath signals and can therefore combine them.

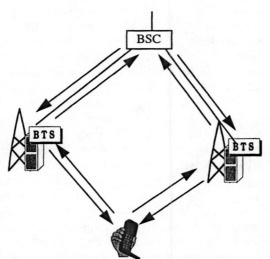

Figure 2.14   Principle of soft handover with two base station transceivers (BTS).

2.3.4   Interfrequency Handover

The third generation CDMA networks will have multiple frequency carriers in each cell, and a hot-spot cell could have a larger number of frequencies than neigboring cells. Furthermore, in hierarchical cell structures, micro cells will have a different frequency than the macro cell overlaying the micro cells. Therefore, an efficient procedure is needed for a handover between different frequencies. A blind handover used by second generation CDMA does not result in an adequate call quality. Instead, the mobile station has to be able to measure the signal strength and quality of an another carrier frequency, while still maintaining the connection in the current carrier frequency. Since a CDMA transmission is continuous, there are no idle slots for the interfrequency measurements as in the TDMA-based systems. Therefore, compressed mode and dual receiver have been proposed as a solution to interfrequency handover [12] (see Section 6.3.7). In the compressed mode, measurements slots are created by transmitting the data of a frame, for example, with a lower spreading ratio during a shorter period, and the rest of the frame is utilized for the measurements on other carriers. The dual receiver can measure other frequencies without affecting the reception of the current frequency.

2.3.5   Multiuser Detection

The current CDMA receivers are based on the RAKE receiver principle, which considers other users' signals as interference. However, in an optimum receiver all signals would be detected jointly or interference from other signals would be removed by subtracting them from the desired signal. This is possible because the correlation properties between signals are known (i.e., the interference is deterministic not random).

The capacity of a direct sequence CDMA system using RAKE receiver is interference limited. In practice this means that when a new user, or interferer, enters the network, other users' service quality will degrade. The more the network can resist interference the more users can be served. Multiple access interference that disturbs a base or mobile station is a sum of both intra- and intercell interference.

Multiuser detection (MUD), also called joint detection and interference cancellation (IC), provides means of reducing the effect of multiple access interference, and hence increases the system capacity. In the first place, MUD is considered to cancel only the intra-cell interference, meaning that in a practical system the capacity will be limited by the efficiency of the algorithm and the intercell interference.

In addition to capacity improvement, MUD alleviates the near/far problem typical to DS-CDMA systems. A mobile station close to a base station may block the whole cell traffic by using too high a transmission power. If this user is detected first and subtracted from the input signal, the other users do not see the interference.

Since optimal multiuser detection is very complex and in practice impossible to implement for any reasonable number of users, a number of suboptimum multiuser and interference cancellation receivers have been developed. The suboptimum receivers can be divided into two main categories: linear detectors and interference cancellation. Linear detectors apply a linear transform into the outputs of the matched filters that are trying to remove the multiple access interference (i.e., the interference due to correlations between user codes). Examples of linear detectors are decorrelator and linear minimum mean square error (LMMSE) detectors. In interference cancellation, multiple access interference is first estimated and then subtracted from the received signal. Parallel interference cancellation (PIC) and successive (serial) interference cancellation (SIC) are examples of interference cancellation.

For a more detailed treatment of multiuser detection and interference cancellation, refer to Chapters 5 and 10 as well as [13–16].

2.4   IS-95 CDMA

In this section, we describe the features of the IS-95 air interface according to the new IS-95B standard, with a focus on the new downlink and uplink channel structure [17].[1] Main air interface parameters, downlink and uplink channel structures, power control principles, and speech coding are discussed. For a more detailed treatment of the IS-95A standard, refer to [18], and for a theoretical analysis of IS-95 air interface solutions, refer to [8].

The IS-95 air interface standard, after the first revision in 1995, was termed IS-95A [19]; it specifies the air interface for cellular, 800-MHz frequency band. ANSI J-STD-008 specifies the PCS version (i.e., the air interface for 1900 MHz). It differs from IS-95A primarily in the frequency plan and in call processing related to subscriber station identity, such as paging and call origination. TSB74 specifies the Rate Set 2 (14.4 Kbps) standard. IS-95B merges the IS-95A, ANSI J-STD-008 [20] and TSB74 standards, and, in addition, it specifies the high speed data operation using up to eight parallel codes, resulting in a maximum bit rate of 115.2 Kbps. In addition to these air interface specifications, the IS-97 [21] and IS-98 [22] standards specify the minimum performance specifications for the mobile and base station, respectively.

Table 2.1 lists the main parameters of the IS-95 air interface. Carrier spacing of the system is 1.25 MHz. Practical deployment has shown that 3 CDMA carriers can be fitted into 5 MHz bandwidth due to required guard bands. Network is synchronous within few microseconds. This facilitates use of the same long code sequence with different phase offsets as pilot sequences. However, an external reference signal such as Global Positioning System (GPS) is needed.

2.4.1   Downlink Channel Structure

Figure 2.15 shows the downlink physical channel structure. The pilot channel, the paging channel, and the synchronization channel[2] are common control channels and traffic channels are dedicated channels. A common channel is a shared channel, and a dedicated channel is solely allocated for the use of a single user. Data to be transmitted on synchronization, paging, and traffic channels are first grouped into 20-ms frames, convolutionally encoded, repeated to adjust the data rate, and interleaved. Then the signal is spread with an orthogonal Walsh code at a rate of 1.2288 Mcps, split into the I and Q channels, and, prior to baseband filtering, spread with long PN sequences at a rate of 1.2288 Mcps.

---

[1] In this book, the terms uplink and downlink are used instead of the forward and reverse link, which are used in the IS-95 standard.

[2] In the IS-95 standard the synchronization channel is actually termed "Sync Channel."

Table 2.1
IS-95 Air Interface Parameters

Bandwidth	1.25 MHz
Chip Rate	1.2288 Mcps
Frequency band uplink	869 - 894 MHz 1930 - 1980 MHz
Frequency band downlink	824 - 849 MHz 1850 - 1910 MHz
Frame length	20 ms
Bit rates	Rate Set 1: 9.6 Kbps Rate Set 2: 14.4 Kbps IS-95B: 115.2 Kbps
Speech codec	QCELP 8 Kbps EVRC 8 Kbps ACELP 13 Kbps
Soft handover	Yes
Power control	Uplink: Open loop + fast closed loop Downlink: Slow quality loop
Number of RAKE fingers	4
Spreading codes	Walsh+Long M-sequence

A mobile station uses the pilot channel for coherent demodulation, acquisition, time delay tracking, power control measurements, and as an aid for the handover. In order to obtain a reliable phase reference for coherent demodulation, the pilot channel is transmitted with higher power than the traffic channels. Typically about 20% of the radiated power on the downlink is dedicated to the pilot signal. After obtaining phase and code synchronization, the mobile station acquires synchronization information (data rate of the paging channel, time of the base station's pilot PN sequence with respect to the system time) from the synchronization channel. Since the synchronization channel frame has the same length as the pilot sequence, acquisition of the synchronization channel takes place easily. The synchronization channel operates at a fixed rate of 1.2 Kbps. The paging channel is used to page a mobile station. The paging channel has a fixed data rate of 9.6 or 4.8 Kbps.

Each downlink traffic channel contains one fundamental code channel and may contain one to seven supplemental code channels. The traffic channel has two different rate sets. The rate set 1 supports data rates of 9.6, 4.8, 2.4, and 1.2 Kbps (Figure 2.15(a)) and the rate set 2 supports 14.4, 7.2, 3.6, and 1.8 Kbps (Figure 2.15(b)). Only the full rate (9.6 or 14.4 Kbps) may be utilized on the supplemental code channels. The mobile station always supports the rate set 1 and it may support the rate set 2. To achieve equal power levels at the base station receiver, the base station measures the received signal and adjusts each mobile station's power levels accordingly. The 20-ms frame is divided into 16 power control groups with a duration of 1.25 ms. One power control bit is multiplexed in for the fundamental code channel for each power control group.

The transmitted data is encoded by a convolutional code with a constraint length of 9. The generator functions for this code are 753 (octal) and 561 (octal). For the synchronization channel, the paging channels, and rate set 1 on the traffic channel, a convolutional code with a rate of 1/2 is used. For the rate set 2, an effective code rate of 3/4 is achieved by puncturing two out of every six symbols after the symbol repetition.

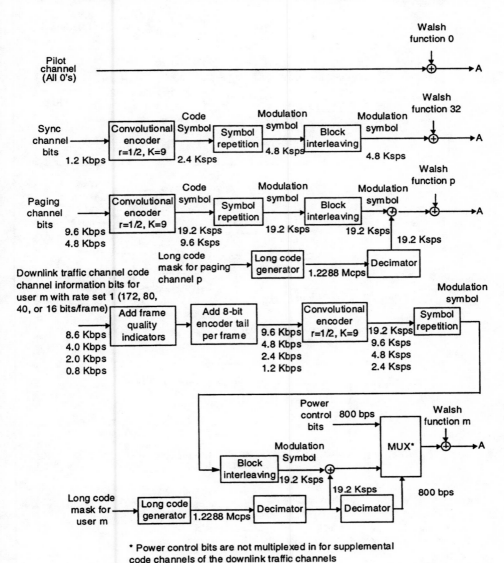

Figure 2.15(a)   Rate set 1 downlink CDMA channel structure. (*Source*: [17], reproduced under written permission of the copyright holder (Telecommunications Industry Association). At the time of the publication, the standard which contains this figure was not finalized, please check with TIA for the correct version.)

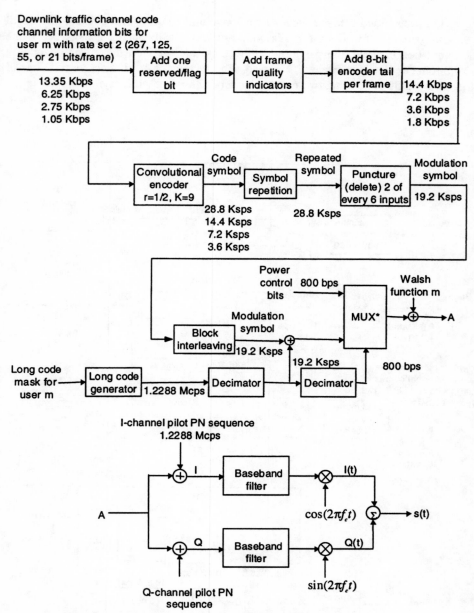

Figure 2.15(b) Rate set 2 downlink CDMA channel structure. (*Source*: [17], reproduced under written permission of the copyright holder (Telecommunications Industry Association). At the time of the publication, the standard which contains this figure was not finalized, please check with TIA for the correct version.)

Since the data rate on different channels varies, symbol repetition is used to achieve a fixed data rate prior to interleaving. For the synchronization channel, each convolutionally encoded symbol shall be repeated once (i.e., each symbol occurs two consecutive times). For the paging channel, each code symbol at the 4800-bps rate shall be repeated once. The code symbol repetition rate on the forward traffic channels varies with data rate. Code symbols are not repeated for the 14.4- and 9.6-Kbps data rates. Each code symbol at the 7.2- and 4.8-Kbps data rates is repeated once, at the 3.6 and 2.4 Kbps data rates three times, and at the 1.8- and 1.2-Kbps data rates seven times.

In the downlink, three types of spreading codes are used. Walsh codes of length 64 at a fixed chip rate of 1.2288 Mcps separate the physical channels. The Walsh function consisting of all zeros (W0, Walsh code number 0) is used for the pilot channel, W1-W7 are used for paging channels (unused paging channel codes can be used for traffic channels). The synchronization channel is W32, and traffic channels are W8 to W31 and W33 to W63. A pair of long M-sequences of length 16,767 ($2^{15}-1$) is used for quadrature spreading, one for the I channel and one for the Q channel. Quaternary spreading is used to obtain better interference averaging. Since the pilot channel Walsh function is all zeros, this pair of M-sequences also forms the pilot code. Different cells and sectors are distinguished with the different phase offsets of this code.

A long pseudo random sequence with a period of $2^{42}-1$ is used for base band data scrambling (i.e., to encrypt the signal on the paging and traffic channels). It is decimated from a 1.2288-Mcps rate down to 19.2 Kbps. The long pseudo noise sequence is the same used in the uplink for user separation, and it is generated by a modulo-2 inner product of a 42-bit mask and the 42-bit state vector of the sequence generator.

2.4.2   Uplink Channel Structure

As depicted in Figures 2.16 to 2.19, the uplink has two physical channels: a traffic channel, which is a dedicated channel, and a common access channel. A traffic channel consists of a single fundamental code channel and zero through seven supplemental code channels. Similar to the downlink, traffic channels always support the rate set 1 data rates and may support the rate set 2 data rates. The supplemental code channel can only use the full rates (9.6 or 14.4 Kbps). Data transmitted on the uplink channels are grouped into 20-ms frames, convolutionally encoded, block interleaved, and modulated by 64-ary orthogonal modulation. Then, prior to baseband filtering, the signal is spread with a long PN sequence at a rate of 1.2288 Mcps, split into the I and Q channels, and spread with in-phase and quadrature spreading sequences.

The access channel is used by a mobile station to initiate a call, to respond to a paging channel message from the base station, and for a location update. Each access channel is associated with a downlink paging channel, and consequently there can be up to seven access channels. The access channel supports fixed data rate operation at 4.8 Kbps.

The transmitted information is encoded using a convolutional code with constraint length 9 and the same generator polynomials as in the downlink. For the access channel and rate set 1 on the traffic channels, the convolutional code rate is 1/3. For rate set 2 on the traffic channels, a code rate of 1/2 is used. Similar to the downlink, code symbols output from the convolutional encoder are repeated before being interleaved when the data rate is lower than 9.6 Kbps for rate set 1 and 14.4 Kbps for rate set 2. However, the repeated symbols are not actually transmitted. They are masked out according to a masking pattern generated by the data burst randomizer to save transmission power. For the access channel, which has a fixed data rate of 4.8 Kbps, each code symbol is repeated once. In contrast to the traffic channel, the repeated code symbols are transmitted.

The coded symbols are grouped into 6-symbol groups. These groups are then used to select one of 64 possible Walsh symbols (i.e., a 64-ary orthogonal modulation is carried out to obtain good performance for noncoherent modulation). After the orthogonal modulation, the transmission rate is 307.2 Kbps. The reason to use the noncoherent modulation is the difficulty in obtaining good phase reference for coherent

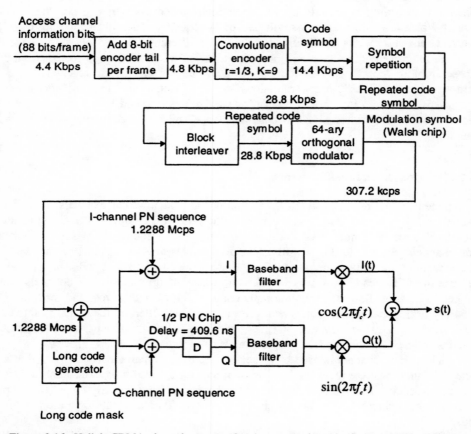

Figure 2.16 Uplink CDMA channel structure for the access channel. (*Source*: [17], reproduced under written permission of the copyright holder (Telecommunications Industry Association). At the time of the publication, the standard which contains this figure was not finalized, please check with TIA for the correct version.)

demodulation in the uplink. It should be noted how the Walsh codes are used differently in the uplink and downlink. In the downlink, they were used for channelization, while in the uplink they are used for orthogonal modulation.

Each code channel in a traffic channel and each access channel are identified by a different phase of a pseudo-random M-sequence with a length of 2^{42} (see 5.6.2). The in-phase and quadrature spreading is performed by the same pair of M-sequences (length 2^{15}) as in the downlink (now augmented by one chip).

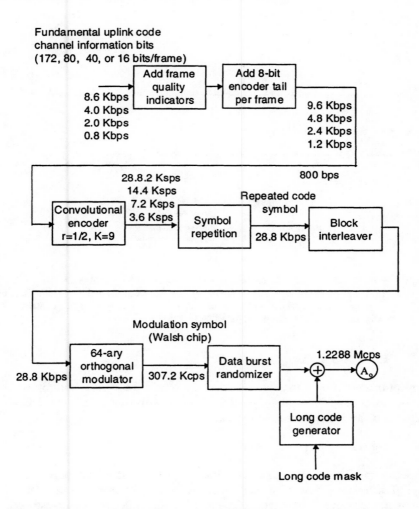

Figure 2.17 Uplink CDMA channel structure for fundamental code channels with rate set 1. (*Source*: [17], reproduced under written permission of the copyright holder (Telecommunications Industry Association). At the time of the publication, the standard which contains this figure was not finalized, please check with TIA for the correct version.)

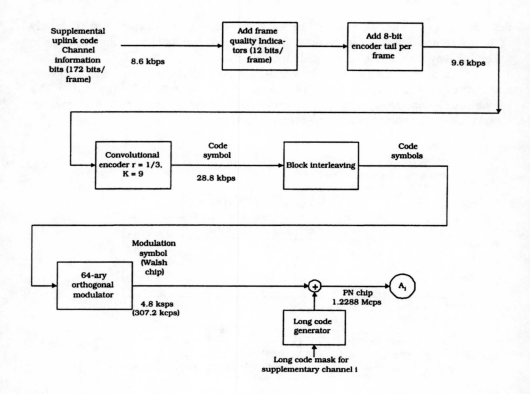

Figure 2.18 Uplink CDMA channel structure for supplemental code channels with rate set 1. (*Source*: [17], reproduced under written permission of the copyright holder (Telecommunications Industry Association). At the time of the publication, the standard which contains this figure was not finalized, please check with TIA for the correct version.)

2.4.3   Power Control

IS-95 has three different power control mechanisms. In the uplink, both open loop and fast closed loop power control are employed. In the downlink, a relatively slow power control loop controls the transmission power.

2.4.3.1   Open Loop Power Control

The open loop power control has two main functions: it adjusts the initial access channel transmission power of the mobile station and compensates large abrupt variations in the pathloss attenuation. The mobile station determines an estimate of the pathloss between the base station and mobile station by measuring the received signal strength at the mobile using an automatic gain control (AGC) circuitry, which gives a rough estimate of the propagation loss for each user. The smaller the received power, the larger the propagation loss, and vice-versa. The transmit power of the mobile station is determined from the equation:

mean output power (dBm) = −mean input power (dBm) + offset power + parameters   (2.2)

The offset power for the 800-MHz band mobiles (band class 0) is −73 dB and for the 1900-MHz band mobiles (band class 1) −76 dB [17]. The parameters are used to adjust the open-loop power control for different cell sizes and different cell effective radiated powers (ERP) and receiver sensitivities [23]. These parameters are initially transmitted on the synchronization channel.

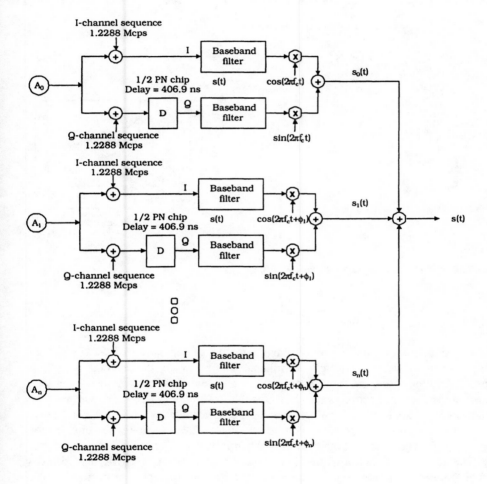

Figure 2.19 Uplink traffic channel structure including fundamental code channel and multiple supplemental code channels with rate set 1. (*Source*: [17], reproduced under written permission of the copyright holder (Telecommunications Industry Association). At the time of the publication, the standard which contains this figure was not finalized, please check with TIA for the correct version.)

The open loop power control principle is described in Figure 2.20. Since the distance (d1) of mobile station 1 to the base station (BTS) is shorter than the distance of mobile station 2 (d2) to the BTS, the signal received by the mobile station 1 has a smaller propagation loss. Assume that the mean input power of the mobile station 1 is −70 dBm (100 pW)[3] and the mean input power of the mobile station 2 is −90 dBm (1 pW). For band class 0 mobiles with no correction parameters, the mobile station transmission power to achieve equal received powers at the base station can be calculated from (2.2) to be 17 dBm (50 mW) and −7 dBm (200 μW), respectively.

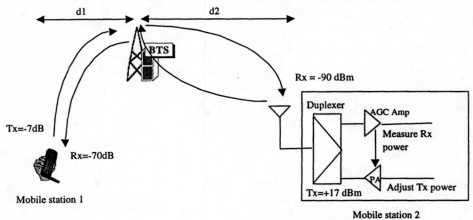

Figure 2.20 Uplink open loop power control principle.

2.4.3.2 Closed Loop Power Control

Since the IS-95 uplink and downlink have a frequency separation of 20 MHz, their fading processes are not strongly correlated. Even though the average power is approximately the same, the short term power is different, and therefore, the open loop power control cannot compensate for the uplink fading. To account for the independence of the Rayleigh fading in the uplink and downlink, the base station also controls the mobile station transmission power. Figure 2.21 illustrates the closed loop power control. The base station measures the received SIR[4] over a 1.25-ms period, equivalent to six modulation symbols, compares that to the target SIR, and decides whether the mobile station transmission power needs to be increased or decreased. The power control bits are transmitted on the downlink fundamental code channel every

---

[3] 1dBm means 1dB over 1 mW. For example, −70 dBm is 70 dB (10 million times) less than 1 mW (i.e., 1E-12 W = 1 picoWatt).

[4] The IS-95 standard suggests that the received signal strength should be measured. However, in practice usually the SIR or the received bit energy to noise density (E_b/I_0) are used, since they have direct impact on the bit error rate (BER).

1.25 ms (i.e., with a transmission rate of 800 Hz) by puncturing the data symbols. The placement of a power control bit is randomized within the 1.25-ms power control group. The transmission occurs in the second power control group following the corresponding uplink traffic channel power control group in which the SIR was estimated.[5]

Since the power control commands are transmitted uncoded, their error ratio is fairly high, on the order of 5%. However, since the loop is of delta modulation type (i.e., power is adjusted continuously up or down) this is tolerable. The mobile station extracts the power control bits commands and adjusts its transmission power accordingly. The adjustment step is a system parameter and can be 0.25, 0.5, or 1.0 dB. The dynamic range for the closed loop power control is ±24 dB. The composite dynamic range for open and closed loop power control is ±32 dB for mobile stations operating in band class 0, and ±40 dB for mobile stations operating in band class 1 [17]. The typical standard deviation of the power control error due to the closed loop is on the order of 1.1 to 1.5 dB [8]

The SIR required to produce a certain bit error rate varies according to radio environment and depends on the amount and type of multipath. Therefore, IS-95 employs an outer loop that adjusts the target SIR. The base station measures the signal quality (bit error rate), and based on that determines the target SIR. However, this outer loop will increase the power control error, resulting in a total standard deviation of 1.5 to 2.5 dB [8].

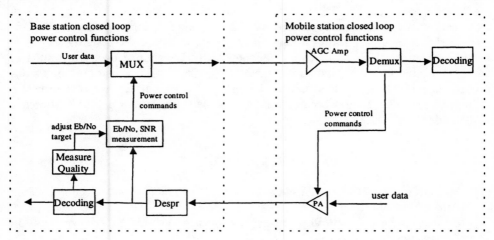

Figure 2.21   Closed loop power control principle.

---

[5] For instance, if the signal is received on the reverse traffic channel in power control group number 5, and the corresponding power control bit is transmitted on the forward traffic channel during power control group number 5 + 2 = 7 [17].

2.4.3.3   Downlink Slow Power Control

The base station controls its transmission power to a given mobile station according to the pathloss and interference situation. The main purpose of the slow downlink power control is to improve the performance of mobile stations at a cell edge where the signal is weak and the interfering base station signals are strong. The downlink power control mechanism is as follows. The base station periodically reduces the transmitted power to the mobile station. The mobile station measures the frame error ratio (FER). When the FER exceeds a predefined limit, typically 1%, the mobile station requests additional power from the base station. This adjustment occurs every 15 to 20 ms. The dynamic range of the downlink power control is only ±6 dB.

   Both periodic and threshold reporting may be enabled simultaneously, either one of them may be enabled, or both forms of reporting may be disabled at any given time.

2.4.4   Speech Codecs and Discontinuous Transmission

IS-95 has three speech codecs, 8-Kbps QCELP, 8-Kbps EVRC, and 13 Kbps. The higher rate codec was developed to provide better voice quality, but due to its higher bit rate it reduces the system capacity. Therefore, the enhanced variable rate codec (EVRC) operating at 8 Kbps was developed. Speech codecs are four-rate code excited linear prediction codecs (CELP). The vocoder rates of the 8-Kbps codecs are 1, 2, 4, and 8 Kbps corresponding to channel rates of 1.2, 2.4, 4.8, and 9.6 Kbps (rate set 1). The 13-Kbps codec uses a 14.4 Kbps channel rate (rate set 2). Since the system capacity is directly proportional to interference, reduction of the transmitted data rate results in better capacity. In IS-95, data rate reduction is implemented with discontinuous transmission (DTX), which is realized by gating the transmitter in pseudo-random fashion on and off. The drawback of this approach is that it creates pulsed interference in the uplink.

REFERENCES

[1] Prasad, R., *CDMA for Wireless Personal Communications*, Boston-London: Artech House, 1996.

[2] Special issue on spread-spectrum communication, *IEEE Transactions on Communications*, Vol. COM-30, May 1982.

[3] Simon, M. K., J. K. Omura, R. A. Scholtz, and B. K. Levitt, *Spread-Spectrum Communications*, Vol. I, II, III, Comp. Sci., 1985.

[4] Scholtz, R. A., "The spread-spectrum concept," *IEEE Transactions on Communications*, Vol. COM-25, August 1977, pp. 74–55.

[5] Torrieri, D. J., *Principles of Secure Communication Systems*, Norwood, Mass.: Artech House, 1985.

[6] Cooper, G. R., and C. D. McGillem, *Modern Communication and Spread-Spectrum*, New York: McGraw-Hill Book Company, 1986.

[7] Glisic, S. G., and P. A. Leppanen (eds.), *Code Divison Multiple Access*

Communications, Boston: Kluwer Academic Publishers, 1995.

[8]   Viterbi, A. J., *CDMA Principles of Spread-Spectrum Communications*, Reading, Mass.: Addison-Wesley Publishing Company, 1995.

[9]   Dixon, R. C., *Spread-Spectrum Systems*, New York: John Wiley & Sons, 1984.

[10] Glisic, S., and B. Vucetic, *Spread Spectrum CDMA Systems for Wireless Communications*, Boston: Artech House, 1997.

[11] Glisic, S., and P. Leppänen, *Wireless Communications: TDMA versus CDMA*, Boston: Kluwer Academic Publishers, 1997.

[12] Gustafsson, M., K. Jamal, and E. Dahlman, "Compressed Mode Techniques for Interfrequency Measurements in a Wide-band DS-CDMA System," *Proceedings of PIMRC'97*, Helsinki, Finland, September 1997, pp. 23–35.

[13] Verdú, S., "Adaptive Multiuser Detection," in *Code Division Multiple Access Communications*, S. G. Glisic and P. A. Leppänen (eds.), Boston: Kluwer Academic Publishers, 1994, pp. 97–116.

[14] Moshavi, S., "Multiuser Detection for DS-CDMA Communications," *IEEE Communications Magazine*, October 1996, pp. 12–36.

[15] Duel-Hallen, A., J. Holtzman, and Z. Zvonar, "Multiuser Detection for CDMA Systems," *IEEE Personal Commun.*, April 1995, pp. 4–8.

[16] Juntti, M., and S. Glisic, "Advanced CDMA for Wireless Communications," in *Wireless Communications TDMA versus CDMA*, S. Glisic and P. Leppanen (eds.), Boston: Kluwer Academic Publishers, 1997, pp. 447–490.

[17] TIA/EIA/SP-3693, "Mobile Station-Base Station Compatibility Standard for Dual-Mode Wideband Spread Spectrum Cellular Systems," to be published as ANSI/TIA/EIA-95B.[6]

[18] Garg, V. K., K. Smolik, and J. E. Wilkes, *Applications of CDMA in Wireless/Personal Communications*, Upper Saddle River, NJ: Prentice-Hall, 1997.

[19] TIA/EIA-95A, "Mobile Station-Base Station Compatibility Standard for Dual-Mode Wideband Spread Spectrum Cellular Systems," 1995.

[20] ANSI J-STD-008, "Personal Station-Base Station Compatibility Requirements for 1.8 to 2.0 GHz Code Division Multiple Access (CDMA) Personal Communications Systems," 1995.

[21] TIA/EIA/IS-97, "Recommended Minimum Performance Standards for Base Stations Supporting Dual-Mode Wideband Spread Spectrum Cellular Mobile Stations," December 1994.

[22] TIA/EIA/IS-98, "Recommended Minimum Performance Standards for Dual-Mode Wideband Spread Spectrum Cellular Mobile Stations," December 1994.

[23] Ross, A. H. M., and K. L. Gilhousen, "CDMA Technology and the IS-95 North American Standard," in *The Mobile Communications Handbook*, J. D. Gibson (ed.), Boca Raton, Florida: CRC Press, 1996, pp. 43–48.

---

[6] To purchase complete version of any TIA document, call Global Engineering Documents at 1800-854-7179 or send a facsimile to 303-397-2740.

Chapter 3

IMT-2000 APPLICATIONS

3.1 INTRODUCTION

The applications of a new system play a major role in capturing the attention of users who are already satisfied with existing systems. Third generation applications can not only offer totally new services, but it can also improve the existing services with its new capabilities such as the provision of simultaneous services, dynamic variable rate services and smaller delay. IMT-2000 service targets are the provision of at least 144 Kbps (preferably 384 Kbps) for wide area coverage with high mobility and 2 Mbps for local coverage with low mobility.

Different applications generate different traffic characteristics and thus have a different impact on the system design and on the network capacity. Service parameters include granularity, minimum transmission delay, discontinuous transmission, packet sizes, and packet distributions. The wide variety of services expected for third generation systems will make the network planning process much more challenging compared to the second generation systems. A video or speech call requires guaranteed quality of service, while non real-time packet data can be offered on a best effort basis (i.e., when there is bandwidth available, the user receives it).

Section 3.2 describes third generation application scenarios to motivate the development of third generation wireless systems with high data rates. In Sections 3.3 and 3.4, the Internet and wireless video traffic and their characteristics are discussed, respectively. Section 3.5 discusses multimedia services, and Section 3.6 describes generic IMT-2000 bearer service requirements. Example traffic models for the analysis and simulation of a wireless system are presented in Section 3.7.

3.2 THIRD GENERATION APPLICATION SCENARIO

As an example of third generation services, we can imagine an IMT-2000 user, Mike, in an outdoors scenario. Mike would like to go to the movies and needs to find out which movies are being shown in town. Voice recognition software translates his query into a format his wireless device can understand. An intelligent search agent initiates a search on the World Wide Web for appropriate information about movies. When the information is found, it is downloaded into Mike's terminal and shown on the terminal screen. After looking at movie titles, he decides to see a preview of a film, which is downloaded to his terminal. Next, he would like to know how to get to the movie theater and downloads a digital map and a picture of the movie theater. It is clear that he could get the idea to access this kind of service anywhere and does not understand the service availability limitations imposed by the system.

In the above example, several features of IMT-2000 were used. To set up the connection, the application negotiated bearer service attributes such as bearer type, bit rate, delay, bit error rate (BER), up/down link symmetry, and error protection including no protection or unequal protection. During the session, parallel bearer services (service mix) were active (i.e., IMT-2000 is able to associate several bearers with a call, and bearers can be added and released independently during a call). Non real-time communication mode was used. Browsing WWW and downloading video or other information are examples of typical asymmetric, non real-time applications. Videotelephony would be an example of a symmetric, real-time application. IMT-2000 is also able to adapt the bearer service bit rate, quality, and radio link parameters depending on the link quality, traffic, network load, and radio conditions within the negotiated service limits.

The mobile office is an IMT-2000 application offering e-mail, WWW access, fax, and file transfer services. These services can already be provided by the second generation networks. However, the higher bit rate of IMT-2000 makes them more attractive to the end-user. A typical engineering document might exceed 1 Mbyte (8 Mbit) if it contains a large number of pictures. It would thus take approximately 4 seconds to transfer it with IMT-2000 but 80 seconds with a 100-Kbps GPRS connection. The higher user bit rate of IMT-2000 will increase the total traffic because users will now use applications that they would not have if only lower bit rates were available. A higher carrier bit rate will enable more efficient sharing of resources between the users and higher multiplexing gains.

Applications adapt their bit rates according to the information that needs to be transmitted. In the second generation systems, speech service uses variable bit rate transmission. In the third generation networks, many more services, such as video and packet data, will produce variable rate data. Consequently, the network will need the capability to take advantage of this variability to utilize the radio spectrum more efficiently.

3.3 INTERNET

The use of the Internet is increasing very rapidly. Figure 3.1 shows the increase in the number of Internet hosts. One reason for the increased Internet usage is the emergence of the World Wide Web (WWW), which can be accessed through a Web browser supported by virtually every personal computer. The WWW provides a common platform for application development, thus stimulating further increases in Internet traffic. Today, the Web provides access to services such as on-line shopping, location finding, banking, on-line brokering, and electronic newspapers. Merging the Internet with mobile radio systems will facilitate access to the Internet anywhere irrespective of the user's location. For example, a user who has lost his way could access one of the location services on the Web through his wireless terminal to get driving instructions.

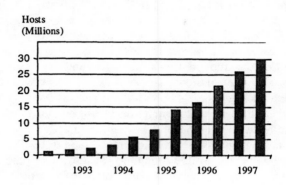

Figure 3.1   Growth of Internet hosts (*Source*: Network Wizards, http//:www.nw.com).

A typical Word Wide Web session could be like the one shown in Figure 3.2 [1]. After the session has started, a user browses Web pages by clicking one page at a time when it is downloaded to his terminal. One page (file) corresponds to a burst of packets, which can be of variable length. Note that these packets are upper-layer packets, which can be segmented for the transmission over the air interface. Between clicking the pages, the user might pause to read a page. The reading time corresponds to the longer delays between bursts of packets. Typically, the World Wide Web traffic is highly asymmetric (i.e., the user is downloading data from the network, and only some acknowledgments or commands are transmitted from the terminal to the network).

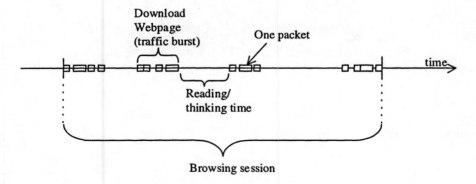

Figure 3.2   Word Wide Web traffic characteristic.

While browsing the Web, a user could download a file using the file transfer protocol (FTP). This would establish a parallel session with the Web browser but with different traffic characteristics. A file transfer can be modeled as one long burst of packets. Transferring e-mail generates a similar type of traffic.

Understanding the nature of the traffic is critical for a good system design. Between the transmission pauses, the system can either keep the radio connection, or release it and then establish it again when there is new data. This is a trade-off between spending radio resources to keep the connection alive and spending radio resources to re-establish the connection. Furthermore, the access time will be longer if the connection is released and then reconnected. The question is how the timers should be set to minimize the consumption of radio resources and the access delay for a user.

The average transmission rate depends on the packet length and the time intervals between packets (see Table 3.2). If the packets are large and the time interval between packets is short, capacity requirements will be much higher. The same applies for reading time between Web pages. The reading time depends on the contents of the page. For example, a page that needs some information to be filled in (e.g., address and credit card information for on-line shopping) requires more time than a page that just contains a simple greeting and a command to continue to the next page.

The WWW traffic and file transfer are examples of packet services that do not need a specific minimum bit rate; rather they can utilize the spare capacity of the network. This is illustrated in Figure 3.3. However, the available bit rate might have an effect on the traffic characteristics of a WWW user. In case the available bit rate is very small, the user might strip-off unnecessary features. For example, he could disable the retrieval of pictures and download only text files, thus reducing the required bit rate.

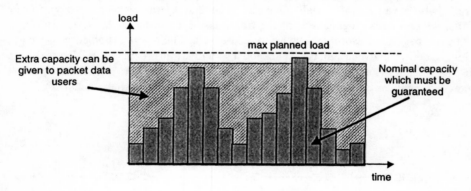

Figure 3.3   Guaranteed versus spare capacity.

3.4 WIRELESS VIDEO

Figure 3.4 illustrates a wireless video concept, which is one of the predicted IMT-2000 applications. In the beginning of this chapter, we described a user watching a preview of a movie, which is a *retrieval-type* service. Other examples of such services are video messaging (photos, drawings) and news delivery through video [2]. *Conversational-type* services (e.g., videoconferencing and videotelephony) involve interaction between two or more end-users. The retrieval-type and conversational-type video services have very different transmission requirements. In the retrieval-type service, transmission delay may be longer, on the order of 0.5 to 1 seconds, and encoding can be performed off-line, resulting in a higher compression ratio and thus a lower bit rate [2]. For the conversational services, transmission delay should not exceed 200 to 300 ms in order to avoid the annoying effect of users speaking simultaneously. The delay has a direct impact on the required signal-to-interference ratio. A longer delay allows for longer interleaving and retransmission of erroneous frames and, subsequently, smaller signal-to-interference ratio and higher overall capacity. The packet error ratio for an acceptable video quality is on the order of 10^{-3} to 10^{-4}, and the bit error ratio in the order of 10^{-6} to 10^{-7}. The conversational services will require more capacity than retrieval services because of lower delays, resulting in higher signal-to-interference ratio requirements to achieve these error rates.

Figure 3.4   IMT-2000 wireless video terminal (courtesy of Nokia).

ITU-T/SG16 (q11 and q15) and the International Organization for Standardization (ISO) are working to set standards for low bit rate video coding. ITU-T/SG15 has produced recommendation H.263, which specifies a coded presentation for compressing the moving picture component of audio-visual signals at low bit rates [3]. Currently, SG16 is developing H.263+ with more efficient compression technology and H.263L (long term standard) aimed at achieving a dramatic increase in compression efficiency. The ISO-MPEG4 (Motion Picture Experts Group) activity is aimed at a generic audio-visual coding system for multimedia communications. It will focus on providing functions for editing the content depicted in video scenes and for coding scenes containing natural video and computer graphics.

Bit rate requirements for videotelephony depend on the video codec used, picture quality required, and frame rate per second required as well as on the actual video sequence. It has been estimated that using an H.263 video codec, the minimum acceptable quality for low-motion video (e.g., a simple head-and-shoulders scene) requires a transmission bit rate of over 20 Kbps, and for generic video telephony over 40 Kbps [2]. If there is more bandwidth available, a higher quality can be achieved. The H.263 and MPEG use a hierarchical scalable video encoding mechanism where the quality of the compressed video can be traded for the compression ratio.

The instantaneous bit rate of video codecs varies greatly depending on the video sequence complexity. The peak rate requirement compared to the average bit rate can be relatively high. A buffer can be used to smooth the bit rate variability of the video encoder output. This variability of the video bit stream can be used to utilize radio resources more efficiently. The spare capacity can be utilized for services without quality of service guarantees (especially delay guarantees), as shown in Figure 3.3.

Figures 3.5 and 3.6 present examples of the bit rate variability of a low bit rate H.263 video codec. The first sequence is called Claire, which is a good quality video clip with a head and shoulders picture and little movement. From Figure 3.5 it can be seen that total bit rate varies between 10 and 17 Kbps dependent on the picture. The highest change in bit rate/second is about 5 Kbps/s (second 10).

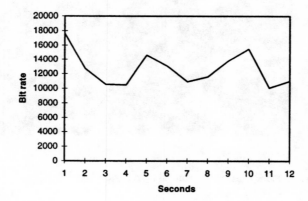

Figure 3.5   Bit rate/second requirement for Claire sequence.

The second example is called Carphone, which is a moving object with a moving background, and has a fair quality. From Figure 3.6 it can be seen that the total bit rate varies between 20 and 80 Kbps. The required bit rate can change almost 50 Kbps during 1 second (from second 8 to 9).

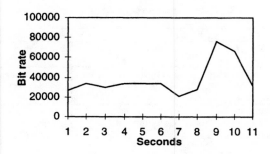

Figure 3.6   Bit rate/second requirement for Carphone sequence.

3.5   MULTIMEDIA SERVICES

Multimedia means transmission of several types of information (e.g., voice, video, and data) simultaneously. Bit rate and quality requirements of these components can be highly variable; for example, BER 10^{-3} and delay of 40 ms for speech and BER 10^{-7} with unconstrained delay for data. Multiplexing of different services is possible using existing multiplexing schemes such as H.223, which is part of the H.324 multimedia communication standard 324 [4], or H.250, which is part of the H.323 multimedia communication standard for packet-based multimedia communications. The H.323 multiplexing scheme converts the different data streams into one single stream for transmission over the network. However, it is more efficient from the radio transmission point of view if the transmission parameters of each service are tailored to the requirements of the service. Therefore, multiplexing of different services should also be possible in the air interface. Third generation networks will offer different transmission bearers for different types of services. The provision of these bearers in wideband CDMA is discussed in Section 5.8.

3.6   IMT-2000 BEARER SERVICES

Since the IMT-2000 applications are not yet known, we cannot specify the supported services based only on today's needs. We need a flexible and broad approach in order to have a future-proof radio interface. Therefore, based on the analysis of a large number of existing services and future needs, more generic radio requirements have been derived (data rate, BER, data rate variation, max delay). These application requirements can be mapped into generic radio bearers as shown in Table 3.1. The exact values are currently under discussion.

Table 3.1
IMT-2000 Bearer Service Requirements

	Real Time/Constant Delay		Non Real –Time/Variable Delay	
Operating environment	Peak Bit Rate	BER / Max Transfer Delay	Peak Bit Rate	BER / Max Transfer Delay
Rural outdoor (terminal speed up to 250 km/h)	at least 144 Kbps (preferably 384 Kbps) granularity appr. 4 Kbps	delay 20 - 300 ms BER 10^{-3} - 10^{-7}	at least 144 Kbps (preferably 384 Kbps)	BER = 10^{-5} to 10^{-8} Max transfer delay 150 ms or more
Urban/ Suburban outdoor (Terminal speed up to 150 km/h)	at least 384 Kbps (preferably 512 Kbps) granularity appr. 10 Kbps	delay 20 - 300 ms BER 10^{-3} - 10^{-7}	at least 384 Kbps (preferably 512 Kbps)	BER = 10^{-5} to 10^{-8} Max transfer delay 150 ms or more
Indoor/ Low range outdoor (Terminal speed up to 10 km/h)	2 Mbps granularity appr. 200 Kbps	delay 20 - 300 ms BER 10^{-3} - 10^{-7}	2 Mbps	BER = 10^{-5} to 10^{-8} Max transfer delay 150 ms or more

Note: Since currently there is no single specification of IMT-2000 or UMTS bearer services or requirements, this table has been composed from various sources and reflects the range of expected IMT-2000 bearer service parameters.

3.7   TRAFFIC MODELS

In order to analyze the capacity and design of access procedures, we need explicit models for the above services. In this section we classify services into real-time and non real-time services. Examples of real-time services are speech and conversational video, and of non real-time services WWW browsing and file transfer.

It should be noted that the models presented here are only one way of modeling these services. As new services emerge and applications develop, traffic characteristics will also change. One example is e-mail: attaching a file to an e-mail message has become very common during recent years, and consequently, the traffic characteristics of e-mail have changed. Therefore, it is important to carefully consider the used traffic model in order to be certain that it represents the main characteristics of the service.

3.7.1   Real-Time Services

For real-time services, the traffic model is a traditional *birth-death process*. Speech and other real-time service users arrive to the system according to a *Poisson process* with an intensity λ_s [users/s].

The departure from the system is modeled by setting the *mean holding time* (*mht* [s]) of the users. That is, the session length for a user is exponentially distributed with an average of *mht*. The average holding time for all real-time users is 120 seconds. A minimum call size of 1 second should be assumed.

The activity of a service can be implemented as an on-off model where the lengths of active and silent bursts are exponentially distributed. The activity factor and the average length of both active and silent periods need to be defined. A typical activity factor for speech is 0.5, and the length of the active and silent periods is 0.7 seconds.

For variable rate video, a more realistic model should consider the different bit rates and their probability.

3.7.2   Non Real-Time Services

As an example of non real-time services, we present a traffic model for a WWW browsing session. The sessions are generated according to a Poisson distribution. A traffic model must take into consideration where in the system we want to model the behavior. The model presented describes how the data packets arrive to a buffer before the air interface. Figure 3.7 illustrates a downlink scenario where the arrow from the base station to the mobile station represents the traffic model. It should be noted that the flow due to one specific session is modeled, not the total flow from all sessions (i.e., there is one buffer per bearer service). Here, only the downlink model is presented. However, this model can be easily generalized for the uplink, reminding that in the uplink, file size and idle period distributions are different.

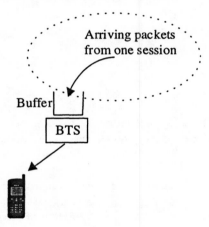

Figure 3.7   The downlink packet flow to a user specific base station (BTS) buffer.

The definition of non real-time traffic models follows [5]. The principles of this model are based on [6]. The following parameters define characteristics of the packet traffic as illustrated in Figure 3.8.

- Session arrival process;
- The number of packet bursts per session (N_{pc});

- The interarrival time between packet bursts (D_{pc}) (i.e., the reading time);
- The number of packets in a packet burst (N_d);
- The interarrival time between packets within a packet burst (D_d);
- The size of a packet (S_d).

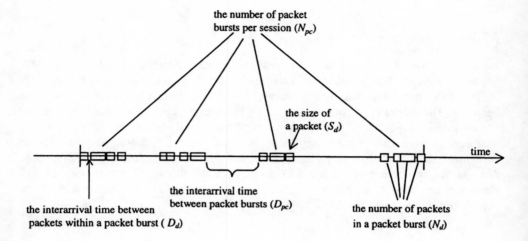

Figure 3.8 Packet service model.

The length of a session is modeled implicitly by the number of events during the session. Session arrival process defines how a session arrives to the system. The arrival of session setups to the network is modeled as a Poisson process. For each service there is a separate process.

N_{pc}, D_{pc}, N_d, and D_d are all geometrically distributed random variables with means μ_{Npc} [packets], μ_{Dpc} [seconds], μ_{Nd} [packets], and μ_{Dd} [seconds], respectively. The geometrical distribution is a discrete representation of the exponential distribution. It is used because the simulations are performed with a discrete time scale. The *geometrical distribution* for a random variable x with mean μ_x is defined by

$$P\{x = k\} = p(1 - p)^k \qquad k = 0, 1, \ldots \qquad (3.1)$$

where $p = \dfrac{1}{1 + \mu_x}$.

According to some studies, the packet size in WWW applications can be modeled as a *Pareto distribution* [7]. The probability density function (pdf) of the Pareto distribution is defined as [8]

$$f_x(x) = \frac{\alpha \cdot k^\alpha}{x^{\alpha+1}} \qquad \alpha, k \geq 0, \ x \geq k \qquad (3.2)$$

and the cumulative distribution function (CDF) is defined as

$$F_x(x) = 1 - \left(\frac{k}{x}\right)^\alpha \qquad x \geq k \qquad (3.3)$$

The average packet length is now obtained from the mean of the Pareto distribution [8]

$$\mu_x = \frac{k \cdot \alpha}{\alpha - 1} \qquad \alpha > 1 \qquad (3.4)$$

Table 3.2 gives example mean values for the distributions of a typical WWW service. The average packet size can be calculated from (3.4) to be 896 bytes. The average requested file size is the average packet size multiplied by the average number of packets: $896 \times 15 = 13.4$ Kbytes. Different values for the average interarrival time between packets correspond to different average source speeds. For example, an average interarrival time of 0.02 seconds between packets corresponds to an average bit rate of 384 kbps. In order to have a finite variance for the distribution, the maximum packet size is limited to 67 Kbyte [5]. Since different parameters of the traffic model will have a different impact on the system performance, it is also important to consider other parameter values as given here.

Table 3.2
Example Mean Values for the Distribution of a Typical WWW Service

Average number of packet bursts within a session (N_{pc})	Average reading time between packet bursts (D_{pc}) [s]	Average number of packets within a packet burst (N_d)	Average interarrival time between packets (D_d) [s]	Average bit rate [Kbps]	Parameters for packet size distribution
5	12	15	0.96	8	k = 81.5
			0.24	32	α = 1.1
			0.12	64	
			0.05	144	
			0.02	384	
			0.004	2048	

REFERENCES

[1] Anderlind E., and J. Zander, "A Traffic Model for Non-Real-Time Data Users in a Wireless Radio Network," *IEEE Communications Letters*, Vol. 1, No. 2, March 1997, pp. 37–39.

[2] Nieweglowski, J., and T. Leskinen, "Video in Mobile Networks," *European Conference on Multimedia Applications, Services and Techniques (ECMAST96)*, Louvain-la-Neuve, Belgium, May 28–30, 1996, pp. 120–133.

[3] Rijkse, K., "The H.263 Video Coding for Low-Bit-Rate Communication," *IEEE Communications Magazine*, Vol. 34, No. 12, Dec. 1996, pp. 42–45.

[4] Lindbergh, D., "The H.324 Multimedia Communication Standard," *IEEE Communications Magazine*, Vol. 34, No. 12, Dec. 1996, pp. 46–51.

[5] ETSI UMTS 30.03, "Selection procedures for the choice of radio transmission technologies of the UMTS," *ETSI Technical Report*, 1997.

[6] Anagnostou, M. E., J.-A. Sanchez-P., and I. S. Venieris, "A Multiservice User Descriptive Traffic Source Model," *IEEE Transactions on Communications*, Vol. 44, No. 10, October 1996, pp. 1243–1246.

[7] Cunha, C. R., A. Bestavros, and M. E. Crovella, "Characteristics of WWW Client-based Traces," *Technical Report TR-95-010*, Boston University Computer Science Department, June 1995.

[8] Mood, A. M., F. A. Graybilland, and D .C. Boes, *Introduction to the Theory of Statistics*, 3rd ed., New York: McGraw-Hill, Inc., 1974.

Chapter 4

RADIO OPERATING ENVIRONMENTS AND
THEIR IMPACT ON SYSTEM DESIGN

4.1   INTRODUCTION

Radio propagation characterization is the bread and butter of communications engineers. Detailed knowledge of radio propagation is essential to develop a successful wireless system. A radio engineer has to acquire full knowledge of the channel to design a good radio communication system [1].

Therefore, knowledge of radio propagation characteristics is a prerequisite for the design of radio communication systems. Many measurements studies have been done to obtain information concerning propagation loss, spatial distribution of power when the environment is physically static, wideband and narrowband statistics concerning the random variables of received signals at a fixed location due to any surrounding movement, and delay spread. References [2–8] provide a good review of these measurements.

A radio channel is a generally hostile medium in nature. It is rather difficult to predict its behavior. Traditionally, radio channels are modeled in a statistical way using real propagation measurement data. In general, the signal fading in a radio environment can be decomposed into a large-scale path loss component together with a medium-scale slow varying component having a log-normal distribution, and a small-scale fast varying component with a Rician or Rayleigh distribution, depending on the presence or absence of the line-of-sight (LOS) situation between the transmitter and the receiver [7–17]. Accordingly, a three-stage propagation model should be appropriate to describe a wireless cellular environment: (1) area mean power depending on the path loss characteristics in the range from the transmitter to the area where the receiver is located; (2) local mean power within that area, which is slow varying, can be very satisfactorily represented by a log-normal distribution; and (3) superimposed fast fading instantaneous power, which follows a Rayleigh or Rician distribution. Figure 4.1 illustrates these three different propagation phenomena. An extreme variation in the transmission path between the transmitter and receiver can be found, ranging from direct LOS to severely obstructed paths due to buildings, mountains, or foliage. The phenomenon of decreasing received power with distance due to reflection, diffraction around structures, and refraction within them is known as path loss. Propagation models have been developed to determine the path loss and are known as large-scale propagation models because they characterize the received signal strength by averaging the power level over large transmitter-receiver separation distances, in the range of hundreds or thousands of meters.

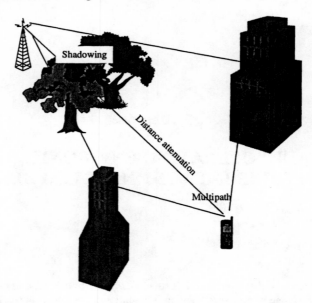

Figure 4.1   Path loss, shadowing, and multipath fading.

Medium-scale propagation models determine the gradual changes in the local-mean power if the receiving antenna is moved over distances larger then a few tens or hundreds of meters. The medium-scale variation of the received signal power is called *shadowing*, and it is caused by obstruction by trees and foliage. The term local-mean power is used to denote the power level averaged over a few tens of wavelengths, typically 40λ [11,13].

Small-scale propagation models characterize the fast variation of the signal strength over a short distance on the order of a few wavelengths or over short time durations on the order of seconds. Small-scale fast fading, also known as short-term fading or multipath fading, is due to multipath reflections of a transmitted wave by local scatterers such as houses, buildings, and man-made structures, or natural objects such as forests surrounding a mobile unit.

The accuracy of the models depends on their purpose. For detailed coverage and capacity analysis in a certain region, detailed models are needed. On the other hand, for rough capacity and range calculations needed in radio interface design, simple and easy-to-use models are sufficient. In the performance assessment, it is important to analyze the sensitivity of the result with respect to the propagation models used.

In addition to the modeling of the propagation environment, the mobile behavior of the wireless terminals needs to be understood in each radio environment. Mobility modeling is involved in the analysis of radio resource management, channel allocation, and handover performance. The accuracy of mobility models becomes essential for evaluation of design alternatives (e.g., different deployment strategies).

In this chapter, we first give an overview of the multipath channel and typical radio environments for third generation wireless systems in Sections 4.2 to 4.4. Thereafter, detailed models for path loss, deployment, shadowing, and wideband radio channels are presented in Sections 4.5 and 4.6. In addition, the impact of two different sets of wideband channel models, ATDMA and CODIT channels, on the performance of narrowband and wideband CDMA system design is discussed. In the last section, mobility models for different radio environments are discussed.

4.2   MULTIPATH CHANNEL

The short-term fluctuations of the received signal caused by multipath propagation are called small-scale fading. The different propagation path lengths of the multipath signal give rise to different propagation time delays. Figure 4.2 shows an example of a *power-delay profile* of a multipath channel consisting of three distinct paths, which are also called *multipath taps*. Depending on the phase of each multipath signal when arriving at the receiver, they sum either constructively or destructively. Consequently, the power of each multipath tap is time varying, resulting in fading dips, as illustrated in Figure 4.3. The depth of the fading dips depends on the channel type. The distribution of the instantaneous power of the channel taps can be described by a distribution function, which depends on the radio environment. A so-called *Rayleigh fading* channel is the most severe mobile radio channel, and accordingly, the fading dips are deep. In a Rayleigh fading channel all multipath taps are independent and there is no dominant path. In a *Rician fading* channel the fading dips are shallower, which is due to a dominant path in addition to the scattered paths. A dominant path typically exists in microcell and picocell environments due to a LOS connection. The impact of different power delay profiles and fading characteristics on DS-CDMA performance will be discussed in Section 4.6.

In order to model the multipath fading channel, its behavior needs to be characterized. The concepts of *delay spread, coherence bandwidth, Doppler spread,* and *coherence time* are used to model the various effects of the multipath channel.

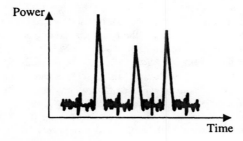

Figure 4.2   An impulse response of a multipath channel.

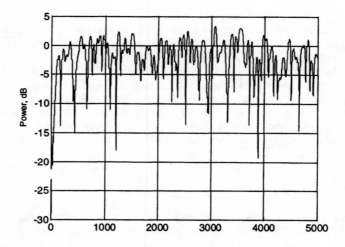

Figure 4.3   Fading channel.

The root-mean-square (rms) value of the delay spread is a statistical measure and tells the spread of the multipaths about the mean delay of the channel (for a mathematical definition of the rms delay spread see Section 7.2.2). The maximum delay spread tells the delay difference between the first and the last multipath components in the power-delay profile. The coherence bandwidth is the maximum frequency difference for which the signals are still strongly correlated. The coherence bandwidth is inversely proportional to the delay spread (i.e., the smaller the delay spread the larger the coherence bandwidth). If the transmission bandwidth of the signal is larger than the coherence bandwidth, the signal will undergo *frequency selective fading*. A *flat fading* channel results if the transmission bandwidth of the signal is smaller than the coherence bandwidth. As can be understood from the above descriptions, the delay spread and the coherence bandwidth are different views of the characteristics of the multipath fading channel. The coherence bandwidth is a measure of the diversity available to a RAKE or equalized receiver. A smaller coherence bandwidth means a higher order diversity. This is illustrated in Figure 4.4, where only part of the frequency carrier is fading. If the coherence bandwidth was as large as the transmission bandwidth, then the entire received spectrum would be observed to fade. The maximum delay spread can be used to calculate how many resolvable paths exist in the channel that could be used in the RAKE receiver (see Section 7.2.2).

The movement of the mobile station gives rise to a Doppler spread, which is the width of the observed spectrum when an unmodulated carrier is transmitted. If there is only one path from the mobile to base station, the base station will observe a zero Doppler spread combined with a simple shift of the carrier frequency (Doppler frequency shift). This is analogous to the change in the frequency of a whistle from a train perceived by a person standing on a railway line when the train is bearing down or receding from that person [6]. The Doppler frequency varies depending on the angle of the mobile station movement relative to the base station. The range of values when the Doppler power spectrum is non-zero is called the Doppler spread.

The reciprocal of the Doppler shift is a measure of the coherence time of the channel. The coherence time is the duration over which the channel characteristics do not change significantly. The coherence time is related to the design of the channel estimation schemes at the receiver (see Section 5.8.1), error correction and interleaving schemes (see Section 5.5), power control schemes (see Section 5.12.), and the frame length of the TDD schemes (see Section 9.4.1).

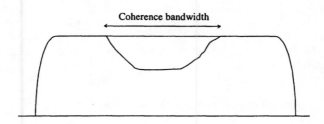

Figure 4.4   Coherence bandwidth and fading.

4.3   RADIO ENVIRONMENTS

There is a large number of environments where third generation mobile radio systems are expected to operate. These include large and small cities, with variations in building construction, as well as tropical, rural, desert, and mountainous areas. In addition, the system construction (e.g., antenna design and height of antennas) impacts the radio environments. Since it is impossible to consider all possible radio environments in the design of a mobile radio system, more general models that will consider the essence of different radio environments are required. Therefore, the large number of possible radio environments has been condensed into three generic radio environments, according to [18–20]:

- Vehicular radio environment;
- Outdoor to indoor and pedestrian radio environment;
- Indoor office radio environment.

These radio environments correspond to the following cell types: macrocell, microcell and picocell, respectively. In this section we describe the characteristics of these environments with respect the path loss attenuation, shadowing, and small-scale fading. Specific path loss attenuation, shadowing and deployment models for each environment will then be presented in Section 4.5. In addition, Section 4.6 presents small-scale fading models. These models will be used in Chapters 7 and 8 for performance simulations. In addition to these models describing single radio environments, a mixed radio environment is described in Chapter 8 in conjunction with the discussion of hierarchical cell structures.

4.3.1 Vehicular Radio Environment

The vehicular environment is characterized by large macrocells and large transmit powers. It is also called the macrocell environment. The mobile speeds are also high because of fast-moving vehicles. There is typically no LOS component, implying that a received signal is composed of reflections. The average power of the received signal decreases with distance raised to some exponent, the *path loss exponent*. The path loss exponent varies depending on the environment but is typically between 3 and 5. The shadowing is caused by obstruction by trees and foliage, and the resulting medium-scale variation in the received signal power can be modeled with a log-normal distribution. The standard deviation varies considerably; for example, 10 dB can be used in urban and suburban areas. In rural and mountainous areas, a lower value can be used. Small-scale fading is Rayleigh. Typical delay spreads are on the order of 0.8 µs and maximum value can be up to tens of microseconds.

4.3.2 Outdoor to Indoor and Pedestrian Radio Environment

Figure 4.5 illustrates the outdoor to indoor and pedestrian radio environment. This radio environment is characterized by small microcells and low transmit powers. The antennas are located below rooftops. Both line-of-sight and no line-of-sight (NLOS) connections exist. Indoor coverage can also be provided from a outdoor base station.

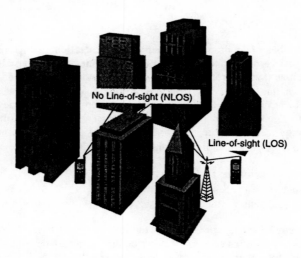

Figure 4.5   Outdoor to indoor environment.

The path loss attenuation in a microcell environment is shown in Figure 4.6. The path loss exponent has large variations, from 2 in areas with LOS, and up to 6 with NLOS due to trees and other obstructions along the path. Furthermore, a mobile station can experience a sudden drop of 15 to 25 dB when it moves around a corner. The standard deviation of the shadowing varies from 10 to 12 dB, and a typical building penetration loss average is 12 dB, with a standard deviation of 8 dB. The small-scale fading is either Rayleigh or Rician with typical delay spreads on the order of 0.2 μs.

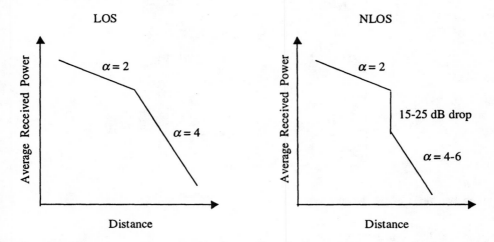

Figure 4.6 Microcell propagation in LOS and NLOS situations (*After*: Wesley, E. K., *Wireless Multimedia Communications*, Reading, MA: Addison-Wesley, 1998).

4.3.3 Indoor Office Radio Environment

In the indoor office radio environment (Figure 4.7), transmit powers are small and base stations and users are located indoors. Path loss attenuation exponent varies from 2 to 5 depending on the scattering and the attenuation by walls, floors, and metallic structures. The wall and floor penetration losses vary according to the material, from 3 dB for light textile to 13 to 20 dB for concrete block walls. A detailed overview of the penetration losses for different types of materials is presented in [7]. Shadowing is log-normal with a typical standard deviation of 12 dB. Fading ranges from Rician to Rayleigh. Rician fading is typically observed if the mobile station is in the same room than the base station and Rayleigh fading is observed when the mobile station is in another room or on a different floor than the base station. Typical average rms delay spreads are on the order of 50 to 250 ns. For a detailed overview, refer to [5]. The mobile station speeds are slow, ranging from a stationary user up to 5 km/h.

4.4   DISTRIBUTION FUNCTIONS

In this section, three important probability density functions (pdf) are given; namely, log-normal pdf used to model the shadowing, and Rayleigh and Rician pdf used to model the distribution of the instantaneous power in small-scale fading models.

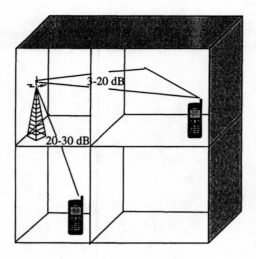

Figure 4.7 Indoor office radio environment.

The log-normal pdf is given by

$$f_{L=}(p_o) = \frac{1}{p_o \sigma \sqrt{2\pi}} \exp\left[-\frac{\left(\ln(p_o) - m^2\right)}{2\sigma^2} \right] \tag{4.1}$$

where p_o is the local mean power, and σ and $p_a = exp(m)$ are the logarithmic standard deviation of the shadowing and the mean, respectively.

The Rician pdf is given by

$$f_R(r) = \frac{r}{p_o'} \exp\left[-\frac{r^2 + s^2}{2p_o'} \right] I_0\left(\frac{r^2}{p_o'} \right) \qquad \text{for } s \geq 0, r \geq 0 \tag{4.2}$$

$$= 0 \qquad \text{for } r \leq 0$$

where r is the amplitude of the received signal, $I_0(\ )$ is the modified Bessel function of the first kind and zeroth order, s is the peak value of the specular radio signal, p_o' is the average power of the signal, $p_o' = p_o/(k+1)$, p_o is the total received local-mean power, and k is the Rician factor, which is defined as the ratio of the average specular power and the average fading power received over specular paths

$$k = \frac{s^2}{2p_o'} \tag{4.3}$$

When the direct signal does not exist (i.e., when s becomes zero $(k = 0)$), (4.2) reduces to the Rayleigh pdf

$$f_R(r) = \frac{r}{p_o} \exp\left[\frac{-r^2}{2p_o}\right] \qquad \text{for } 0 \leq r \leq \infty \qquad (4.4)$$
$$= 0 \qquad \text{for } r < 0$$

4.5 PATH LOSS AND DEPLOYMENT MODELS

In this section, we describe path loss models for the radio environments presented previously. Typically, path loss models are derived using a combination of analytical and empirical methods. In the empirical approach, the measured data is modeled using curve fitting or analytical expressions. The validity of empirical models in other environments and frequencies can only be validated by comparing the model to data measured from the specific area and for the specific frequency. It should be noted that these models present only a snapshot of the real radio environment and are provided only to illustrate the modeling principles of the radio channel. They will be used in Chapters 7 and 8 for the system level simulations of a wideband CDMA system. As will be discussed in Chapter 11, for detailed network planning purposes, the propagation models need to be adjusted according to the real deployment environment.

4.5.1 Vehicular Radio Environment

The vehicular radio environment corresponds to a macrocell. The macrocell layout is typically modeled with a homogenous hexagonal grid, as depicted in Figure 4.8.

We use the path loss model presented in [19] where a Hata-like propagation model was proposed. The path losses are computed as a distance-dependent function $x^{-\alpha}$. Values ranging from 3.0 to 5.0 have been reported for the path loss exponent α depending on the environment.

In addition to this distance-dependent loss, there is the shadowing effect, which is modeled as a zero mean Gaussian random variable, ξ (dB). The standard deviation of the shadowing is assumed to be 10 dB, which is also used for the performance simulations in Chapter 7. The resulting path loss is computed as

$$L_{\text{macro}} = \xi + \alpha \cdot 10 \cdot \log(x) \quad \text{(dB)} \qquad (4.5)$$

where x is distance in kilometers.

4.5.2 Outdoor to Indoor and Pedestrian Radio Environment

In this radio environment, the cell deployment model is of the Manhattan microcell type, as depicted in Figure 4.9. The base station locations can be on street corners, as in Figure 4.9, or in the middle of street canyons. The number of base stations also has to be determined.

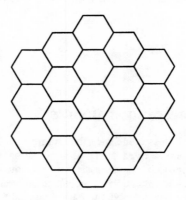

Figure 4.8 Macrocell layout.

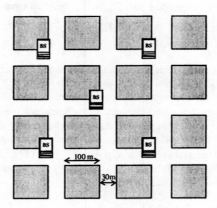

Figure 4.9 Manhattan microcell deployment model.

A three-slope model has been adopted for microcells [19]. A simplifying view has been adopted in the definition of this model, which is not suited for coverage analysis but rather for cellular capacity estimation.

A break point – located at a distance R_b from the transmitter, where the losses equal L_b (dB) – marks the separation between two LOS segments. The second of these LOS segments has a higher slope and predicts larger losses than the first one as illustrated in Figure 4.6. On the other hand, turning a corner causes an additional loss, L_{corner}, in just a few meters but still causes an increase in the slope of this third segment. The model is mathematically given by the following expressions, where loss magnitudes are given in dB [18]

$$
\left.\begin{aligned}
L_{\text{LOS}_1} &= L_b + 20 \cdot n_{\text{LOS}_1} \cdot \log\left(\frac{x}{R_b}\right) & x \leq R_b, \text{ LOS} \\
L_{\text{LOS}_2} &= L_b + 40 \cdot n_{\text{LOS}_2} \cdot \log\left(\frac{x}{R_b}\right) & x > R_b, \text{ LOS}
\end{aligned}\right\} \tag{4.6}
$$

$$
L_{\text{NLOS}} = L_{\text{LOS}}(x_{\text{corner}}) + L_{\text{corner}} + 10 \cdot n_{\text{NLOS}} \cdot \log\left(\frac{x}{x_{\text{corner}}}\right) \qquad \text{NLOS}
$$

where n_{LOS1}, n_{LOS2} and n_{NLOS} denote each segment slope, and the receiver position is defined by the distance from the transmitter to the receiver measured along the street path, x, and, if in a NLOS situation, the distance from the transmitter to the corner, x_{corner}.

The parameters are derived from the expressions in [19]

$$
\begin{aligned}
R_b &= \frac{4 \cdot h_b \cdot h_m}{\lambda} \\
L_b &= \left|20 \cdot \log\left(\frac{\lambda^2}{8 \cdot \pi \cdot h_b \cdot h_m}\right)\right| \\
L_{\text{corner}} &= -0.1 \cdot w_s + 0.05 \cdot x_{\text{corner}} + 20 \\
n_{\text{NLOS}} &= -0.05 \cdot w_s + 0.02 \cdot x_{\text{corner}} + 4
\end{aligned} \tag{4.7}
$$

For the microcell scenario $w_s = 30$m and assuming a wavelength of $\lambda = 0.15$m and the MS and BS antenna heights to be in the ratio $h_b \cdot h_m = 11.25$m^2, the following are obtained

$$
\begin{aligned}
R_b &= 300\text{m} \\
L_b &= 82\text{dB} \\
L_{\text{corner}} &= 17 + 0.05 \cdot x_{\text{corner}} \\
n_{\text{NLOS}} &= 2.5 + 0.02 \cdot x_{\text{corner}} \\
n_{\text{LOS}_1} &= 1 \\
n_{\text{LOS}_2} &= 2
\end{aligned} \tag{4.8}
$$

and the resulting expressions for estimating the path losses will be the following

$$
\begin{aligned}
L_{\text{LOS}_1} &= 82 + 20 \cdot \log\left(\frac{x}{300}\right) & x \leq 300, \text{ LOS} \\
L_{\text{LOS}_2} &= 82 + 40 \cdot \log\left(\frac{x}{300}\right) & x > 300, \text{ LOS}
\end{aligned} \tag{4.9}
$$

$$
L_{\text{NLOS}} = L_{\text{LOS}}(x_{\text{corner}}) + 17 + 0.05 x_{\text{corner}} + (25 + 0.2 x_{\text{corner}})\log\left(\frac{x}{x_{\text{corner}}}\right) \qquad \text{NLOS}
$$

As in the case of macrocells, the shadowing effect is modeled by $10^{\xi/10}$, where ξ is again a Gaussian random variable with the following first order statistics:

$$<\xi> = 0\,dB$$
$$\sigma_\xi = 4\,dB$$

(4.10)

4.5.3 Indoor Office Radio Environment

An indoor office is modeled by a homogenous isolated building with a several floors. Typically no stairs or elevators are modeled. Other relevant parameters are the location and number of base stations, but also the penetration loss of walls and floors. An example of an indoor office model is given in Figure 4.10.

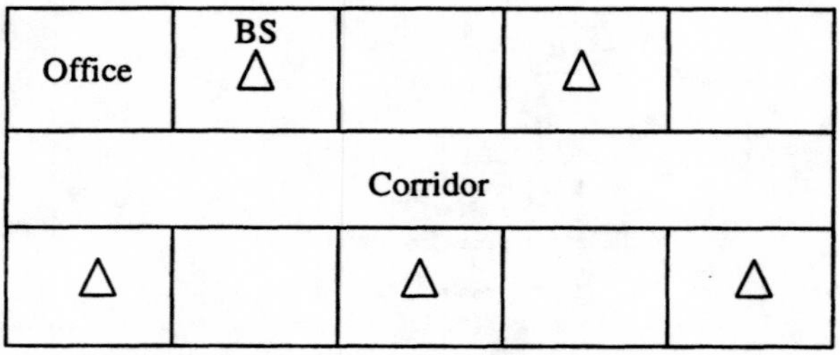

Figure 4.10 Indoor office model example.

The Motley-Keenan model is used to model indoor path losses. This is an empirical model that takes into account the attenuation due to walls and floors on the way from the transmitter to the receiver. The path loss (dB) predicted by this model is given by [19]

$$L_{\text{pico}} = L_0 + 10 \cdot n \cdot \log(x) + \sum_{j=1}^{J} N_{w_j} \cdot L_{w_j} + \sum_{i=1}^{I} N_{w_i} \cdot L_{w_i}$$

(4.11)

where L_0 denotes the loss at the reference point (at 1m), n is the power decay index, and x represents the transmitter to receiver path length. N_{w_j} and N_{f_i} denote the number of walls and floors of different kinds that are traversed by the transmitted signal and L_{f_i} (dB) and L_{w_j} (dB) represent their corresponding loss factors. The values proposed for these parameters are the following

$$L_0 = 37\,dB$$
$$L_f = 20\,dB$$

$$L_w = 3 \text{ dB}$$
$$n = 2 \qquad (4.12)$$

This yields to the following formulation for the loss model:

$$L_{pico} = 37 + 20 \cdot \log(x) + 3 \cdot N_w + 20 \cdot N_f \quad (\text{dB}) \qquad (4.13)$$

Unlike in other radio environments, no shadowing is modeled in the indoor office radio environment.

4.6 SMALL-SCALE MODELS

The small-scale fading is included in wideband channel models. There has been a vast amount of literature about channel modeling. During recent years, ATDMA and CODIT projects have performed wideband channel modeling specifically for third generation systems and their expected radio environments. The outcome of the wideband propagation activities performed by ATDMA is reported in [21]. A complete and detailed study of the work carried out on propagation matters within the CODIT project can be found in [22,23]. The ATDMA models together with models from joint technical committee (JTC) in the United States have been used as a basis for the ITU channel models [24,25]. As will be noticed from the following, channel models intended to model the same environment can differ significantly.

A comparison of ATDMA and CODIT channel models was carried out by SIG5 [19]. In the following we use this comparison to highlight differences in propagation environments and to analyze the impact of these channel on the performance of narrowband and wideband CDMA air interfaces assuming 1.25- and 5-MHz channel bandwidths, respectively.

The approach taken by CODIT for wideband modeling is statistical-physical modeling [23], while ATDMA uses stored channels, which are sets of complex impulse responses corresponding to different radio environments [21]. For SIG5, the channels were transformed into the well-known COST 207 format, which is used here as well.

4.6.1 Vehicular Radio Environment

In the vehicular environment, macrocell channel models are used. Table 4.1 presents the CODIT channel model in COST 207 format and Table 4.2 presents the ATDMA macrocell channel model. To illustrate the differences in the channel models, the impulse responses of both channels are given in Figure 4.11.

It can be observed from Figure 4.11 that in the ATDMA wideband model, the multipath spread is much wider than in the CODIT channel model. Also, in the ATDMA model the average amplitude values differ more than in the CODIT model. In the ATDMA the second path is already attenuated by 10 dB, which leads to a difficult channel since the first tap is also Rayleigh fading [19]. So, the ATDMA channel is almost equally difficult for narrowband and wideband CDMA, even though wideband CDMA can resolve more taps and thus obtains somewhat more diversity gain.

Table 4.1
CODIT Macrocell Channel Model

Tap	Relative Delay (ns)	Relative Power (dB)	Doppler Spectra
1	100	−3.2	CLASS
2	200	−5.0	CLASS
3	500	−4.5	CLASS
4	600	−3.6	CLASS
5	850	−3.9	CLASS
6	900	0.0	CLASS
7	1050	−3.0	CLASS
8	1350	−1.2	CLASS
9	1450	−5.0	CLASS
10	1500	−3.5	CLASS

Table 4.2
ATDMA macrocell channel model

Tap	Relative Delay (ns)	Relative Power (dB)	Doppler Spectra
1	0	0	CLASS
2	380	−10.0	CLASS
3	930	−22.7	CLASS
4	1940	−24.7	CLASS
5	2290	−20.7	CLASS
6	2910	−22.1	CLASS

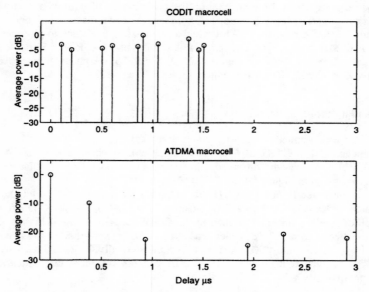

Figure 4.11 Impulse responses of the CODIT and ATDMA macrocell channel models.

In the CODIT channel, the delay spread is more than 1 μs. The delayed paths are attenuated less than 5 dB relative to the strongest path since the delay spread is large enough for some diversity and there is enough energy in the delayed paths. In this channel, narrowband CDMA cannot resolve the channel taps in contrast to wideband CDMA. Thus, wideband CDMA has larger diversity gain.

4.6.2   Outdoor to Indoor and Pedestrian Radio Environment

The channel parameters for the microcell are given in Table 4.3 for CODIT and in Table 4.4 for ATDMA. The impulse responses for both channel models are illustrated in Figure 4.12.

Table 4.3
CODIT Microcell Channel Model

Tap	Delay (ns)	Power (dB)	Doppler Spectrum	Ricean Factor (dB)	Doppler Shift (fi/fdop)
1	0	−2.3	RICE	−7.3	0.6066
2	0	0.0	RICE	−3.5	0.6066
3	0	−13.6	CLASS	−	−
4	50	−3.6	RICE	−3.5	0.6066
5	50	−8.1	CLASS	−	−
6	100	−10.0	CLASS	−	−
7	1700	−12.6	RICE	−2.2	0.6066

Table 4.4
ATDMA Microcell Channel Model

Tap	Delay (ns)	Power (dB)	Doppler Spectra
1	0	−4.9	CLASS
2	230	0	CLASS
3	630	−11.4	CLASS
4	1110	−13.9	CLASS
5	1440	−16.1	CLASS
6	1840	−23.5	CLASS

From Figure 4.12 it can be observed that the first six taps in the CODIT microcell channel are very close to each other, and the receivers cannot resolve these paths. Even though some of these taps are Rice distributed, the total signal will be close to Rayleigh distributed, since there are two other paths with strong average power (−2.3 and −3.6 dB). This leads to almost the same situation as with a typical ATDMA macrocell, except that now there is one path at 1.7-μs relative delay, which provides some diversity. Both narrowband and wideband CDMA can only resolve the last tap, while the first six taps cannot be resolved.

The ATDMA microcell channel is also a rather harsh communications channel. The taps are now wider apart, but all taps are Rayleigh fading and the second strongest path is already −4.9 dB compared to the strongest one [19]. Wideband CDMA with 5-MHz bandwidth can resolve all six taps, while the one for narrowband CDMA can only resolve two groups of three taps each.

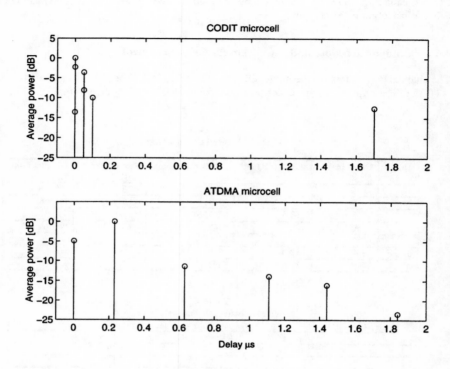

Figure 4.12 Impulse responses of the CODIT and ATDMA microcell channel models.

4.6.3 Indoor Office Radio Environment

The tap settings for CODIT picocell are given in Table 4.5, and for ATDMA picocell in Table 4.6. The impulse responses of both channels are given in Figure 4.13.

We notice from Figure 4.13 that the delay spread of CODIT picocell channel is very small. This, together with the fact that all the taps are Rayleigh fading, will lead to difficult flat fading channel [19]. The ATDMA indoor channel is much easier. The first strong Rice distributed tap should ensure better amplitude distribution than Rayleigh. Also, when using wideband CDMA, all the different paths should be resolvable.

Table 4.5
CODIT picocell channel model

Tap	Delay (ns)	Power (dB)	Doppler Spectra
1	0	−3.6	CLASS
2	50	0.0	CLASS
3	100	−3.2	CLASS

Table 4.6
ATDMA picocell channel model

Tap	Delay (ns)	Power (dB)	Doppler Spectra	Ricean Factor (dB)	Doppler Shift
1	0	0	RICE	−10.0	−
2	77	−3.3	RICE	−4.0	−
3	186	−9.3	RICE	−4.0	−
4	299	−14.3	RICE	−4.0	−
5	404	−20.3	RICE	−4.0	−
6	513	−26.8	CLASS	−	−

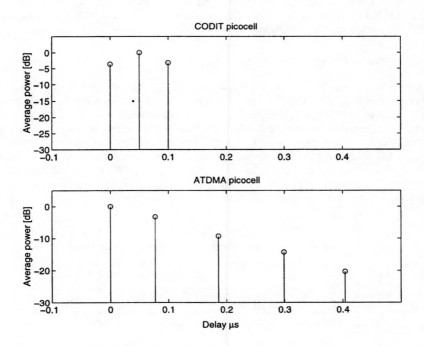

Figure 4.13 Impulse responses of the CODIT and ATDMA picocell channel models.

4.6.4 Spatial Channel Models

Deployment of adaptive/smart antenna techniques requires an understanding of the spatial properties of the wireless channel. Modern spatial channel models build on the classical understanding of fading and Doppler spread and incorporate additional concepts such as time delay spread, angle of arrival, and adaptive antenna geometries. For a more detailed treatment of spatial channel models, refer to [26] and references therein.

4.7 MOBILITY MODELS

Mobility modeling is involved with the analysis of aspects related to location management (location area planning, paging strategies), radio resource management (multiple access techniques, channel allocation strategies, handover rates), network signaling loads, and propagation (handover decision) [26–27]. Different purposes require different types of mobility models. For example, location area planning requires the knowledge of the user location with an accuracy of a large-scale area, while analysis of radio resource schemes such as handover require medium-scale area accuracy (cell area) [27]. Mobility modeling can be based on simulation or analytical approaches. Typically, analytical models with simplified assumptions are used for analysis of critical network dimensioning parameters [27].

In this section we present analytical mobility models for a medium-scale area. The models are based on the SIG5 models [19], which were also used, with some changes, in the ETSI UMTS air interface evaluation process [20].

4.7.1 Vehicular Radio Environment

In the vehicular radio environment, users are completely free to move with a constant speed within the whole service area. The mobility model is a pseudo-random mobility model with semidirected trajectories [20]. Mobiles are uniformly distributed on the map, and their direction is randomly chosen at initialization. After every *decorrelation length,* there will be an update for the direction of the movement. The probability for the direction update is given in Table 4.7, and it is independent of the previous direction update event. If the direction update takes place, there is a uniform probability to change the direction of the movement between -45 and 45 degrees. The direction can also be changed within a given sector to simulate a semi-directed trajectory.

Table 4.7
Parameters of the Vehicular Radio Environment Mobility Model

Parameter	Example Value
Speed value	120 km/h
Probability to change direction at position update	0.2
Maximal angle for direction update	45°
Decorrelation length	20 m

95

4.7.2 Outdoor to Indoor and Pedestrian Radio Environment

The deployment area is a regular grid of streets and buildings (see Section 4.5.2). The simulation area has a finite size with borders (the outer street tier). All buildings and streets have equal size. All users move only in the middle of the streets.

New users will be uniformly distributed over the possible coordinates in the service area. Each possible coordinate has an equal chance of being chosen. Mobiles move along streets (always in the middle of the streets) and may turn at crossings with a given probability (turning probability, TP), as indicated in Figure 4.14. The mobile's position is updated every 5m, and speed can be changed at each position update according to a given probability. The mobility model is described by the parameters shown in Table 4.8.

Table 4.8
Parameters of the Outdoor to Indoor and Pedestrian Radio Environment Mobility Model

Parameter	Example Value
Mean speed	3 km/h
Standard deviation for speed	0.3 km/h
Probability to speed change at position update	0.2
Probability to turn at cross street	0.5

Mobiles are uniformly distributed in the streets, and their direction is randomly chosen at initialization. Users leaving the simulation area via the borders are lost. This loss influences the traffic density: it becomes inhomogeneous as it decreases from the middle to the borders. To keep the traffic density homogenous for the whole border, there must be an incoming probability equal to the leaving probability. In this model, this is achieved simply by generating a new subscriber when one leaves the area. This new subscriber is placed exactly at the border with the same probability of each point of the border. Since it is assumed that the leaving probability is the same for each part of the border, the resulting traffic density will be homogenous.

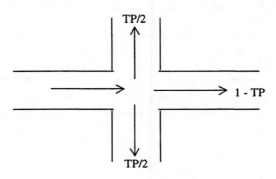

Figure 4.14 Turning point probability.

4.7.3 Indoor Office Radio Environment

In the indoor office radio environment, only one mobility class is assumed. The movement of the users is characterized as follows [20]:

- There is no mobility between the floors.
- Mobiles are either stationary or moving with constant speed from an office room to corridor or vice versa.
- If a mobile is in an office room, it has higher probability of being stationary.
- If a mobile is in the corridor, it has lower probability of being stationary.

Each mobile is either in the stationary or the moving state. The transition from the stationary state to the moving state is a random process. The time duration each mobile spends in the stationary state is drawn from the geometric (discrete exponential) distribution with different mean values depending on whether the mobile is in an office room or in the corridor. The transition from the moving state to the stationary state takes place when mobile reaches its destination.

When a mobile is in an office room and it is switched to the moving state, it moves to the corridor (see Figure 4.15), according to the following procedure:

1. Select the destination coordinates in the corridor with uniform distribution. (Each place in the corridor has equal probability of becoming the destination point.)
2. The mobile "walks" from its current location to the destination location so that first the vertical (y) coordinate is matched with the new coordinate, and next the horizontal (x) coordinate is matched with the destination coordinate. The speed is constant during the "walking."
3. When the mobile reaches the destination point, it is transferred into the stationary state.

By letting mobiles simply walk straight out from the office room, it is assumed that the door dividing each office room and corridor is as wide as the office room itself. When a mobile

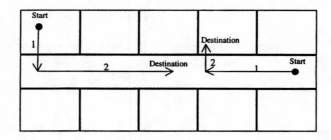

Figure 4.15 The mobility model. Two example routes are drawn. On the left: from an office room to the corridor; and on the right: from the corridor to an office room.

is in a corridor and it is switched to the moving state, it moves to any of the office rooms with equal probability. The following four-step procedure defines the movement along the corridor and from the corridor to an office room:

1. Select the destination office room by using discrete uniform distribution.
2. Select the destination coordinates with uniform distribution. (Each place in the corridor or in an office room has equal probability of becoming the destination point.)
3. The mobile "walks" from its current location so that first the horizontal (x) coordinate is matched with the new coordinate and next the vertical (y) coordinate is matched with the destination coordinate. The speed is constant during the "'walking."
4. When the mobile reaches the destination point, it is transferred into the stationary state. At the stationary state, mobiles do not move at all.

Figure 4.16 represents the mobile station's movement with a state model. P(S,S) denotes the transition probability from the stationary state into the stationary state, P(S,M) denotes the transition probability from the stationary state into the moving state, P(M,S) denotes the transition probability from the moving state into the stationary state, and P(M,M) denotes the transition probability from the moving state into the moving state. To derive transition probabilities from the stationary state to the moving state, the following parameters must be set: ratio of mobiles at office rooms (r), mean office room stationary time (mr) and iteration time step (Δt). With these parameters, the transition probabilities per iteration time step ($1-\Delta t/mr$, $1-\Delta t/mc$) and mean corridor stationary time (mc) can be derived so that flow to the office rooms equals to the flow from the office rooms:

$$r \cdot \frac{\Delta t}{mr} = (1-r) \cdot \frac{\Delta t}{mc} \tag{4.14}$$

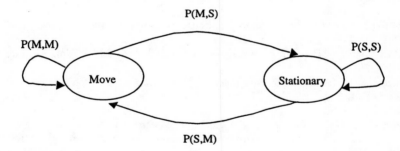

Figure 4.16 State automate presentation of the mobile station movement.

With the parameters listed in Table 4.9, the following values are obtained:

- P(S,S) in an office room = 1–0.005/30 = 0.999833;

- P(S,M) in an office room = 0.005/30 = 0.0001667;
- P(S,S) in corridor room = 1−0.0009444 = 0.9990556;
- P(S,M) in corridor room = 0.005×85/(30×15) = 0.0009444;
- The average stationary time in the corridor becomes $\Delta t/P(S,M)$ = 5.294 seconds.

Table 4.9
Parameters of the Indoor Office Mobility Model

Ratio of mobiles at office rooms	85%
Mean office room stationary time	30 seconds
Simulation time step	0.005 seconds
Number of office rooms	12
Mobile speed	1 m/s

P(M,S) and P(M,M) are not random. P(M,M) equals 1 if the destination is not reached, 0 otherwise; P(M,S) equals 0 if the destination is not reached, 1 otherwise.

REFERENCES

[1] Prasad, R., "European radio propagation and subsystems research for the evolution of mobile communications," *IEEE Communication Magazine*, Vol. 34, February 1996, p. 58.

[2] Anderson, J. B., T. S. Rappaport, and S. Yoshida, "Propagation measurements for wireless communications channels," *IEEE Communication Magazine*, Vol. 33, January 1995, pp. 42–44.

[3] Fleury, B. H., and P. E. Leuthold, "Radiowave propagation in mobile communications: an overview of European research," *IEEE Communication Magazine*, Vol. 34, February 1996, pp. 70–81.

[4] Molkdar, D., "Review on radio propagation into and within buildings," *IEE Proc. –H*, Vol. 138, February 1991, pp. 61–73.

[5] Hashemi, H., "The indoor propagation channel," *Proc. IEEE*, Vol. 81, July 1993, pp. 943–968.

[6] Steele, R., (ed.), *Mobile Radio Communications*, London: Pentech Press Publishers, 1992.

[7] Rappaport, T. S., *Wireless Communication Principles & Practice*, New Jersey: Prentice Hall PTR, 1996.

[8] Prasad, R., *Universal Wireless Personal Communications*, Boston: Artech House, 1998.

[9] Jakes, W. C., Jr. (ed.), *Microwave Mobile Communications*, New York: John Wiley & Sons, 1974.

[10] Lee, W. C. Y., *Mobile Communications Engineering*, New York: McGraw-Hill, 1982.

[11] Linnartz, J. P., *Narrowband Land-Mobile Radio Networks*, Boston: Artech House, 1993.

[12] Lee, W. C. Y., *Mobile Cellular Telecommunications Systems*, New York: McGraw-Hill, 1989.

[13] Lee, W. C. Y., *Mobile Communications Design Fundamentals*, New York: John Wiley & Sons, 1993.

[14] Macario, R. C. V., (ed.), *Personal & Mobile Radio Systems*, London: Peter Peregrinus,

1991.

[15] Stuber, G. L., *Principles of Mobile Communications*, Boston: Kluwer Academic Publishers, 1996.

[16] Mehrotra, A., *Cellular Radio Performance Engineering*, Boston: Artech House, 1994.

[17] Hess, G. C., *Land-Mobile Radio System Engineering*, Boston: Artech House, 1993.

[18] Recommendation ITU-R M.[FPLMTS.REVAL], "Guidelines for Evaluation of Radio Transmission Technologies for IMT-2000/FPLMTS."

[19] Pizarroso, M., and J. Jiménez (eds.), "Common Basis for Evaluation of ATDMA and CODIT System Concepts," *MPLA/TDE/SIG5/DS/P/001/b1*, MPLA SIG 5, September 1995.

[20] ETSI UMTS 30.03, "Selection procedures for the choice of radio transmission technologies of the UMTS," *ETSI Technical Report*, 1997.

[21] Strasser, G., (ed.), "Propagation Models Issue 1," *R2084/ESG/CC3/DS/P/012/b1*, April 1993.

[22] Jiménez, J., (ed.), "Final Propagation Model," *CODIT Deliverable, R2020/TDE/PS/ P/040/b1*, June 1994.

[23] Braun, W., and U. Dersch, "A Physical Mobile Radio Channel," *IEEE Trans. on Vehicular Technology*, Vol. 40, No. 2, May 1991, pp. 472–482.

[24] Xia, H. H., "Reference Models for Evaluation of Third Generation Radio Transmission Technologies," *Proceedings of ACTS Summit*, Aalborg, Denmark, October 1997, pp. 235–240.

[25] Recommendation ITU-R M.1225, "Guidelines for Evaluation of Radio Transmission Technologies for IMT-2000," *Question ITU-R 39/8*, 1997.

[26] Special issue on smart antennas, IEEE Personal Communications, Vol. 5, No. 1, February 1998.

[27] Markoulidakis, J. G., G. L. Lyberopoulos, D. F. Tsirkas, and E. D. Sykas, "Mobility Modeling in Third-Generation Mobile Telecommunications Systems," *IEEE Personal Communications*, Vol. 4, No. 4, August 1997, pp. 41–54.

[28] Jabbari, B., "Teletraffic Aspects of Evolving and Next-Generation Wireless Communications," *IEEE Personal Communications*, Vol. 3, No. 6, December 1996, pp. 4–9.

Chapter 5

CDMA AIR INTERFACE DESIGN

5.1   INTRODUCTION

Chapter 2 has presented the basic elements of DS-CDMA: RAKE receiver, power control, soft handover, and multiuser detection. This chapter discusses these elements in greater depth and presents the most important aspects to be considered when designing a wideband CDMA air interface. It starts by highlighting the system requirements and assumptions in Section 5.2, and their impact on the wideband CDMA air interface design. Secondly, logical and physical channel design are discussed in Sections 5.3 to 5.5. Thereafter, spreading codes and their properties are described in Section 5.6, along with the relevant criteria for selection and optimization. Section 5.7 on modulation covers not only data modulation but also spreading modulation and different spreading circuit types. Section 5.8 on control schemes gives the principles of *forward error control* (FEC) and *automatic repeat request* (ARQ). Next, the different data rates and quality of service requirements are described in Section 5.9, along with multirate scheme implementation. In addition, requirements and solutions for efficient transmission of packet data are given in Section 5.10. Transceiver design is illustrated from the receiver and transmitter perspective in Section 5.11. In addition to the RAKE receiver, an advanced receiver scheme, namely multiuser detection, is introduced in Section 5.12. Random access procedure is briefly discussed in Section 5.13. Section 5.14 describes handover and covers handover procedure, design requirements, and also interfrequency handover. Design principles of power control are presented in Section 5.15. Finally, admission and load control are introduced in Section 5.16.

5.2   DEFINITION   OF   REQUIREMENTS   AND   GENERAL   DESIGN PROCESS

The starting point for the air interface design are the system requirements. Understanding these is critical for a good end result of the design process. Basic system requirements determine data rates, *bit error rate* (BER), and delay, as well as channel models, which define the radio environment. In Chapter 3, UMTS bearer service requirements were derived based on some service examples. In Chapter 4, pathloss, shadowing, and wideband channel models for different radio environments were introduced. The radio environment is not in itself a system requirement, but it places constraints on the system design and therefore needs to be carefully modeled. A number of other issues, such as available frequencies, synchronization constraints, signaling requirements, and complexity aspects, have to be considered. The IMT-2000 system requirements have been specified in [1], and more detailed radio requirements are given in [2]. Corresponding ETSI documents for UMTS are [3,4].

CDMA radio interface design involves several areas, as illustrated in Figure 5.1. Each area itself is very broad, and different parameters can be optimized alone. However, a good design is always a trade-off between several and often contradicting requirements. Good radio performance, for example, requires more advanced receivers resulting in higher costs and complexity, while another system requirement might be precisely the minimization of equipment cost. Spreading codes can be optimized alone or together with channel coding. The design of the multirate scheme for variable data rates requires careful joint-optimization of coding, modulation, and spreading codes. The handover synchronization scheme depends on the network synchronization (i.e., whether synchronous or asynchronous base stations are used). The preceding examples are just a few of the very large number of interdependencies among the different areas. Moreover, the final design depends on the way each optimization criteria is weighted. In the following sections, we highlight each area and their inter-relationship in more detail.

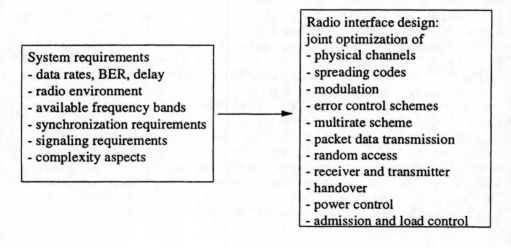

Figure 5.1 CDMA air interface design process.

5.3 LAYERED AIR INTERFACE STRUCTURE

As shown in Figure 5.2, the air interface functions are structured into protocol layers according to principles presented in [5]. The physical channel performs coding, modulation, and spreading for the physical channels. As shown in Figure 5.2, the link layer is further divided into two sublayers: *medium access control* (MAC) and *link access control* (LAC). The medium access layer coordinates the resources offered by the physical layer. The link access control performs the functions essential to set up, maintain, and release a logical link connection. The network layer contains call control, mobility management, and radio resource management functions. In this chapter we will restrict our discussion to the air interface dependent functions (i.e., to those that function in different protocol layers that depend on the underlying multiple access method, in this case, wideband CDMA). These layers include the physical layer, the MAC layer, and radio resource management in the network layer. The other functions are air interface independent and could be used for other types of multiple access schemes such as TDMA or FDMA. For a more detailed discussion on the layered structure, refer to Chapter 12.

The physical layer-related functions are discussed in Sections 5.5 to 5.12. The MAC layer is discussed as a part of packet data transmission in Section 5.10. Radio resource management schemes are discussed in Section 5.13 (random access), in Section 5.14 (handover), in Section 5.15 (power control), and in Section 5.16 (admission and load control). Some radio resource management functions such as handover, power control, and random access have an impact on the physical layer (measurement capabilities, impact on synchronization channels). Logical channel structure is an air interface independent function, but we discuss it in Section 5.4 in order to give background for the physical channel design.

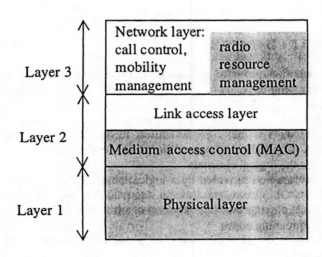

Figure 5.2   The layered structure of the air interface.

5.4   LOGICAL CHANNELS

Certain functions are needed in every cellular system to set up, release, and maintain connections. First, the mobile station acquires *synchronization* from the network. Thereafter, it obtains system information, such as cell identities, spreading codes, access channel, and neighbouring cell lists, from the network. In case the call originates from the network, the mobile station is *paged*. After synchronization, the mobile station initiates connection setup by sending a *random access* message, and the network responds by sending an *access grant* message. Connection is established on a *traffic channel*. During the connection, control information carrying measurement and signaling data is transmitted.

The logical channels can be divided into control channels and traffic channels. A control channel is either common or dedicated. A *common control channel* is a *point-to-multi-point* control channel that carries connectionless messages, which are primarily intended to carry signaling information necessary for access management functions broadcast information (access request, access grant, paging messages, and user packet data) [5]. A *dedicated control channel* is a *point-to-point* bidirectional control channel, intended to carry signaling information such as handover measurements, service adaptation information, and power control information. Traffic channels carry a wide variety of user information streams. The procedures described lead to the following logical channel structure:

- Synchronization channel;
- Random access channel;
- Broadcast channel;
- Paging channel;
- Dedicated control channel;
- Traffic channel.

5.5   PHYSICAL CHANNELS

In general, the functions provided by logical channels need to be mapped into physical channels. The mapping depends on several aspects such as frame design, modulation method, and code design. In this section we discuss general principles for the design of different channels. These are discussed in detail later in this chapter and with example designs in Chapter 6. The naming conventions for the physical channels of different wideband CDMA proposals in Chapter 6 vary considerably. Here we primarily use the same names as for the logical channels.

A specific function provided by a logical channel may be mapped into several physical channels. Or, vice versa, several fuctionalities may be mapped using, for example, time multiplexing into one physical channel. The different physical channels are separated by spreading codes.

A synchronization channel is used to provide the receiver with chip, bit, and frame synchronization. A pilot signal can be used as a reference signal for chip level synchronization and coherent detection, as discussed below. The mobile station uses the downlink synchronization channel for handover measurements and synchronization, as discussed in Section 5.14.3.

The broadcast channel transmits system-specific information. The system designer needs to determine the type and rate of information to be transmitted so that the broadcast channel data rate can be set. For the paging channel, the number of paging channels and the data rate need to be determined.

The random access channel structure depends on how fast the synchronization needs to be established and on the selected access strategy (see Section 5.13). In addition, the expected number of access attempts and synchronization time requirements determine the number of random access channels. Also, the amount of data transmitted during the access attempt is an important design aspect.

Typically, a dedicated control channel is bundled together with a traffic channel. Traffic and dedicated control channels can be time multiplexed, code multiplexed or I&Q multiplexed. The last alternative means that the user data is transmitted on the I channel and the control information on the Q channel. A pure code multiplexing without I&Q multiplexing leads to multicode transmission, which slightly increases the envelope variations of the transmitted signal, requiring a more linear power amplifier. Furthermore, with a code multiplexed dedicated control channel, the mobile station needs to receive two code channels instead of one and thus, complexity is increased.

5.5.1 Frame Length Design

The frame length of the traffic channel depends on the service requirements and desired radio performance. In order to obtain good performance in a fading channel, the frame length of the traffic channel has to be long enough to support a reasonable interleaving depth. On the other hand, it cannot be too long and exceed the transmission delay requirement of service. Since the speech service is typically a dominant service, frame length is tailored according to it. As seen in Chapter 3, for third generation wireless systems, the delay requirement for speech is specified to be 30 to 40 ms. Thus, a frame length of 10 ms is short enough, assuming interleaving is spanning over one frame. Some speech codecs introduce a 10-ms delay, and processing adds a few milliseconds. Thus, the total delay budget is between 20 and 25 ms. If a longer delay is allowed, a 20-ms frame could be used to improve the performance.

For data services, a longer interleaving period than for speech is often tolerable. Therefore, we would like to extend the interleaving period over one frame. This can be performed in multiples of the frame period and is called *inter-frame interleaving*. For packet services, a shorter frame length than 10 ms might be desirable to minimize the retransmission delay of erroneous packets.

The frame length control channel is typically assumed to be the same as for the traffic channel. However, the introduction of packet data might motivate a smaller frame size for the control channel to decrease the channel setup time and to increase the traffic channel efficiency.

5.5.2 Measurement Signaling

Signaling requirements depend on the measurement needs (i.e., how often measurements for power control, handover, and load control are performed and how they need to be transmitted between mobile stations and the base station). When the rough transmission requirements for signaling have been defined, appropriate channel structures can be designed. The signaling traffic can either be transmitted on a dedicated control channel (outband signaling) or on a traffic channel by puncturing the user data (inband signaling). The selected method depends on the quality requirements for user data and mobile station complexity. Inband signaling degrades the transmission quality since user data bits are deleted. However, if the number of deleted bits is small, a power increase (with the same amount as for outband signaling) will compensate for the loss of quality.

5.5.3 Pilot Signals

Because of fading channels, it is hard to obtain a phase reference for the coherent detection of data modulated signal. Therefore, it is beneficial to have a separate pilot channel. Typically, a channel estimate for coherent detection is obtained from a common pilot channel. However, a common pilot channel transmitted with an omnidirectional antenna experiences a different radio channel than a traffic channel signal transmitted through a narrow beam. It has been noticed in second generation systems that common control channels are often problematic in the downlink when adaptive antennas are used. The problem can be circumvented by user dedicated pilot symbols, which are used as a reference signal for channel estimation. In this case, a common pilot is only used for handover measurements and for channel estimation when the downlink beamforming is not used.

The user dedicated pilot symbols can either be time or code multiplexed. Figures 5.3 and 5.4 depict block diagrams of a transmitter and receiver for time multiplexed pilot symbols and, a code multiplexed, parallel pilot channel, respectively. A drawback of the pilot channel transmitted on a separate code is that it requires extra correlators for despreading. The performance of both approaches is the same if the same amount of power is used for the pilot signals, assuming that optimal least mean squares channel recovery techniques are employed.

For time multiplexed pilot symbols, the ratio between the number of data symbols (L_d) and the number of pilot symbols (L_p) needs to be determined. For the parallel pilot channel, the power ratio of pilot and data channels needs to be determined. In the receiver, pilot symbols are averaged over a certain time period. The averaging length varies according to Doppler frequency. One possibility is to find a compromise by simulating the performance in different scenarios and by selecting a number of pilot symbols and an averaging period that results into reasonable performance over different operating conditions [6]. Another possibility is to have more flexibility and to have a variable number of pilot symbols and a variable averaging period. For a code multiplexed pilot channel, it is easier to adjust the pilot power.

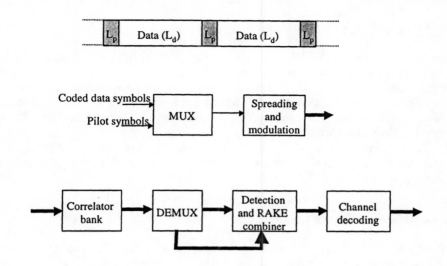

Figure 5.3   Time multiplexed pilot symbols.

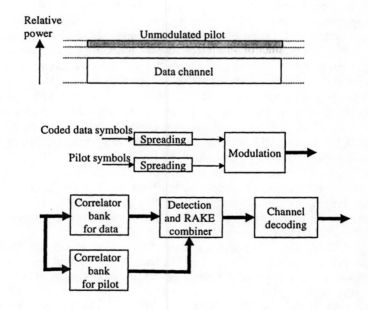

Figure 5.4   Code multiplexed, parallel pilot channel.

5.6   SPREADING CODES

Spreading codes[1] can be divided into *pseudo-noise* (PN) codes and *orthogonal* codes. PN codes are pseudo-random codes generated by a feedback shift register. The most commonly considered PN codes for DS-CDMA systems are generated using linear shift registers. The cross-correlation between orthogonal codes is zero for a synchronous transmission. Orthogonal codes such as Walsh sequences are typically used for channel separation in DS-CDMA systems.

This section provides an engineer's perspective on the design and selection of spreading codes for DS-CDMA systems. For discussion of the fundamental properties of PN sequences and linear shift registers refer to [7, 8].

5.6.1   Basic Properties of Spreading Codes

In a DS-CDMA transmitter, the information signal is modulated by a spreading code, and in the receiver it is correlated with a replica of the same code. Thus, low cross-correlation between the desired and interfering users is important to suppress the multiple access interference. Good autocorrelation properties are required for reliable initial synchronization, since large sidelobes of the autocorrelation function might lead to erroneous code synchronization decisions. Furthermore, good autocorrelation properties are important to reliably separate the multipath components. Since the autocorrelation function of a spreading code should resemble, as much as possible, the autocorrelation function of white Gaussian noise, the DS code sequences are also called pseudo-noise (PN) sequences. The autocorrelation and cross-correlation functions are connected in such a way that it is not possible to achieve good autocorrelation and cross-correlation values simultaneously. This can be intuitively explained by noting that having good autocorrelation properties is also an indication of good randomness of a sequence. Random codes exhibit worse cross-correlation properties than deterministic codes [9]. Figure 5.5 illustrates auto and cross-correlation functions for a 31 chip length M-sequence.

There exist two basic correlation functions: even and odd. For an *even* correlation function, the consequent information data bits are the same; and for an *odd* correlation function, consequent data bits are different. In an asynchronous CDMA system both functions exist with equal probability. Typically, however, only even cross-correlation properties can be optimized.

Typically, PN sequences are generated with a linear feedback shift register generator, depicted in Figure 5.6. The outputs of the shift register cells are connected through a linear function formed by exclusive-or (XOR)-type logic gates into the input of the shift register. Nonlinear codes are used for military and cryptographic applications. Nonlinearity can be implemented, for example, with the aid of AND gates [9].

---

[1] Sometimes spreading codes are called spreading sequences. In this book, the terms spreading code and spreading sequence are used interchangeably.

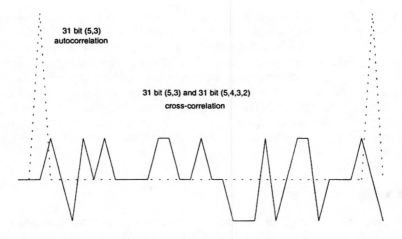

Figure 5.5   Auto and cross-correlation functions.

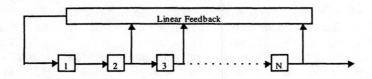

Figure 5.6   Feedback shift register.

The symbol of a DS code can be classified as binary or nonbinary [9]. Nonbinary sequences are typically complex valued having equal signal amplitude. Binary or quaternary (4-phase) sequences are preferred in DS systems, since they fit in with BPSK, QPSK, and O-QPSK modulation methods typically used in DS-CDMA systems [9].

A spreading code can be either short or long.[2] Short codes span one symbol period, while long codes span several symbol periods. The reason to use short codes is either to attempt to control the correlation properties through a suitable selection of a short spreading code set or to minimize the complexity of multiuser detectors. Furthermore, the signal obtained by spreading with short codes is cyclostationary (periodic), which can be utilized in the design of the multiuser detection algorithm (see Section 5.12). Long codes facilitate a large number of spreading sequences, and they aim to randomize the interference.

---

[2] It should be noted that in the IS-95 system [10], the term *short code* is used (differently than in this book) for a relatively long sequence with length of 2^{15} used to spread several symbols, and the term *long code* is used for a code with length of 2^{42}.

A DS-CDMA system can be classified as synchronous or asynchronous[3]. In a synchronous system the transmission times of spreading codes are the same, whereas in an asynchronous system no time-control of transmission times between users exists. In a synchronous system (e.g., the downlink of a cellular system) completely orthogonal codes can be used. However, multipath will result in a partially non-orthogonal system. The uplink of a cellular system is asynchronous, although synchronization has been proposed for a wideband CDMA system in [11]. Short codes in an asynchronous system require a special synchronization scheme for handover since the time difference between base stations cannot be measured using spreading codes, unlike in a long code system [12].

5.6.2 Pseudo-Noise Sequences

Table 5.1 presents the number of M-sequences for different sequence lengths. As can be seen, the number of codes is rather limited. However, M-sequences possess a certain property that has been used in the design of the IS-95 air interface. For each M-sequence, the time-shifted version of the code is a new code. Thus, the number of codes is the same as the length of the code. However, the propagation delay limits the minimum time separation between two codes, and therefore, long spreading sequences are needed. Assuming a chip rate of 3.6864 Mcps, one chip equals 0.27 µs. In IS-95 the minimum PN-offset is 64 chips corresponding to 15.6 km. To maintain the same PN-offset in kilometers between different pilot sequences as in IS-95, a PN-offset of 192 chips is required, and a code length of 2^{15} provides now 170 different pilots. Therefore, the use of the same M-sequence with different phase-offsets might lead to code planning [13] (see Chapter 11). An advantage of this code design is that cross-correlation properties between the codes are now actually determined by the autocorrelation functions. The autocorrelation properties of M-sequences are very good, in contrast to their cross-correlation properties. However, since partial correlations are used (i.e., correlation is performed over one symbol, not over the whole sequence), the good autocorrelation sequences of the full-length sequence are not fully utilized.

Table 5.1
M-Sequences

Sequence Length	Number of Codes
7	2
15	2
31	6
63	6
127	18
255	16
511	48

---

[3] It should be noted that synchronous/asynchronous CDMA in this context is different from the network synchronous/asynchronous scheme, which is related to synchronization between base stations.

In Table 5.2, a number of other well-known PN code families are presented. The number of Gold codes is approximately the same as the code length. Gold codes have two variations: even and odd degree codes. With an odd degree code (code lengths of 31, 127, 511, 2047, ...), a smaller value of maximum cross-correlation relative to the code period is achieved [9]. Furthermore, there exists a code family called Gold-like sequences whose maximum cross-correlation is slightly smaller than normal Gold codes. However, in practice, this reduction is so small that there is no performance difference.

There exist several Kasami families: small (S-Kasami), large (L-Kasami), and very large (VL-Kasami). In addition, Kasami sets can be further extended into the very very large family and so on. However, the larger the number of codes, the larger the maximum cross-correlation value; thus, the overall performance of the system degrades.

For BCH dual-codes, the number of codes and maximum cross-correlation value depends on the error correction capability of the corresponding block code [9].

Recently, complex spreading sequences, called 4-phase sequences, have raised interest because of their possible improvement to the maximum cross-correlation value [14]. Complex spreading sequences are suitable for quadrature modulation methods such as QPSK and O-QPSK. The most interesting code family is so-called Set A [14]. The code set size is the same as for the Gold sequences.

Table 5.2
Spreading Code Families

	Degree[a]	Number of Codes	Maximum Cross-correlation
Gold	$n = 1 \bmod 2$	$2^n + 1$	$2^{\frac{n+1}{2}} + 1$
Gold	$n = 2 \bmod 4$	$2^n + 1$	$2^{\frac{n+2}{2}} + 1$
Gold like	$n = 0 \bmod 4$	2^n	$2^{\frac{n+1}{2}} - 1$
Gold like	$n = 0 \bmod 4$	$2^n + 1$	$2^{\frac{n+2}{2}} - 1$
S-Kasami	$n = 2 \bmod 4$	$2^{\frac{n}{2}}$	$2^{\frac{n}{2}} + 1$
L-Kasami	$n = 1 \bmod 2$	$2^{\frac{n}{2}}(2^n + 1)$	$2^{\frac{n+2}{2}} + 1$
	$n = 0 \bmod 4$	$2^{\frac{n}{2}}(2^n + 1) - 1$	$2^{\frac{n+2}{2}} + 1$
VL-Kasami	$n = 2 \bmod 4$	$2^{\frac{n}{2}}(2^n + 1)^2$	$2^{\frac{n+4}{2}} + 1$
Dual-BCH	$n = 0 \bmod 2$	$2^{\frac{n}{2}}$	$2^{\frac{n+2}{2}} + 1$
4-phase Set A	$n > 0$	$2^n + 1$	$2^{\frac{n}{2}} + 1$

[a] Code length is $2^n - 1$. *Source*: [9].

5.6.3 Orthogonal Codes

Orthogonal sequences are completely orthogonal for zero delay. For other delays, they have very bad cross-correlation properties, and thus they are suitable only for

112

synchronous applications. The downlink of a cellular system is synchronous without multipath. However, due to multipath, an overlay code is required to suppress the multipath interference. Furthermore, an overlay code is required due to bad auto-correlation properties. The performance gain of the orthogonal Walsh codes depends on the channel profile, delay spread, and path loss. The better the isolation between cells and the smaller the multipath spread, the higher the performance gain [15].

The Walsh codes used in IS-95 are an example of orthogonal codes. Walsh code can be easily despread by a Walsh-Hadamard transform. The length of a Walsh code is an even number of chips, and the number of codes is equal to the number of chips (e.g., there are 128 codes of length 128). A variable spreading factor scheme is one way of transmitting variable data rates (see Section 5.9). A Walsh code of length n can be divided into two codes of length $n/2$. All $n/2$ length codes generated from length n codes are orthogonal against each other. Furthermore, any two length $n/2$ and n codes are orthogonal except when one of the two codes is a mother code of the other. Walsh codes can be constructed according to (5.1).

$$H_0 = [0] \qquad H_n = \begin{bmatrix} H_{n-1} & H_{n-1} \\ H_{n-1} & \overline{H}_{n-1} \end{bmatrix} \tag{5.1}$$

Another example of orthogonal codes are tree-structured orthogonal codes as proposed in [16]. Orthogonality between the different spreading factors can be achieved by tree-structured orthogonal codes whose construction is illustrated in Figure 5.7.

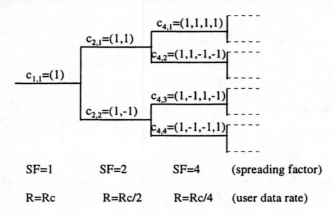

Figure 5.7 Construction of orthogonal spreading codes for different spreading factors.

The tree-structured codes are generated recursively according to the following equation:

$$c_{2n} = \begin{pmatrix} c_{2n,1} \\ c_{2n,2} \\ \vdots \\ c_{2n,2n} \end{pmatrix} = \begin{pmatrix} \begin{pmatrix} c_{n,1} & c_{n,1} \\ c_{n,1} & -c_{n,1} \end{pmatrix} \\ \vdots \\ \begin{pmatrix} c_{n,n} & c_{n,n} \\ c_{n,n} & -c_{n,n} \end{pmatrix} \end{pmatrix} \qquad (5.2)$$

where c_{2n} is an orthogonal code set of size $2n$. The orthogonality properties of these codes are similar to the variable length Walsh codes. Actually, this is just another way to generate Walsh functions. The order of the functions in the matrix is not the same as in the Hadamard matrix, but the functions themselves are the same. The generated codes within the same layer constitute a set of orthogonal functions and are thus orthogonal. Furthermore, any two codes of different layers are also orthogonal except when one of the two codes is a mother code of the other. For example, code $c_{4,4}$ is not orthogonal with codes $c_{1,1}$ and $c_{2,2}$.

5.6.4 Selection Criteria

The correlation properties and the number of codes are two important selection criteria when designing or selecting the most appropriate code family for a specific air interface design.

5.6.4.1 Correlation Properties

Since in an asynchronous system the codes can have any phase shift relative to each other, cross-correlation properties of PN codes must be optimized. In addition, the optimization of the autocorrelation properties is important for acquisition performance. In what follows, we discuss different optimization criteria used for code design and their impact on system performance.

Several optimization criteria for spreading code design have been presented in the literature [9]. Typically, different variations of correlation optimization are used. However, quite often they lack a link to the final system performance, which is measured in terms of bit error rate(BER) and average signal-to-noise ratio (SNR) [9].

The code design criterion used most often is minimization of the maximum peak even cross-correlation value. However, for reasonably large code lengths (255), it has not been shown to give any difference between code families since both the odd and the even cross-correlations impact the performance [9]. In addition, the minimization of the maximum correlation value does not minimize the overall multiple access interference, since in an asynchronous system all phase shifts impact the system's performance. Therefore, the distribution of cross-correlation values is more important than the peak cross-correlation value.

The maximum correlation value has more meaning when the SNR is large, the number of users is small, or when the codes are very short. This is the case in a variable spreading factor system for high data rates. However, optimization of codes for this kind of situation is difficult since the codes should be optimized against both codes of the same length and other spreading factors with different code lengths. If so-called subcode modulation is used [17], where the length of the spreading code is equal to the longest symbol corresponding to the lowest symbol rate, integration is not done over the full length of the sequence for higher symbol rates. Thus, the partial cross-correlation properties determine the performance. In practice, the optimization of the partial crosscorrelations is, however, difficult. Of course, in a situation when two high bit rate users with the same data rate co-exist within the same bandwidth, a suitable selection of spreading codes would improve the performance.

Furthermore, the maximum correlation value has significance when users have a delay difference that produces this value. In this case, if the relative positions of the users do not change, then the observed performance degrades drastically. So, even if the system performs well on average, some users can experience low quality. This problem does not occur with long spreading codes in spite of occasional large cross-correlation values, since correlation changes with every data bit. However, for spreading factors of 127 and 255, short spreading codes have a performance advantage of 0.5 to 1.0 dB over long spreading codes [9].

A more suitable optimization criterion than the maximum correlation value is the optimization of mean-square (MSQ) value of the cross-correlation values. The MSQ value of even (periodic) and odd cross-correlation values is almost equal for Gold, Kasami, and M-sequences [9].

From the performance point of view, it seems that any of the pseudo-noise codes presented in Section 5.6.2 could be used as the spreading codes for an asynchronous DS-CDMA system. Short spreading codes do have a small performance advantage over long codes [9].

5.6.4.2   Number of Codes

In the case of long codes, the number of codes is sufficient to accommodate enough users in a cell. On the other hand, one might sometimes run out of short spreading codes. In this section, different factors influencing the required number of spreading codes are discussed.

The required number of codes depends on the expected traffic load and spectrum efficiency. A large number of codes is required to support low bit rate speech users and also when a multicode transmission method is used (see Section 5.9.1). As an example, assume a spectrum efficiency of approximately 100 Kbps/MHz/cell for 5-MHz wideband CDMA [18]. With voice activity of 50% and voice codec rate of 8 Kbps, this would result into 125 users per cell (= 5·100/8·0.5).

In the downlink, orthogonal codes are usually used for channel separation. For example, there are 256 codes of length 256, and thus, even with a dedicated signaling channel using a different Walsh code than the traffic channel for every user, the number of codes would be enough in the above-described example. However, if the neighboring cells are empty, then one cell can accommodate a much higher number of users and more codes are required. Furthermore, mobiles in soft handover consume codes from at least two base stations. One way to increase the number of codes is to allocate the same set of orthogonal codes with different pilot codes for the same sector. However, this would degrade the system performance since the two sets of codes were no longer orthogonal. Also, quasi-orthogonal codes have been proposed to increase the number of orthogonal codes available [19].

In a variable data rate system, code allocation for different connections can be either static or dynamic [20]. The dynamic code allocation releases the spreading codes when the user is inactive. In static allocation, a connection is allocated a number of codes to support the maximum data rate. If the activity factor were very low, each connection would only use a very small part of the allocated codes. If there are several such users, then the base station might run out of codes. Therefore, the dynamic code allocation is the preferred method, especially in a packet data system with highly variable data rates.

In the uplink, each user has its own individual spreading code. If short codes and QPSK modulation are used, 250 codes are required for each cell assuming the same spectrum efficiency of 100 Kbps/MHz/cell as in the downlink example. Since the same codes cannot be reused in the neighboring cells, a much larger number of codes is required. Thus, it seems that for length 255 codes only the VL-Kasami set has enough codes (over one million codes). The L-Kasami set could be used as well, but then the code reuse factor is 16 (= 4112/250), which might not be enough.

5.6.4.3   Other Code Design Requirements

Orthogonal codes have an even number of chips in contrast to PN-sequences which have an odd number of chips. If both types of codes are used in the system, it is beneficial for the spreading code length to be a multiple of a power of 2 in order to facilitate synchronization. Therefore, PN-sequences should be extended by one chip. This extra chip should be selected in such a way that the number of ones and zeros in the extended sequence is equal. In general, this extension of the code sequence does not change the correlation properties significantly.

For the access and synchronization channel, synchronization requirements might impact the spreading code selection. Acquisition time should be short, and thus, the number of codes cannot be too large or the codes too long. However, if long codes are used, then special synchronization schemes might be needed to reduce uncertainty in the code phase (see Section 6.3.4).

Receiver complexity impacts the selection of spreading codes. Multiuser detection, for example, might require the use of short spreading codes to reduce implementation complexity.

5.7   MODULATION

Figure 5.8 illustrates DS-CDMA transmission system functions related to modulation. The *data modulator* maps the m incoming coded data bits into one of M (= 2^m) possible real or complex valued transmitted data symbols. The M-ary data symbols are fed to the *spreading circuit,* where the resulting signal is filtered and mixed with the carrier signal (i.e., spreading modulation is performed). Typical data modulation schemes are BPSK and QPSK. Spreading circuits can be binary, balanced quaternary, dual-channel quaternary, or complex DS-spreading. Typical spreading modulation schemes are

BPSK used with a binary spreading circuit, QPSK, and O-QPSK used with a quaternary or complex spreading circuit.

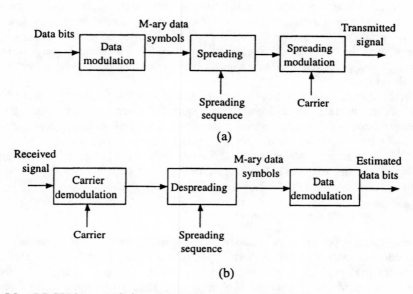

Figure 5.8 DS-CDMA transmission system: (a) transmitter and (b) receiver.

5.7.1 Data Modulation

Binary data modulation maps incoming data bits into transmitted data symbols with a mapping rule: $(1 \Rightarrow +1)$, $(0 \Rightarrow -1)$. Quaternary data modulation converts two consecutive data bits into one complex data symbol. Typically, Gray encoding is used, which has the following mapping rule: $(00 \Rightarrow +1+j)$, $(01 \Rightarrow -1+j)$, $(11 \Rightarrow -1+j)$, $(10 \Rightarrow +1-j)$. The resulting complex data symbol can be transmitted with quaternary or complex spreading. It should be noted that the quaternary data modulation does not give any advantage in a DS-CDMA system, as compared to binary data modulation. Actually, it is slightly more sensitive to phase errors due to cross coupling of the I and Q channels.

Even higher order modulation methods such as 16-QAM have been proposed for CDMA to increase data throughput. However, the increase in data rate does not offset the higher carrier-to-interference ratio (C/I) requirement. With multiuser detection the difference is smaller since most of the interference is removed.

Demodulation can be noncoherent, differentially coherent, or coherent. With an ideal synchronization and without an external reference signal, the coherent reception gives a 3-dB advantage over noncoherent reception in an additive white Gaussian noise

(AWGN) channel. Coherent demodulation requires a phase reference, which can be obtained either from the data bearing signal or from an auxiliary source, such as external reference symbols or the pilot signal. Thus, the actual gain achieved against non-coherent detection depends on the amount of radio resources spent for reference symbols/pilot signal and performance loss due to errors in phase estimation.

5.7.1.1   M-ary Modulation

In addition to the above-described conventional M-ary data modulation, which converts incoming data symbols into complex valued symbols, the M-ary data modulator can convert m incoming data symbols into one of M ($M = 2^m$) possible data symbols represented by K channel symbols. These can then be transmitted using binary, balanced quaternary, quaternary, or complex spreading schemes. The M-ary modulator spreads the transmitted signal in the frequency domain. An example of such a scheme is the IS-95 uplink, where orthogonal 64-ary modulation is used (i.e., six consequent data symbols are used to select one Walsh-Hadamard symbol length of 64) [10]. In this way, the energy of six symbols is collected before a quadratic combiner at the receiver and the performance is improved when compared to binary modulation. The choice of the code rate depends on implementation complexity and channel coherence time, and performance improves as long as the carrier phase remains constant over the duration of a given M-ary symbol. Implementation of coherent detection is more difficult for this kind of scheme but has been proposed in [21–23] and is discussed further in Section 5.8.7.

5.7.2   Spreading Circuits

The binary DS-spreading circuit has a simple transmitter structure, and only one PN-sequence is needed for spreading. Its drawback is that it is more sensitive to multiple

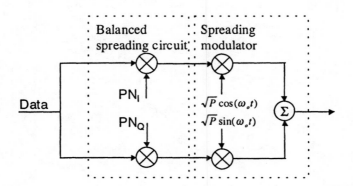

Figure 5.9   Balanced quaternary transmitter.

access interference than balanced quaternary spreading [24].

Balanced quaternary DS-spreading spreads the same data signal into I and Q channels using two spreading sequences. Figure 5.9 illustrates the balanced quaternary DS transmitter.

In the dual-channel QPSK spreading circuit depicted in Figure 5.10, the symbol streams on the I and Q channels are independent of each other. They can even have different powers. Similar to the balanced quaternary spreading circuit, two spreading sequences are needed.

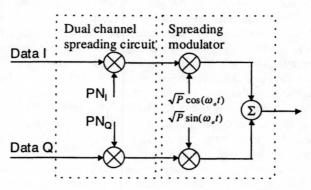

Figure 5.10   Dual-channel quaternary transmitter.

Figure 5.11 depicts a complex spreading scheme. Complex spreading uses either a complex spreading sequence or two real spreading sequences. Complex spreading reduces the peak to average ratio of the signal. It should be noted that with complex spreading, offset QPSK should not be used, since it will actually increase the peak-to-average power ratio. For despreading architectures see Section 10.4.1.3.

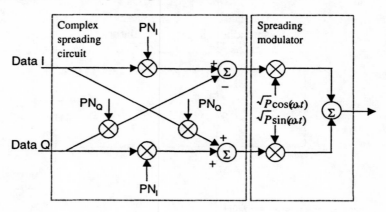

Figure 5.11   Complex spreading transmitter.

5.7.3   Spreading Modulation

Typically, linear modulation methods (BPSK, QPSK, and offset QPSK) have been proposed for wideband CDMA because they offer good modulation efficiency. Band-limited linear modulation methods with nonlinear power amplifiers result in spectrum regrowth (i.e., spectrum leakage to adjacent carriers). Thus, especially in the uplink, the modulated signal should tolerate nonlinear power amplifier effects as much as possible, since a nonlinear power amplifier is more power efficient than a linear one. In practice, after the modulation method has been fixed, pulse shaping determines the final spectrum properties of the modulation scheme. As discussed in Chapter 8, the roll-off factor depends on power amplifier linearity and adjacent channel attenuation requirements.

BPSK spreading modulation can be used with the binary spreading circuit. Non-filtered BPSK has a constant envelope and an infinite spectrum. However, in practice the BPSK signal is filtered, resulting in a nonconstant envelope and higher linearity requirements.

QPSK and offset QPSK are quadrature types of modulation methods. The offset QPSK has a one half chip delay in the Q channel, which prevents phase transitions via zero, as can be observed in Figure 5.12. Therefore, filtered O-QPSK has fewer envelope variations than QPSK, thus reducing the linearity requirements on the power amplifier. This is especially important in the uplink. However, if complex spreading is used, then QPSK should be used instead of offset QPSK.

The scattering diagrams of BPSK, QPSK, and offset QPSK modulation methods are shown in Figure 5.12.

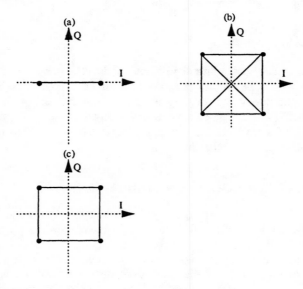

Figure 5.12   Scattering diagrams: (a) BPSK, (b) QPSK, and (c) O-QPSK.

5.8   ERROR CONTROL SCHEMES

Error control schemes can be classified in two basic categories: forward error control (FEC) and automatic repeat request (ARQ) schemes [25]. In addition, a combination of FEC and ARQ, a hybrid scheme, can be used [25]. In FEC, an error correcting code, sometimes referred to as the channel code, is used to combat transmission errors in a fading radio channel. In ARQ, an error detecting code together with a retransmission protocol is used. In the hybrid scheme, FEC reduces the need for retransmission by improving the error rate before ARQ.

5.8.1   Selection of Error Control Scheme

The choice of the error control scheme depends on the required quality of service (data rate, delay and BER) and the radio channel. Raw channel error rates for transmission in a fading channel are typically on the order of 10^{-1} to 10^{-2}. For speech, the resulting error rate after demodulation must be on the order of 10^{-2} to 10^{-3} or better. This can be achieved, for example, with convolutional codes, which is the standard coding for all cellular systems. For data transmission, a bit error rate of 10^{-6} or better is required. This can be achieved with a concatenated coding scheme using a convolutional and a block code — typically a Reed-Solomon code [26]. Furthermore, nontransparent data services can utilize an additional ARQ scheme to improve the error performance. Maximum transmission delay of a service determines the maximum interleaving depth. Since the application requirements (BER, bit rate, and delay) vary greatly, the coding scheme should be adaptable.

5.8.2   Convolutional Codes

For convolutional codes, constraint length and code rate need to be selected. The constraint length should be as large as possible to obtain good performance [27]. However, the complexity of the decoder increases with increasing constraint length. State-of-the-art VLSI (very large scale integration) implementations have been obtained for constraint length 9 convolutional codes. The code rate depends on the interleaving depth and the coherence time of the channel.

5.8.3   Concatenated Reed-Solomon/Convolutional Coding

This coding scheme employs an outer Reed-Solomon (RS) code and an inner convolutional code in conjunction with outer and inner interleavers [28]. The use of an inner bit interleaver/deinterleaver pair renders the fading channel memoryless and allows the convolutional code to mitigate multiuser interference more effectively [26]. The burst errors at the output of the Viterbi decoder are randomized by the outer symbol interleaver. The concatenated coding for mobile radio systems is designed in a way that the convolutional code produces an error rate on the order of 10^{-3}, and the RS code then improve this to a level of 10^{-6}. Of course, the actual improvement depends on the channel and on interleaving lengths. In addition, the error detection capability of the RS code can be used to provide an indication for automatic retransmission request for nontransparent data services.

5.8.4 Turbo Codes

Recently, turbo coding has received attention as a potential FEC coding scheme for cellular applications [29–32]. Turbo codes are parallel or serial concatenated recursive convolutional codes, whose decoding is carried out iteratively [29]. They have been demonstrated to closely approach the Shannon capacity limit on both AWGN and Rayleigh fading channels. The turbo code encoding structure is based on a combination of two or more weak error control codes. The information bits are interleaved between two encoders, generating two parity streams. These streams are then multiplexed and possibly punctured. The whole process results in a code that has powerful error correction properties.

5.8.5 Hybrid ARQ Schemes

The type-I hybrid ARQ scheme uses a code designed for error detection and correction [25]. Thus, it requires more parity check bits (overhead) than a code used for error detection only. Type-II hybrid ARQ codes the first transmission of a message with parity check bits used solely for error detection [25]. If an error is detected, the erroneous word is saved in the receiver buffer and a retransmission is required. The retransmission is a block of parity check bits formed from the original message and an error-correcting code. These parity check bits are then applied to the stored code word in the receiver. In case there is still an error, the next retransmission might contain either the original code word or an additional block of parity check bits.

5.8.6 Interleaving Schemes

A radio channel produces bursty errors. However, since convolutional codes are most effective against random errors, interleaving is used to randomize the bursty errors. The interleaving scheme can be either block interleaving or convolutional interleaving [27]. Typically, block interleaving is used in cellular applications.

   The performance improvement due to interleaving depends on the diversity order of the channel and the average fade duration of the channel. The interleaving length is determined by the delay requirements of the service. Speech service typically requires a shorter delay than data services. Thus, it should be possible to match the interleaving depth to different services.

5.8.7 Combined Channel Coding and Spreading

Very low rate coding schemes, which involve a combination of channel coding and spectrum spreading, have been presented by Hui and Viterbi [21-22]. In these schemes, signal spreading is performed using a very low rate convolutional code, and users are separated with PN sequences, which have the same rate as the channel symbol rate (i.e., spectra is not spread by the PN sequence at all but it is used to randomize the users' signals against each other). Hui actually proposed a combination of a very low rate convolutional code and short spreading sequences. The encoder consists of a binary convolutional encoder, a simple shift register of length k (k = constraint length of the code), and a M-ary orthogonal signal mapper [23]. The k code bits are used to select one of the M orthogonal code words (e.g., Walsh symbol set). The code rate is now $1/M$. It is also possible to use bi-orthogonal signal sets, as presented in [23].

5.9 MULTIRATE SCHEMES

Multirate design means that different services with different quality of service requirements are multiplexed together in a flexible and spectrum-efficient way. The provision of variable data rates with different quality of service requirements can be divided into three sub-problems:

- How to map different bit rates into the allocated bandwidth;
- How to provide the desired quality of service;
- How to inform the receiver about the characteristics of the received signal.

The first problem concerns issues like multicode transmission and variable spreading. The second problem concerns coding schemes, and the third problem concerns control channel multiplexing and coding.

In addition to the above basic requirements, the design of a multirate solution needs to consider a number of other requirements and constraints. The uplink and downlink directions set different requirements for the multirate solution. Mobile station power amplifier requirements should not be too stringent, in order to facilitate use of power-efficient power amplifiers. In the downlink, the multirate solution should allow the mobile station receiver to save processing power.

5.9.1 Accommodation of Higher Data Rates

The increase of data rate in a CDMA system can be implemented by two basic schemes: *variable spreading factor* (VSF) and *multicode*. In a VSF scheme, the spreading ratio is reduced as the data rate increases. In a multicode scheme, additional parallel codes are allocated as the data rate increases. It is also possible to have combinations of these. Figure 5.13 illustrates the multicode transmission scheme. The high bit rate data stream is split into N parallel channels, each with data rate R_b. With multicode transmission, the mapping of simultaneously transmitted services into frames can be performed in two different ways. In the first way, services are transmitted simultaneously in different frames and in different codes. In the second way, services are mapped into the same frame, whose bits are then mapped into different codes. The advantage of this flexibility in service mapping will be discussed further in conjunction with power control (see Section 5.9.1.4).

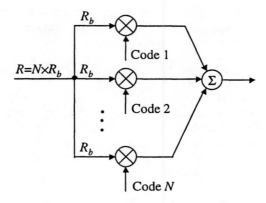

Figure 5.13 Multicode transmission scheme.

In the selection of a multirate transmission scheme, the following criteria need to be considered:

- Performance in the presence of multipath;
- Multiple access interference characteristics;
- Forward link orthogonality;
- Power control requirements;
- Complexity of the receiver and transmitter.

5.9.1.1 Performance in the Presence of Multipath

The performance of multicode and VSF schemes in the presence of multipath is considered. In the case of no multipath, there is no self-interference, assuming fully orthogonal spreading sequences [20]. When multipath exists, the self-interference is the same since in the case of multicode, there is N times less interference from other codes due to the higher spreading factor but there are N interfering codes instead of only one, as in the single code approach. Thus, in this respect, there seems to be only a small difference in the performance, assuming RAKE detection for BER greater than 10^{-6}. A drawback of the VSF scheme is that a very short code duration leads to intersymbol interference (ISI). For example, a data rate of 500 Kbps corresponds to a symbol duration of 2 μs. Thus, with a delay spread of 10 μs, there is ISI over five symbols.

5.9.1.2 Multiple Access Interference Characteristics

A fundamental assumption behind a matched filter detection approach is that the multiple access interference is Gaussian. The VSF approach may violate this assumption since there may be a high power user degrading the performance of the low rate users. In addition, for the highest data rates the spreading factor of the VSF scheme is very small; thus, a high maximum cross-correlation relative to autocorrelation peak might degrade the performance. Therefore, in some instances it might be better to implement higher bit rates with a multicode scheme.

5.9.1.3 Forward Link Orthogonality

Since the forward link is synchronous, it is possible to maintain orthogonality between the codes. For multicode transmission we can select a set of orthogonal codes. With VSF, it is also possible to maintain orthogonality between different spreading factors if we impose the following constraint between different rates:

$$R = R_c/2^n \qquad n = 0, 1, 2, \ldots \tag{5.3}$$

This can be achieved with the variable length Walsh sequences or with the three structured orthogonal codes [16] discussed in Section 5.6.3.

5.9.1.4 Power Control Requirements

In order for the receiver to be able to estimate the path loss for power control purposes, transmission power should not vary. Otherwise, the receiver has to first know the rate of a traffic channel. Another possibility is to use a fixed rate control channel for the power measurements. With multicode transmission, each of the parallel channels has a fixed power. For VSF, the power varies according to transmission rate, and thus, explicit transmission power needs to be signaled to the receiver. One further advantage of multicode transmission is that in case of several simultaneous services, the QoS can be adjusted using power control. Parallel service with similar power requirements can be time multiplexed, while code multiplexing is used for parallel services with different power requirements.

5.9.1.5 Complexity

Power amplifier linearity requirements should be as low as possible to allow the use of power-efficient power amplifiers. The multicode scheme results in larger envelope variations than the single code transmission. An additional drawback of the multicode scheme is that it requires as many RAKE receivers as there are codes. However, each RAKE receiver may be less complex, since a higher spreading factor might facilitate the use of fewer bits for quantization [20].

5.9.2 Granularity of Data Rates

Granularity means the minimum possible bit rate change. In order to maintain high flexibility, it is desirable to have as fine of granularity as possible. For multicode, granularity is R/M Kbps. If one code has the capability of smaller quantization of data rates, then better granularity can be achieved. For the VSF scheme, granularity varies according to the spreading factor and is finest for the low bit rates [20]:

$$\Delta R = \frac{R_c}{SF-1} - \frac{R_c}{SF} = \frac{R_c}{\frac{R_c}{R}-1} - R = \frac{R^2}{R_c - R} \qquad (5.4)$$

where SF is the spreading factor, R_c is the chip rate, and R is the user bit rate.

For low bit rate services, this would most likely be sufficiently fine granularity. However, if we restrict the spreading factors according to (5.2), then the granularity is too coarse. Furthermore, if we assume for multicode transmission that one code carries 10 Kbps of traffic, the basic granularity is 10 Kbps. This is too large for some services such as low rate speech codecs, which typically require better granularity. The problem of granularity can also be considered from the viewpoint that, given the coded user bit rate, it is possible to match it with the given chip rate. For example, if one code carries 32 ksymbol/s and the source data rate is 9.6 Kbps coded with a 1/3 convolutional code, it results in a gross bit rate of 28.8 Kbps. Thus we need to match the 28.8 Kbps to 32 Kbps.

One alternative to matching the bit and chip rate is the use of rate compatible punctured codes (RCPC) [33]. The basic idea of RCPC coding is to divide the bit stream of the mother code into blocks whose bits are either punctured or repeated according to a perforation/repetition matrix. The drawback of this approach is that a fixed number of rates have to be selected, since not all service rates can be directly matched to the available chip rate. Furthermore, for RCPC, good codes are only known up to the constraint length of 7.

Repetition coding, or puncturing, is another possibility for rate matching. Even unequal repetition coding (i.e., only some symbols in the frame are repeated to implement the desired symbol rate within the frame) can be used [17].

5.9.3 Transmission of Control Information

In order to vary the data rate or other service parameters, the receiver needs to know the structure of the received signal. This can be achieved either by transmitting explicit control information or by blind rate detection, for example, from the CRC information [34]. In case explicit control information is transmitted, the following issues need to be solved:

- Coding of the control information to achieve desired quality of service;
- Multiplexing of the control information;
- Position of the control information.

The control bits have to have a considerably lower error rate than the information bits, since, if a control word has an error, the whole frame is lost. Control information can either be coded together with the user data or independently from the user data. Furthermore, control information can be transmitted either code multiplexed or time multiplexed.

There are two possibilities for the position of control information: in the previous frame or in the same frame as the user data, as illustrated in Figure 5.14. Since the receiver is informed in advance of the transmission parameters, the processing of the data can be done "on-line." However, for services with a short delay requirement the additional delay of one frame might be too long. A further drawback is that erroneous control information results in loss of the previous and the next frame, except if the next frame is transmitted with the same parameters as the previous frame.

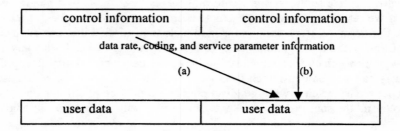

Figure 5.14 Control information transmitted (a) in the previous frame, (b) in the same frame as user data.

In case the control information is transmitted within the same frame as the user data, then the receiver needs to decode the control information first. Thus, the data need to be buffered. Memory requirements for the buffer depend on the spreading code solution. In case the spreading ratios are multiples of each other, it is possible to despread with the lowest spreading ratio and to buffer the sub-symbols after despreading, instead of the samples before the despreading, which need to be used if arbitrary spreading ratios are used. However, for high data rates buffering might be still a problem.

5.10   PACKET DATA

Since non real-time packet data services are not delay sensitive, they use the retransmission principle implemented with ARQ protocol to improve the error rate. The retransmission protocol can be either implemented in layer 2 as part of MAC and RLP or in the physical layer (layer 1). If packet data retransmission is implemented as part of the layer 2, the transmission of packet data in the physical layer does not differ from the transmission of circuit switched data. So the multirate aspects discussed above also apply to the transmission of packet data. If the physical layer ARQ is used, then the physical layer is modified depending on the ARQ scheme used (see Section 5.8 about ARQ schemes). In both cases, the access procedure and handover for packet data services have certain special implications, which are discussed in the next subsections.

5.10.1 Packet Access Procedure

The packet access procedure in CDMA should minimize the interference to other users. Since there is no connection between the base station and the mobile station before the access procedure, initial access is not power controlled and thus the information transmitted during this period should be minimized. There are three scenarios for packet access:

- Infrequent transmission of short packets containing little information;
- Transmission of long packets;
- Frequent transmission of short packets.

Since the establishment of a traffic channel itself requires signaling and thus consumes radio resources, it is better to transmit small packets within the random access message without power control. For long and frequent short packets, a dedicated traffic channel should be allocated.

If a dedicated channel has been reserved and there is nothing to send, the mobile station either cuts off the transmission or keeps the physical connection by transmitting power control and reference symbols only. In the former case, a virtual connection (higher layer protocols) is retained in order to rapidly re-establish the link in case of a new transmission. Selection between these two alternatives is a trade-off between resources spent for synchronization and power control information, and resources spent for random access.

5.10.2 MAC Protocol

The task of the medium access protocol is to share the transmission medium with different users in a fair and efficient way. Sometimes, multiple access protocols such as FDMA, CDMA, and TDMA are also classified as medium access protocols. However, as already discussed in the beginning of this chapter, the medium access protocol is part of the link layer, while the multiple access scheme is part of the physical layer. The medium access protocol has to resolve contention between users accessing the same physical resource. Thus, it also manages the packet access procedure described in the previous section. Since the third generation systems offer a multitude of services to customers at widely varying quality of service requirements, the MAC needs to offer capabilities to manage the access demands of different users and different service classes. This can be performed using reservation and priority schemes. Services with delay constraints can use a reservation scheme to reserve capacity to guarantee the quality of service. Priority schemes can be used to prioritize the requests from different services.

5.10.3 Packet Data Handover

Since CDMA operates with a reuse factor of one, it needs efficient and fast handover in order to avoid excessive interference with the other cells. This has been realized with soft handover in the case of circuit switched connections. Soft handover also improves performance through increased diversity. For packet connections, and especially for short packets, there may be no need to establish soft handover even if the user is at the cell edge. However, there is still the need to route packets via the base station that provides the best connection. This is more important with frequent packet transmissions than with infrequent transmission. This can be implemented with frequent rescheduling of packet data transmission (i.e, if the transmission time exceeds a predefined timer, rescheduling is performed and the base station offering the best connection is selected).

If soft handover is used, then the implementation of ARQ depends on where it is terminated (i.e., where in the network the retransmission decision is made). If the ARQ is terminated after the soft handover combining device typically situated in the BSC, then ARQ is implemented as without soft handover. The only drawback of this solution is that it increases the delay for retransmission because of the fixed network transmission between the BTS and the BSC.

If ARQ is terminated in the BTS before the handover combining unit, a more complicated protocol is required. The mobile station transmits a packet in the uplink direction. Since all base stations involved in the handover receive the packet and each of them makes an independent decision whether there was an error or not in the packet, the problem is how to transmit and coordinate the retransmission requirements. The mobile station should retransmit only if both cells require retransmission. However, a return channel would be required for both links. Furthermore, it would not be possible to combine the information from the two links like the power control information, since acknowledgments are coded. Therefore, this scheme looks impractical.

5.11   TRANSCEIVER

In this section, the receiver and transmitter functions of the mobile and base stations and their relation to the overall system design are covered. Receiver algorithms for detection, code tracking, and acquisition are introduced and design examples are given. Based on this, their impact on different radio interface parameters is discussed. On the transmitter side especially, the power amplifier design and its impact on system parameters are explained.

5.11.1   Receiver

Figure 5.15 depicts the block diagram of a RAKE receiver. A RAKE finger despreads the received signal with a correlator. For coherent demodulation, the despread signal is multiplied by a complex amplitude to correct phase error and to weight each finger according to the selected combining strategy (maximal ratio or equal gain combining). The impulse response (IR) measurement block continuously measures the multipath profile. Whenever the delays of the impulse response change, the IR measurement block allocates new code phases for the code tracking block that tracks the small changes. After combining the signals from different RAKE fingers, deinterleaving and decoding of the channel code is performed. In addition, a searcher, not shown in the figure, is continuously scanning the neighboring cell pilots in order to provide pilot measurements for handover.

The number of RAKE fingers depends on the channel profile and the chip rate. The higher the chip rate, the more resolvable the paths. However, more RAKE fingers are needed to catch all the energy from the channel to maintain good performance (see

Section 7.2.2). A very large number of RAKE fingers leads to combining losses. Practical implementation architecture for a RAKE receiver is discussed in Chapter 10.

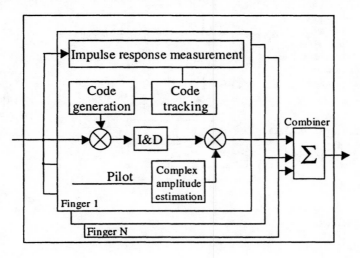

Figure 5.15 Block diagram of CDMA RAKE receiver.

5.11.1.1 *Impulse Response Measurement*

The impulse response measurement correlates the received signal with different phases of the pilot code to find the multipath components. The required measurement speed for IR measurement depends on the mobile speed and the radio environment. The faster the mobile station is moving, the faster the measurements need to be performed in order to catch the best multipath components for the RAKE fingers. Furthermore, in a long delay spread environment, the scanning window needs to be wider (see Chapter 11). In addition to measurement functions, this block performs RAKE finger allocation (i.e., assigning multipath components for RAKE fingers). For the code allocation, different strategies can be applied. Allocation can be done after the whole impulse response has been measured or immediately after a strong enough multipath component has been found.

5.11.1.2 *Code Acquisition*

Code acquisition is performed prior to the acquisition of system synchronization. The mobile station scans through the pilot signals. A priority order for the pilot signals may be set based upon the last pilot and neighboring pilots. If the connection is lost for some reason, scanning starts from the highest priority pilots. In the case of high interference, code acquisition could become a system bottleneck. A matched filter depicted in Figure 5.16 can be used for fast code acquisition. The predefined parallel data corresponds to the spreading code chips.

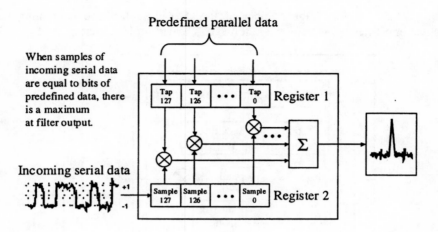

Figure 5.16 Matched filter.

5.11.1.3 Code Tracking

Typical implementation of a code tracking loop is an early late delay locked loop. It consists of two correlators (early and late), which are allocated half a chip apart from the on-time correlator. Depending on the correlation result, the code phase is adjusted. Tracking loop performance depends on loop bandwidth. If the update is faster than a multipath component moves in delay, then synchronization errors are negligible, but, on the other hand, loop noise increases. The requirements also depend on the detection strategy (i.e., whether conventional or multiuser detection is applied). The impact of code tracking errors on conventional detectors (matched filters) have been studied in [35, 36]. A comprehensive study of code tracking and acquisition can be found in [37].

5.11.1.4 Complex Amplitude Estimation

The complex amplitude estimation includes both amplitude and phase estimation. In the case of maximal ratio combining, the signal is weighted with the complex conjugate of the complex amplitude. If equal gain combining is performed, only the phase error is corrected, and each RAKE finger is considered with equal weight. The complex amplitude estimate needs to be averaged over a reasonably long period, while coherence time sets the upper limit for the averaging time (i.e., the channel should not change during the estimation).

5.11.1.5 Searcher

A searcher scans other cell pilots. During a call, the mobile station is scanning through the pilot signals, as well as measuring the downlink interference and possibly receiving uplink interference results. Since the number of pilots is very large, it might take a long time before a rising pilot in a neighboring set is noticed. Thus, searcher time can limit the system performance especially in a micro cellular environment, where a new base station can become active very fast due to the corner effect (see Section 11.5.1). One possibility of reducing the required hardware is to have flexible allocation between RAKE and searcher fingers. This would increase the scanning effectiveness in a low multipath environment. The number of fingers for searching depends on the desired speed for pilot scanning. Search window planning is discussed in Chapter 11 and in [38].

5.11.2 Transmitter

Figure 5.17 is a block diagram of a DS-CDMA transmitter. The data coming from the source is channel encoded, interleaved, and possibly different services are multiplexed. The resulting frames are passed through spreading and filtering, D-A converted, and passed trough IF and RF parts. The transmission parameters are controlled by the TX control block.

The most important component from the point of view of complexity and cost is the power amplifier in the RF section, especially for the mobile station. A nonlinear power amplifier can be operated near the compression point, and thus, it has good power efficiency resulting in low battery consumption. The linearity requirements are set by the sensitivity of the transmitted signal waveform for the nonlinearities and adjacent channel requirements. Multicode transmission is more sensitive to power amplifier nonlinearities than single code transmission since its peak-to-average ratio is higher because of the summing of several signals. If orthogonal Walsh sequences are used, the signal should be scrambled to avoid a high peak-to-average ratio. This is because if Walsh functions modulated with all ones or zeros are summed, then they form an impulse, which places high linearity requirements on the power amplifier. Nonlinear power amplifiers are discussed in more detail in Chapters 8 and 10.

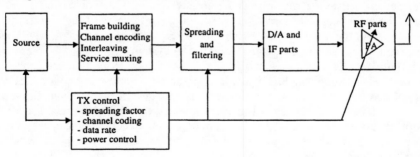

Figure 5.17 Transmitter block diagram.

5.11.2.1   Transmit Diversity

The forward link performance can be improved by transmit diversity. So far, in CDMA systems, delay transmit diversity has been applied. It has, however, two main drawbacks: the loss of orthogonality that results in self-interference, and the requirement for additional RAKE fingers in the mobile station. Therefore, several more advanced transmit diversity concepts have been developed for third generation CDMA systems.

   Time division transmit diversity (TDTD) can be implemented by time-orthogonal signalling, by using a pseudo-random antenna hopping sequence. In a more optimal implementation, the transmit antennas are determined using closed loop antenna selection (CLAS). This results in the selective transmit diversity (STD) concept [39]. In order to be effective, the STD concept requires that the control loop delay is kept within the channel coherence time.

   In code division transmit diversity (CDTD), the coded bits are distributed to more than one antenna, and all antennas transmit simultaneously to a given user. In orthogonal transmit diversity (OTD), different orthogonal codes for each antenna are used for spreading eliminating self-interference in flat fading.

   The performance of TSD, TDTD, and OTD schemes has been compared in [39]. Power control with TDTD schemes has been studied in [40].

   In multi carrier CDMA, transmitter diversity can be implemented by transmitting the carriers on multiple antennas. One or more carriers are mapped to one antenna. Since signals transmitted through different antennas experience independent fading, the performance is improved.

5.12   MULTIUSER DETECTION

Multiuser detection (MUD) and interference cancellation (IC) seek to improve performance by canceling the intra-cell interference and thus increasing the system capacity. The actual capacity increase depends on the efficiency of the algorithm, radio environment, and the system load. In addition to capacity improvement, MUD and IC alleviate the near/far problem typical to DS-CDMA systems.

   Figure 5.18 depicts a system using multiuser detection/interference cancellation. Each user is transmitting data bits, which are spread by the spreading codes. The signals are transmitted over a Gaussian multiple access channel. In the receiver, the received signal is correlated with replicas of the user spreading codes. The correlator consists of a multiplier and an integrate and dump function. A matched filter can also be used. Multiuser detection processes the signal from the correlators jointly to remove the unwanted multiple access interference from the desired signal. The output of a multiuser detection block are the estimated data bits.

5.12.1   Capacity and Coverage Improvement

The performance of multiuser detection depends on the captured energy and impact of phase and code tracking errors [41]. The fraction of captured energy is a function of

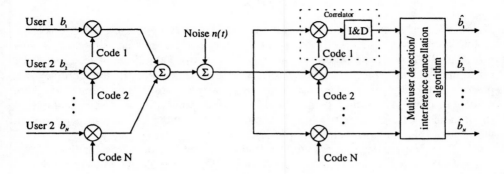

Figure 5.18 System model of multiuser detection.

chip duration multiplied by the number of RAKE branches divided by the delay spread. Furthermore, since typically the interference from a single cell (i.e., intracell interference) is canceled, the intercell interference limits the achievable capacity. If only intracell interference is canceled and intercell interference amounts to a fraction f of the intracell interference, then the boundary for capacity increase is $(1+f)/f$. With a propagation power law of 4, intercell interference is 55% of intracell interference; therefore, ideally, MUD can improve the capacity of the system by a factor of 2.8, compared to a system without MUD [42]. However, in practice, the MUD efficiency is not 100% but depends on the detection scheme, channel estimation, delay estimation, and power control error.

Typically, MUD is studied mainly for the uplink. One of the main motivations, in addition to complexity considerations, to justify the study of multiuser detection for the uplink has been the claim that the capacity of DS-CDMA systems is uplink limited. However, this claim comes from the studies for IS-95. The noncoherent uplink of IS-95 with antenna diversity has normally worse performance than the coherent downlink with orthogonal codes. Even for IS-95, however, the downlink is sometimes the limiting direction [43]. All third generation wideband CDMA systems use coherent detection also in the uplink and thus, together with antenna diversity, the uplink has better performance than the downlink [11,44]. One way of improving downlink performance would be canceling the worst interferers from other cells.

As will be shown in Chapter 7, multiuser detection improves coverage and makes it more independent of the system load. Since MUD efficiency varies in different radio environments, this improvement is not fixed. This impact of MUD on coverage introduces a new variable to the network planning process, since MUD efficiency needs to be taken into account in the coverage design. On the other hand, the load factor is no longer crucial for the coverage area design because cell size does not shrink as fast with an increasing load as in the conventional detector. However, due to orthogonal codes, the impact might be different in the uplink and downlink, especially if MUD is applied only in the uplink. This is because in low delay spread radio environments, forward link coverage does not depend heavily on load.

5.12.2 Development of Multiuser Detection

The idea of multiuser detection was first mentioned in 1979 by Schneider [45]. In 1983, Kohno et. al. published a study on multiple access interference cancellation receivers [46]. In 1984, Verdú proposed and analyzed the optimal multiuser detector and the maximum likelihood sequence detector, which, unfortunately, is too complex for practical implementation since its complexity grows exponentially as a function of the number of users [47]. Consequently, Verdú's work inspired multiple researchers to find suboptimal multiuser detectors, which could achieve a close to optimal performance with reasonable implementation complexity.

The first studies concentrated on finding suboptimal detectors for the AWGN channel [48]. The following wave in the research of multiuser detection were studies on suitable detectors for fading multipath channels, first in slowly fading channels and then in relatively fast fading channels. Both flat fading and multipath channels have been considered. Since the multiuser detector parameters such as amplitude, phase, and cross-correlations between users change over time, studies of adaptive multiuser detectors that self-tune the detector's parameters based on the received signal were started [49]. In the real world, the parameters required for signal detection are never perfect. Consequently, studies taking into account the impact of nonideal estimation of phase, amplitude, and delays have been initiated [35,41]. Channel coding is an essential part of all proposed wideband CDMA air interfaces. However, so far, relatively few studies have considered multiuser detection together with channel coding [50–53]. MUD with channel coding can be either partitioned or integrated. Partitioned approaches treat the multiuser interference equalization problem and decoding problem separately. Integrated approaches perform both the equalization and decoding operations together [50]. Partitioned approaches can be further divided into hard and soft decision approaches. In the soft decision approach a multiuser detection algorithm feeds the soft symbols after detection into a channel decoder. As discussed in Section 5.9, third generation wideband CDMA systems will have multiple data rates. Recently, multiuser detectors for multirate CDMA systems have been proposed in [53-55].

5.12.3 Multiuser Detection Algorithms

Multiuser algorithm classification is depicted in Figure 5.19. Multiuser detection can be either optimal or suboptimal. Suboptimal multiuser detection algorithms can be classified into linear and interference cancellation type algorithms. Some detectors can also be classified into both categories. Furthermore, few other schemes not fitting this classification have been proposed. These include partial trellis-search algorithms [51,52] and neural networks proposed for AWGN channels [56,57].

Another proposed way to classify multiuser detection algorithms is by linear and nonlinear algorithms [58]. However, in this classification the same algorithm can belong to both categories depending on the implementation.

Optimal multiuser detection consists of a matched filter followed by a maximum likelihood sequence detector implemented via a dynamic programming algorithm (e.g., Viterbi algorithm). The formulation of optimal multiuser detection for an AWGN channel, the maximum likelihood sequence estimator (MLSE), was published in 1986 by Verdú. Presentations of an optimal detector in a Rayleigh fading channel can be found in [59,60]. Since the complexity of the optimal multiuser detector is exponential as a function of the number of users, a large number of studies to find suboptimal detectors have been carried out.

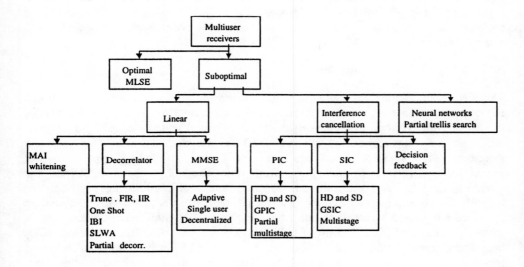

Figure 5.19 Multiuser detection classification.

5.12.3.1 Linear Detectors

The decorrelator (also called zero-forcing detector) proposed by Lupas and Verdú in [48] multiplies the matched filter outputs by the inverse of the cross-correlation matrix. An advantage of this detector is that the received signal amplitudes do not have to be known. On the other hand, the received noise process is also filtered with the inverse matrix and hence increases the noise power. This is proportional to the users' mutual crosscorrelations. Since the decorrelator is a sequence detector, the detection process cannot be started until the whole transmitted sequence is received at the receiver. In practice this is not feasible and would result in a very long delay. Therefore, several finite delay decorrelator schemes have been proposed: isolation bit insertion (IBI), multishot sliding window algorithm (SLWA) and hard decision method (HDM) [61,62], improved one-shot detection scheme [63], and finite length FIR and IIR schemes [59]. The partial decorrelator forces the MAI due to past symbols to zero, while the MAI due to future symbols might be suppressed, for example, by interference cancellation. The partial decorrelator belongs to the family of nonlinear decision feedback (DF) detectors, which are typically characterized by a linear feedforward filter and a nonlinear feedback filter [64].

Another linear receiver is the linear minimum mean square error (LMMSE) [65], which, unlike the decorrelator, doesn't enhance the noise. Even though LMMSE has also been proposed for centralized receivers in AWGN and known fading channels, the main interest has been in decentralized adaptive implementations, which are especially attractive for implementation in mobile stations [66]. A modified LMMSE receiver for fading channels has been presented in [67]. MAI-whitening filters modeling the multiple access interference as colored noise are also examples of decentralized linear receivers [68].

5.12.3.2   *Interference Cancellation*

The idea of interference cancellation is to estimate the multiple access and multipath induced interference and then to subtract the interference estimate [69]. Successive interference cancellation (SIC) cancels the interference estimate on a user by user basis, while parallel interference cancellation (PIC) cancels the interference of all users simultaneously [70]. Group-wise algorithms detect symbols within a given group and cancel the interference on that group from other users. In wideband CDMA, the grouping can be performed based on data rates [53]. Multistage interference cancellation algorithms improve the interference estimates iteratively. If tentative data decisions are used, the scheme is called hard decision (HD) interference cancellation. If tentative data decisions are not used, the scheme is called soft decision (SD) interference cancellation. Complete subtraction of the MAI in PIC has been shown to result in biased decision statistics [71]. Therefore, partial cancellation has been proposed to reduce the effect of bias in SD-PIC [71] and HD-PIC [72]. A sliding window PIC receiver has been presented in [73].

5.12.4 System Model and MUD Algorithm Formulation

In the following paragraphs, we present a standard model for the DS-CDMA uplink with K users. BPSK modulated data of each user are spread by multiplying the data modulated signal by a pseudo-random sequence. The composite signal after passing through the channel and entering the base station can be modeled as

$$r(t) = \sum_{k=1}^{K} s_k(t - \tau_k) + n(t) \tag{5.5}$$

where τ_k is the delay of kth user's transmitted signal, and $n(t)$ is the complex AWGN with two-sided power spectral density σ^2. The kth user's signal is given as

$$s_k(t - \tau_k) = \sqrt{2P_k} \sum_i b_k(i) a_k(t - iT - \tau_k) e^{j\phi_k} \tag{5.6}$$

where P_k is the power of the kth user, $b_k(i)$ is the BPSK modulated data for the kth user transmitted at time i, ϕ_k is the received signal phase of the kth user, and $a_k(t)$ is the spreading waveform given by

$$c_k(t) = \sum_{i=1}^{G} c_k^{(i)} p(t - iT_c) \tag{5.7}$$

where G is the number of chips, $c_k^{(i)} \in \pm 1$, and $p(t)$ is the pulse shape of a chip with duation T_c.

The conventional detector (matched filter or correlator receiver) can be modeled as follows. The output of the matched filter for user k is

$$y_k(i) = \int \mathrm{Re}\{r(t)e^{-i\hat{\phi}_k}\} c_k(t - iT - \hat{\tau}_k) dt \tag{5.8}$$

where $\hat{\phi}_k$ and $\hat{\tau}_k$ are the estimated time delay and phase, respectively.

The estimated bits are then given as

$$\hat{b}_k(i) = \mathrm{sgn}(y_k(i)) \tag{5.9}$$

5.12.4.1   Decorrelator

For the decorrelator, the output of the matched filters can be expressed in matrix form using the block based model [48]:

$$\mathbf{y} = \mathbf{R}\mathbf{W}\mathbf{b} + \mathbf{n} \tag{5.10}$$

where

$$\mathbf{y} = \left[\mathbf{y}^{\mathrm{T}}(1), \mathbf{y}^{\mathrm{T}}(2), ..., \mathbf{y}^{\mathrm{T}}(M)\right]^T \tag{5.11}$$

$$\mathbf{y}(i) = \left[y_1(i), y_2(i), ..., y_{K1}(i)\right]^T \tag{5.12}$$

$$\mathbf{b} = \left[\mathbf{b}^{\mathrm{T}}(1), \mathbf{b}^{\mathrm{T}}(2), ..., \mathbf{b}^{\mathrm{T}}(M)\right]^T \tag{5.13}$$

$$\mathbf{b}(i) = \left[b_1(i), b_2(i), ..., b_{K1}(i)\right]^T \tag{5.14}$$

where M is the number of symbols in the block. The correlation matrix is given by

$$\mathbf{R} = \begin{bmatrix} \mathbf{R}(0) & \mathbf{R}(-1) & 0 & \cdots & 0 & 0 \\ \mathbf{R}(1) & \mathbf{R}(0) & \mathbf{R}(-1) & 0 & \cdots & 0 \\ \vdots & \ddots & \ddots & \ddots & \ddots & \vdots \\ 0 & \cdots & 0 & \mathbf{R}(1) & \mathbf{R}(0) & \mathbf{R}(-1) \\ 0 & 0 & \cdots & 0 & \mathbf{R}(1) & \mathbf{R}(0) \end{bmatrix} \tag{5.15}$$

where the entries in the correlation matrix $R(i)$ are given by

$$R_{uk}(i) = \cos(\phi_k - \phi_l) \int a_k(t - \tau_k) a_l(t + iT + \tau_l) dt \qquad (5.16)$$

The noise vector **n** is additive Gaussian and colored. The amplitude matrix **W** is a diagonal matrix containing the amplitudes of all users. Now the decorrelator output decision is given by

$$\hat{\mathbf{b}} = \text{sgn}(\hat{\mathbf{R}}^{-1}\mathbf{y}) \qquad (5.17)$$

5.12.4.2 MMSE

The MMSE receiver performs a linear transformation on the matched filter outputs that minimizes the mean square error (MSE). The detected bits are obtained from

$$\hat{\mathbf{b}}(i) = \text{sgn}\left(\left(\hat{\mathbf{R}} + \sigma^2 \mathbf{W}^{-2} \right)^{-1} \mathbf{y} \right) \qquad (5.18)$$

5.12.4.3 Parallel Interference Cancellation

The PIC receiver detects all users at the same time and then cancels interference simultaneously. It applies the multistage principle, and the received signal at stage n is given by

$$r_k^{(n)}(t) = r^{(n-1)}(t) - \sum_{l=1}^{K} \hat{s}_l(t - \hat{\tau}_l) \qquad (5.19)$$

5.12.4.4 Serial Interference Cancellation

The SIC receiver cancels one user at a time. It ranks the users according to their powers cancelling first the higher power users. The received signal for user k' after cancelling k'-1 users is given by

$$r_{k'}(t) = r(t) - \sum_{l=1}^{k} \hat{s}_l(t - \hat{\tau}_l) \qquad (5.20)$$

5.12.5 Design Aspects for Multiuser Detection

5.12.5.1 Uplink Versus Downlink

The uplink and downlink have different characteristics, which impact the design of the MUD scheme. The uplink is asynchronous (i.e., the users' transmission times are independent of each other). Propagation delays for cell ranges of 1 to 30 km are from 3 to 100 μs, introducing a significant amount of intersymbol interference between the symbols of different users. The downlink is synchronous, and, in contrast to base stations, a mobile station needs to detect only its own signal. Typically, orthogonal codes are used in the downlink, but multipath propagation partially destroys the orthogonality. The additional gain from multiuser detection depends on the multipath profile. Multipath spread in outdoor systems ranges from a few microseconds up to 20 μs.

Since a mobile station is only interested in demodulating its own signal, a different multiuser detection strategy than in the base station could be applied to reduce the complexity in the mobile station, while still improving the performance over conventional detection. Many of the proposed multiuser detectors can also be modified for single user detectors.[4] For multirate users, one possibility is to cancel only the self-interference from different parallel codes of the same user. In the downlink, the common control channels also cause interference. Since a mobile station has to detect the pilot signal, it could cancel the pilot signal, as proposed in [74].

5.12.5.2   Multiuser Parameter Estimation

Coherent multiuser detectors require the knowledge of the complex channel coefficients, the carrier phases, and the propagation delays. Even though conventional methods for parameter estimation can be used, better results would be achieved with methods that take interference from other users into account. Large parameter estimation errors might offset the gain from the multiuser detection.

The optimal MLSE receiver estimates the received complex amplitudes and multiplies the matched filter bank output with it [75]. In suboptimal implementations, data detection and complex channel coefficient estimation are decoupled from each other (i.e., multiuser detection is applied to data), while the amplitude and phase for each user are recovered independently from each other [59]. In this case, the optimal channel estimator would be LMMSE [59]. However, since the channel co-variance matrix required by the LMMSE estimator changes according to fading rate and signal-to-noise ratios, it is very difficult to implement in practice. However, it can be approximated by an adaptive channel estimator filter such as FIR or recursive IIR predictors or smoothers [59]. Channel estimation can be either decision feedback (decision directed) or feedforward channel estimation (data aided). The latter uses only so-called pilot or reference symbols to estimate the channel. Since for wideband CDMA either pilot channel or symbols are used, data aided estimation schemes can be used.

Multiuser delay estimation methods include optimal ML estimation [76], subspace methods (MUSIC algorithm) [77], the hierarchical maximum likelihood method [78], and PIC-based delay estimators [79]. However, subspace methods seem not to be suitable for fast fading, multipath channels with an SNR less than 10 dB [77].

---

[4] Multiuser detectors that make a joint detection of the symbols of different users are also called centralized multiuser detectors. Single user detectors, also called decentralized multiuser detectors, demodulate a signal of one desired user only [59].

5.12.5.3   System Design Choices

If MUD is to be part of the next standard, some minimum performance requirements have to be specified. Since multiuser detection is very complex, one might have to make a trade-off between complexity and performance. For example, how many users at maximum are involved in the joint detection process? As we saw in the previous section, MUD research is still in a phase that would not justify to make it a mandatory feature for wideband CDMA standards.

Even though multiuser detection is a receiver technique, it might have an impact on the system design because of its large complexity, while a proper system design might ease the implementation of the multiuser detection. If short codes are used, then the implementation of multiuser detection is easier. For example, the receiver can exploit the cyclostationarity (i.e., the periodic properties) of the signal to suppress interference without knowing the interfering codes. This would be especially useful in the mobile station.

5.12.6 Choice of Multiuser Detector Algorithm

When considering which algorithm is the most appropriate choice for a practical multiuser detection algorithm, the following criteria should be considered:

- Complexity;
- Performance.

5.12.6.1   Complexity

The final implementation complexity of a multiuser detector receiver depends on the selected architecture and cannot be estimated without a detailed analysis of algorithms down to the ASIC and DSP implementation level. However, rough estimates of the implementation complexity can be obtained by estimating the number of arithmetic operations per second and the number of clock cycles required by synchronous DSP [59]. It should also be noted that an arithmetic operation could be implemented using either an integer or a floating point operation depending on the requirements of the selected algorithm. This might cause further significant differences to the implementation complexities of the algorithms. When assessing the required number of arithmetic operations, we follow the methodology in [59] and consider only the dominant term.

The ideal implementation of the linear detectors such as decorrelator and MMSE has cubic dependence on the number of the users times the number of multipath components [58]. Therefore, iterative algorithms such as the conjugate gradient (CG) method have been proposed to implement them [80]. The preconditioned conjugate gradient algorithm (PCG) seems to be one of the simplest algorithms for linear multiuser detection [59].

Tables 5.3 and 5.4 summarize the computational complexities of the algorithms. Table 5.4 has been calculated following the example in [59] where parameters of the FMA2 wideband CDMA scheme have been used. The number of users is assumed to be

$K = 150$. The number of multipaths is $L = 4$, the number of samples per chip is $N_s = 4$, and the code length is $N_c = 256$. The symbol rate after channel coding is 20 ksymbols/s (i.e., the symbol interval is $T = 50$ μs). The number of iterations for the PGC algorithm is $I = 124$, and the observation window size is $N = 13$. The number of cancellation stages in the PIC algorithm is $M = 2$ and in the SIC algorithm $M = 1$.

The iterative PCG algorithm is most complex and hardly feasible from an implementation point of view, especially if long spreading codes are used [81]. An alternative implementation for a decorrelator, an approximate decorrelator, has been proposed to relieve the processing requirement [82]. Even though it does not require the matrix inversion, the cross-correlations between different user codes have to be calculated with every symbol requiring $O[(2(KL)^2 N_c N_s)/T] \approx 15$ Tflops. The symbol level PIC is the least complex multiuser detection algorithm. However, for long spreading codes, the correlation matrix needs to be updated for every symbol, thereby increasing complexity much further. It should be noted that the cross-correlations could be calculated in parallel. The regenerative PIC is only five times more complex than the conventional matched filter (MF) detector and is well suited for long spreading codes. The regenerative SIC algorithm also has low complexity. The high clock rate, however, limits the number of users in the cancellation process. The high clock rate can be reduced by increasing the detection delay. The architecture presented in [83] would, however, introduce one symbol delay for each user in the detection process, making, for example, fast power control impossible for a large number of users. Thus, from a complexity point of view, the PIC receiver seems best suited for systems with long spreading codes.

Table 5.3
Implementation Complexity Requirements

	Operations/s	Clock Cycles
Conventional	$O[(4N_c N_s KL)/T]$	$2/T$
PCG	$O[(12IN(KL)^2)/T]$	$16M/T$
PIC (symbol level)	$O[(4M(KL)^2)/T]$	$4M/T$
PIC (regenerative)	$O[(8MN_c N_s KL)/T]$	$5M/T$
SIC (regenerative)	$O[(8M N_c N_s KL)/T]$	$4MK/T$

Table 5.4
Implementation Complexity

	Operations/s	Clock Cycles
Conventional (MF)	49 Gflops/s	40 kHz
PCG	139 Tflops/s	40 MHz
PIC (symbol level)	57 Gflops/s	160 kHz
PIC (regenerative)	196 Gflops/s	200 kHz
SIC (regenerative)	98 Gflops/s	12 MHz

5.12.6.2 Performance

In this section we discuss the performance of the decorrelator, MMSE, PIC, and SIC detectors in AWGN and Rayleigh fading channels. Furthermore, the impact of imperfect phase and delay estimation is discussed. Special attention is paid to near-far performance.

According to [58,84] the decorrelator, MMSE, and SD-PIC receivers have almost equal performance in AWGN channel with bit energy-to-noise density ratio (E_b/N_0) less than 10 dB. The MMSE detector is slightly better than the decorrelator due to the noise enhancement property of the decorrelator [84]. The SD-SIC is considerably worse than the other schemes. The HD-PIC performs better than the SD-PIC receiver does. This is due to more reliable tentative decision of HD-PIC and thus less residual interference.

The impact of phase errors alone is not very large on any multiuser detection receiver and is much less severe than the delay errors [58]. Furthermore, it seems that the performance of all detectors deteriorates equally as the phase error increases [84].

In flat Rayleigh fading the performance of all detectors is almost equal [58]. The performance of the SIC scheme has improved as compared to the AWGN channel since the instantaneous powers of the users are different. To obtain the performance improvement requires, of course, power ranking of the users on a symbol by symbol basis.

In frequency selective Rayleigh fading the performance of the SD-PIC and SD-SIC receivers is slightly worse than the performance of the linear receivers. The reason is the additional multipath-induced multiple access interference, which degrades the estimation of the channel gains [58]. The decorrelator and MMSE receivers do not need separate estimation of the channel gains. Since the HD-PIC can use separate estimates of the channel gains, its performance is not degraded as much. Actually, for moderate channel gain estimation errors, the hard decision PIC still outperforms the decorrelator receiver [59].

The sensitivity of the decorrelator, MMSE, SD-PIC, and SD-SIC receivers for delay estimation errors has been compared in [58,84]. In an AWGN channel all detectors degrade equally fast as the delay error increases. In a Rayleigh fading channel, however, the performance of the decorrelator and MMSE receivers degrades faster up to the standard deviation of 0.1 chip. Thus, the performance of the receivers becomes nearer and for larger delay errors the performance is almost equal. Sensitivity of the optimum receiver and HD-PIC for channel mismatch has been analyzed in [41].

In order to interpret the sensitivity result, we need to know how large an error the delay tracking scheme will produce. The most commonly used delay tracking device in the conventional CDMA receiver is the delay locked loop (DLL). The variance of delay tracking error can be assumed to be less than 0.01 chip [85]. Of course, in the near-far situation, the variance would be larger. A variance of 0.01 (standard deviation of 0.1 chips) still leads to significant degradation of multiuser detector performance in [58,84]. For the conventional detector the degradation is smaller. This can be easily explained by the fact that the performance of the conventional detector deteriorates only because of reduced energy in the correlation process. In the multiuser detection algorithms, the error also impacts the multiple access interference estimates, and thus, the delay error of one user impacts all other users' signals as well.

There exist two issues concerning the near-far performance of multiuser detectors. For the VSF scheme, different data rates have different powers and thus there exists a constant near-far situation. For the multicode scheme, imperfect power control leads to variations in the received user powers.

Some conclusions for performance of multiuser detectors in the VSF system can be drawn from the results presented in [58,84]. One high power user degrades the performance of the SD-PIC scheme significantly. The reason for this is that the cancellation of the weak user is inaccurate at the first stage due to the interference from the high power user [58]. Juntti observed similar behavior for the HD-PIC receiver in real multirate simulations [53]. The SIC scheme benefits from unequal powers. However, the SIC algorithm is sensitive not to the power of the strongest user but to the power of the second strongest user [58]. This is because if there are two very strong users, the second strongest user degrades the estimation of the first user. This further reduces the attractiveness of the SIC algorithm. The decorrelator is insensitive to user powers as long as there is no delay estimation error. However, with imperfect delay estimation, it also loses its near-far resistance. In the presence of delay estimation errors, the near-far performance degrades and all schemes come closer to each other in performance.

The performance of different multiuser detection schemes for the VSF scheme has been studied in [53]. The most promising scheme was groupwise SIC (GSIC), where users are grouped according to their spreading factors and the PIC or decorrelator is applied within a group. Users with the lowest spreading factor were detected first, and their multiple access interference was subtracted from the MF outputs of the other users, which are then further detected by the PIC or decorrelator. However, the performance of the high bit rate users was not good because the undetected low bit rate users degrade the performance of high bit rate users.

5.12.6.3   MUD and Power Control

Third generation wideband CDMA uses fast closed loop power control both in the uplink and the downlink. Since all multiuser detectors are in practice near-far limited, power control is required even if MUD is used. In the uplink, fast power control improves the performance in three ways: by equalizing the user powers, the detrimental near-far effect is mitigated; by compensating the channel fading, E_b/N_0 performance is improved; and by minimization of the transmission power, the battery life of mobile stations is increased and intercell interference reduced. Thus, even with multiuser detection it is important to use power control.

In the downlink, power control also improves the performance against fading but contrary to the uplink, it increases power differences between the received signals at the mobile station. Multiuser detection can facilitate larger differences in the power levels and thus offers better compensation of deep fades [86].

All wideband CDMA schemes apply fast power control both in the uplink and the downlink. Fast power control tries to compensate for the effect of fast fading. Power control has the largest impact on the performance of the SIC scheme. The SIC scheme performs best when the user powers are different. This is obviously the case for instantaneous powers in Rayleigh fading. By ranking the users on a symbol basis, equal performance compared to the other schemes can be obtained. To relieve the implementation complexity of the SIC receiver with respect to the power ranking, it has been proposed to apply SIR-based power control [87]. The user that is detected first sees the interference from all other users, and the user that is detected last sees only noise and other cell interference. Consequently, given an equal quality goal, the power control will adjust the user powers so that the first user has the largest power and last user the smallest power. The performance of the first user would be equal to the conventional detector. Further gain can be obtained by taking into account the spatial distribution of the users within a cell. In order to minimize the intercell interference, the first user in the detection process should be closest to the base station [87]. For high mobility users this kind of assignment would be difficult.

It is well known that imperfect power control degrades the performance of the conventional detector. Imperfect power control has two effects on the receiver performance. Since fading is not compensated perfectly, the performance is degraded. Furthermore, nonequal powers due to imperfect power control lead to a near-far situation. As was discussed earlier, none of the multiuser detection receivers are near-far resistant when considering imperfect parameter estimation. Thus, imperfect power control will degrade also the performance of multiuser detectors. The power control error can be modeled as a log-normally distributed random variable. The combined power control error in the reverse link caused by the fast power control loop and open loop power control is on the order of 1.5 to 2.5 dB. Based on the studies so far it seems that all detectors will have similar performance degradation under imperfect power control. Furthermore, multiuser detectors require equally stringent power control as does the conventional matched filter detector.

5.12.6.4   *Performance Versus Complexity*

Since performance depends on how complex a multiuser detector we apply, we should weigh the complexity against performance. The 1-stage SIC is not as good as others for equal powers. The 2-stage SIC performs as well as the 2- and 3-stage PIC. The high clock rate and delay make it less attractive. The 2-stage PIC (one cancellation stage) performs in most situations almost as well as the 3-stage PIC, especially in the presence of parameter estimation errors [58,84]. However, given the reasonably low complexity of the PIC algorithm, it might be worthwhile to implement the 3-stage PIC.

5.13   RANDOM ACCESS PROCEDURE

Random access is a process where a mobile station requests an access to the system, and the network answers the request and allocates a traffic channel for the mobile station. Random access is executed when the power of a mobile station is switched on or the synchronization is lost for some reason. Furthermore, random access is carried out for packet data transmission whenever there is something to transmit. As described in Section 5.3, before the random access can be performed, the following steps need to be performed:

- Code and frame synchronization;
- Retrieval of cell parameters such as random access code;

- Estimation of the downlink path loss and the initial power level for the random access.

Optimization criteria for the random access procedure are the speed of the process and low transmission power. Requirements for the speed of the random access procedure depend on the requirements for the initial synchronization time. The number of access channels depends on the anticipated access load. Information transmitted during the random access state will have an effect on this. Since excessive transmission power degrades CDMA system capacity, it is important to minimize the total power transmitted during the random access state. This is especially important because during the random access transmission power cannot be controlled by the fast closed loop power control. If initial transmission happens with a very low power, then access attempt can take a long time. On the other hand, high transmission power during the initial access results in fast synchronization but causes higher interference to other users during the synchronization period.

The minimum information that needs to be transmitted during a random access attempt is some kind of mobile identification. The actual traffic channel negotiation and service contract negotiation can then happen in a nonpublic channel. This can be performed in later state. A typical random access message consists of preamble, synchronization part, and data part, as illustrated in Figure 5.20. The data part contains at least the mobile station identification, while the preamble is an unmodulated wideband spread signal.

Preamble	Synchronization	Data

Figure 5.20 Random access message.

5.14 HANDOVER

In this section, we first define the different handover types that can be used in CDMA systems, including soft handover. In the second subsection, the handover procedure is defined. The implementation of the different phases of the soft handover procedure is studied in the third subsection. The third subsection also discusses macro diversity combining and other factors that impact the performance of soft handover. The fourth and fifth subsections describe softer and interfrequency handover, respectively.

5.14.1 Definitions

In soft handover a mobile station is connected to more than one base station simultaneously. *Idle handover* is a handover that is made when the mobile station is in idle state (i.e., does not have an active connection). *Hard handover* is a handover occurring when only one traffic channel is available at a time. *Softer handover* means a soft handover between sectors in the same cell. *Interfrequency handover* means handover between two different frequencies. *Intra system handover* is a handover between two systems (e.g., from a second generation system to third generation system).

Active set consists of the base stations involved in the soft handover with the given mobile station. If the active set is changed, an *active set update* occurs. *Candidate set* consists of the base stations that fulfil the criteria to be included in the active set but have not yet been included in the active set. *Neighbor set* contains the base stations that are likely candidates for soft handover. These are usually base stations whose geographical coverage areas are near the mobile station. *Remaining set* contains all base stations excluded from the other sets. *Discard set* is defined to be the set of base stations that belongs to the current active set but are going to be dropped from the active set since they no longer fulfill the criteria for the active set.

5.14.2 Handover Procedure

The handover procedure can be divided into three phases: measurement, decision, and execution phase, as illustrated in Figure 5.21. In the handover measurement phase, the typical downlink measurements performed by the mobile station are signal quality and signal strengths of its cell and neighboring cells. In the uplink, the base station measures signal quality. The measurement results are signaled to relevant network elements, to mobile stations, and to the BSC. In the handover decision phase, sometimes also called the evaluation phase, the measurement results are compared against predefined thresholds and it is decided whether handover should be initiated or not. In addition, admission control (see Section 5.16) is performed to verify that the new user can be accommodated into the new cell without degrading the quality of the existing users. In the execution phase, the mobile station enters the soft handover state, a new base station is added or released, or interfrequency handover is performed.

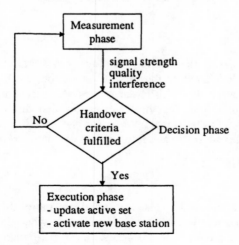

Figure 5.21 Handover phases.

5.14.3 Soft Handover

Soft handover is one of the most discussed features of CDMA. On the one hand it brings increased performance by increasing diversity, but on the other it is a necessity in a CDMA system for avoiding excessive interference from neighboring cells. The increased diversity comes at the expense of increased backhaul connections in the network.

5.14.3.1 Handover Decision

In IS-95, the handover decision is based on the pilot strength measurements of the downlink only. In wideband CDMA for third generation systems with asymmetric traffic, more decision parameters are needed. At least the following parameters can be identified:

- Distance attenuation;
- Uplink interference;
- Downlink interference.

Figure 5.22 illustrates the handover parameters. In reality, there are delays associated with adding and removing the pilots from the active set. When the signal strength of base station 2 exceeds the *add threshold*, the mobile station enters the soft handover state. When the signal strength of base station 1 drops below the *drop threshold* and stays there until the drop timer has expired, base station 1 is removed

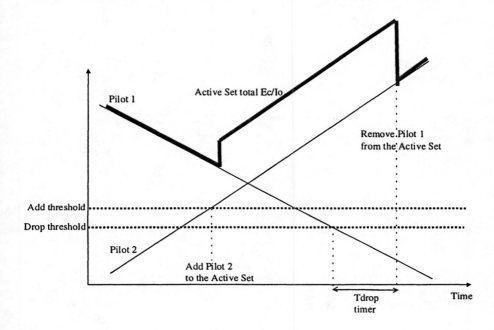

Figure 5.22 Handover parameters.

from the active set.

The add and drop thresholds can also be adjusted dynamically to take into account situations where a weak pilot added to the active set would not bring any benefit. If there are already some dominant strong pilots, adding a new weak pilot does not improve performance. On the other hand, if all pilots are weak then even adding a new, weak pilot is beneficial. This type of *dynamic threshold* is specified in the IS-95B standard.

The selection of a cell with a high uplink load but with a low distance attenuation, instead of a cell with a low uplink load but a slightly higher distance attenuation, will result in a higher mobile station transmission power. Thus, it is important to take the uplink interference situation into account in the handover decision. Figure 5.23 depicts a situation where the high uplink load influences the handover decision. The mobile station is currently connected to BTS1 but should make a handover to BTS2 since the path loss attenuation is lowest because of the LOS connection. However, the high bit rate multimedia user is producing a high interference level in the uplink, and thus, the required transmission power for the mobile station to connect to BTS2 is actually higher than to BTS3. Therefore, handover to BTS3 should be made.

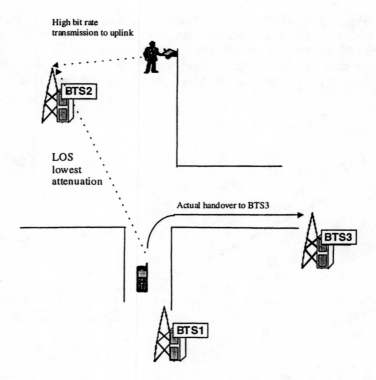

Figure 5.23 High uplink load handover situation.

In every moment, the mobile station tries to keep the best base stations available in the active set. However, the handover procedure should avoid the so-called ping-pong effect, where new base stations have to be added and old ones released too often. This results in an excessive signaling reducing capacity and increased possibility of dropping the call.

The mobile station orders base stations according to the handover decision parameters, signal strength, and interference situation. The best base stations are included in the active set. If a base station is decided to be dropped from the active set, a timer is started. If the quality is still low after the timer expires, then the base station is dropped. The ping-pong effect is avoided by long enough timers. The rate of handover depends on users' mobility and cell size.

From a capacity perspective, it is beneficial if the downlink loads are balanced between the base stations. Thus, a base station that has a high load in the downlink should hand over traffic to other base stations. This can be done by either adjusting the pilot strength according to the downlink load or by the mobile station taking into account the downlink load in the handover decision.

There are a number of alternatives on how the uplink and downlink interference situation can be taken into account in the handover decision. The first and most straightforward alternative is to have constant pilot strength and to signal the uplink interference value to the mobile station. Furthermore, the downlink interference is measured in the mobile station. A drawback of this approach is the high signaling load in the downlink, which reduces the capacity. The second possibility is to adjust the pilot strength according to the uplink load so that the selection of the base station according to pilot strength also takes the interference conditions into account. This method avoids signaling the uplink interference value to the mobile station. The third possibility is that the pilot is adjusted according to downlink load. A fourth possibility is that the pilot is adjusted according to the uplink and downlink loads. A drawback of the pilot adjustment strategies is that they are slower than constant pilot strength plus signaling, and reducing the pilot power impacts the receiver performance of the mobile station. Some of the above strategies can be decided to be used by the network operator. Only the procedures that impact signaling needs and mobile station behavior need to be specified in the standard.

5.14.3.2   Handover Measurements

Handover measurements are performed to gather information for a handover decision. The averaging period for the measurement results depends on the mobile speed. Measurement results should be updated more frequently when the speed of the mobile station increases.

The different pilot sets need to be measured with different frequencies. The active set needs to be measured most frequently. Also the candidate set should be measured as often. The neighbor set is measured less often, and the remaining set is measured least often. On the other hand, it would be good to measure the neighbor set more frequently to avoid problems with rapidly rising pilots creating high interference.

5.14.3.3   Macro Diversity Combining

It has to be decided where and how the macro diversity combining is performed. There are several instants at which the macro diversity combining can be performed:

- In the RAKE receiver;
- In the channel decoder;
- After the channel decoder;
- After the source codec.

Combining in the RAKE receiver requires that the transmissions from different base stations are synchronized. Combining at this point provides the largest gain, and is most suitable for the mobile station, since the RAKE receiver sees the transmission from other base stations as additional multipath signals. In the network side, combining at the RAKE level or channel decoder is impractical. It would then need a large amount of signaling, since the soft bits should be transmitted to the soft handover termination point situated either in the BSC or in the switch. Thus, combining after the channel decoder seems to be the most practical alternative.

Maximal ratio, equal gain, and selection diversity are the principal combining techniques. In maximal ratio combining, the received signals are weighted according to their received SIR, while in equal gain combining, no weighting is performed. With selection diversity, the signal with the highest SNR is selected.

Figure 5.24 illustrates the different possibilities for the soft handover diversity combiner. There exist two basic possibilities: BSC or switch. According to the current UMTS network view, the radio access network comprising the base stations and base station controllers and the core network comprising the switch are separated. Thus, all radio-related functions should be placed in the access network, and consequently the diversity combiner should be situated in the BSC. This will also reduce the signaling load in the network, compared to a situation where the combiner is placed in the switch.

Soft handover between base stations belonging to different BSCs requires special arrangements, (e.g., inter BSC links, as indicated in Figure 5.24). Furthermore, base stations can even belong to different switches.

5.14.3.4   Performance Considerations

Delays and errors in macro diversity measurements and execution lead to performance degradation. Some parameters have more significance in the overall performance.

Errors in SIR and signal strength measurements seem to have little impact on overall performance according to simulations performed in [88]. Standard deviation varied between 0 and 1.25 dB. However, increased handover rate, and thus signaling, was not taken into account in the simulations.

Measurement delay is caused by the scanning procedure of the pilots. Measurement speed depends on the number of pilot signals and the number of searchers in the mobile station receiver. In IS-95, the worst case assumption is 200 ms. In [88], this was assumed to be 250 ms. Since the number of the pilots to be scanned is large, a sophisticated algorithm must be used to decide which base stations to scan, how often

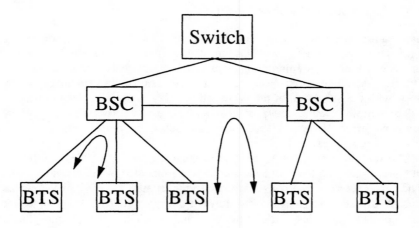

Figure 5.24 Placement of the diversity combiner in the network.

they should be scanned, and how averaging measurements would most likely improve the system performance [88].

After the decision to connect a new base station has been made, some time elapses before the setup of the new connection. This delay consists of signaling to the new base station, acquisition of the mobile station reverse link signal by the new base station, and starting the transmission in a new channel. Furthermore, the new base station has to signal the used traffic channel to the old base station, which in turn signals it to the mobile station. Thereafter, the mobile station receives the message and starts combining the two traffic channels. This process can take up to 200 to 300 ms.

Overall delay exceeding 300 ms starts to degrade the system capacity [88]. Furthermore, the downlink was noticed to be more sensitive to the delay due to the slow power algorithm. For third generation systems using fast downlink power control, the impact of delays in the handover measurement and execution might be different.

The size of the active set depends on the radio conditions and number of RAKE fingers available. Two base stations often already give most of the achievable gain. The number of RAKE fingers decides whether the additional path diversity obtained through macro diversity can be utilized. In the forward link, the addition of a new base station to the active set can even degrade the performance, since transmission of additional signals creates more interference that could not be utilized because of a limited number of RAKE fingers. An active set update rate larger than 0.5 Hz (2 sec), preferably 1 Hz, is needed to not degrade the system quality [88]. Active set threshold (add threshold, see Figure 5.22) means that if the signal strength of a base station is within that window, for example 5 dB, it can be included in the active set. Too large a threshold will degrade the soft handover gain, since combining two base stations with large power differences does not produce very much diversity gain. On the other hand, too small an add threshold combined with a small drop threshold (Figure 5.22) increases the number of the handovers leading to excessive signaling due to the ping-pong effect.

5.14.3.5   Synchronization During Handover

Synchronization requirements and solutions for soft handover depend on the network synchronization and spreading code design. If the network is synchronized at chip level with an accuracy of a few microseconds as in IS-95, the handover procedure does not need to consider time differences between the base stations involved in the soft handover, with the exception of the propagation delay. Since it is desirable not to have external synchronization because of cost and complexity reasons, an asynchronous network is preferred. However, it will place some special requirements on the handover synchronization, as explained below.

The mobile station has to be able to measure the time difference between the current base station and the new base station to be included in the active set. This time difference can exceed, in the worst case, the frame length. In an asynchronous network, code design further impacts the handover synchronization scheme. Long codes facilitate the measurement of this long time difference. With short codes, a special synchronization scheme is required that first measures the difference in frames with the help of, for example, the synchronization channel [12]. After the timing difference has been reduced to the frame level, it can be further reduced to the chip level. After the timing difference has been measured, the new base stations align the transmission of its forward link for the user in soft handover with the first base station.

5.14.4 Softer Handover

Softer handover occurs when the mobile station moves from one sector to another within one cell. The forward link resembles the soft handover situation (i.e., the mobile station receives the signals from the two sectors and combines them in the RAKE receiver). In the reverse link, in contrast to soft handover, the combining can also be performed in the RAKE receiver. In softer handover, there is no need for transactions between the base stations and, for example, the base station controller. Thus, softer handover between sectors can be established much faster than in soft handover, as fixed network signaling is not required. Therefore, it might be useful to implement street microcells as sectored cells to reduce the time for time-critical corner handover (see Chapter 11).

5.14.5 Interfrequency Handover

Seamless interfrequency handover is an important new feature in third generation CDMA networks that have (1) hierarchical cell structures (i.e., microcells overlaid by a macrocell) and (2) hot spot cells with more carriers than surrounding cells, as illustrated in Figure 5.25. In the latter scenario, the interfrequency handover can be performed as an intracell handover within the hotspot cell. First, interfrequency handover is performed to frequency 1, and then a normal soft handover is carried out to the neighboring destination cell. For the first scenario, an efficient method to measure other carrier frequencies is required.

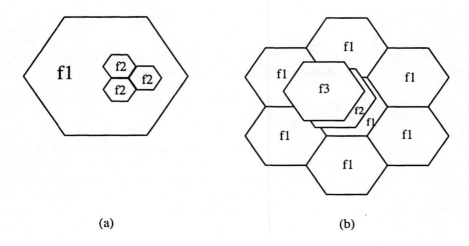

Figure 5.25   Interfrequency handover scenarios: (a) macro cell overlaying micro cells, and (b) hot spot cell with several carriers.

For interfrequency handover, there are two implementation alternatives: compressed mode/slotted mode and dual receiver. The dual receiver has an expensive receiver but simple system operation. However, if there is already a diversity receiver (two receiver chains, and not selection diversity), then it is possible to use the other receiver for measurements on other carriers. Of course, diversity gain is lost during the measurement period. Compressed mode, on the other hand, has a simple receiver but a more complex system operation. The receiver algorithm has to operate with bursty signals during the slotted mode, resulting in poorer performance and more control effort.

For the compressed mode there exist a number of alternatives for implementation. In [89] four different methods are presented: variable spreading factor, code rate increase, multicode, and higher order modulation. Variable rate spreading and coding rate increment (puncturing) cause 1.5 to 2.5 dB loss in E_b/N_0 performance and higher order modulation causes an even higher loss (5 dB). This is due to breaks in power control and less coding. A further drawback of the variable spreading factor is that simple terminals have to be able to operate with different spreading ratios (e.g., a speech terminal that normally operates with a spreading ratio of 256, should now also operate with a spreading ratio of 128).

5.15   POWER CONTROL

In the design of a power control scheme, power control criteria, step size, dynamic range, and command rate need to be considered. Each of these aspects is discussed in the following subsections.

5.15.1 Power Control Criteria

As discussed in Chapter 2, all wideband CDMA proposals use open and fast closed loop power control methods. Depending on the power control criteria, several different algorithms can be derived. Most typical criteria are:

- Path loss based power control;
- Quality-based power control.

Normally, the power control algorithm is a combination of these two basic criteria. Quality can be measured through, for example, SIR. Since different SIRs correspond to the same FER in different radio environments, we need to have a function that maps the desired FER into the required SIR target. This is performed by continuously measuring the FER and SIR, and then adjusting the SIR target to produce the desired FER.

5.15.2 Power Control Step Size

Power control step size defines how much a power control command changes the transmission power. Either a simple up/down adjustment or several power adjustment levels can be used. Typical step sizes are between 0.5 and 1 dB. It should be noted that power control adjustment is relative to the previous power setting, since an absolute power setting would require extremely accurate, and thus expensive, power control circuitry [90].

5.15.3 Dynamic Range Requirements

In the uplink, the open loop power control tries to equalize the powers of all mobile stations received by the base station to avoid the near-far effect. The dynamic range requirement is set by the maximum distance difference between two mobile stations and can be expressed as

$$\frac{P_{rx1}}{P_{rx2}} = \left(\frac{d_2}{d_1}\right)^{\alpha} \tag{5.21}$$

where:
P_{rx1} = received signal power from the mobile station 1;
P_{rx2} = received signal power from the mobile station 2;
d_1 = distance between mobile station 1 and base station;
d_2 = distance between mobile station 2 and base station;
α = path loss attenuation factor.

In an extreme situation, if mobile station 1 is 100m from the base station and the mobile station 2 is 20 km away, and assuming a path loss attenuation factor of 3.5, the required dynamic range is 80 dB. Fast power control compensates slow and fast fading fluctuations. To compensate for these, the fast power control needs a dynamic range on the order of 10 to 30 dB around the setpoint set by the open loop control.

In the downlink, dynamics cannot be as high, since transmissions for different mobile stations come from a single source. Even if code channels, in theory, are orthogonal (when using orthogonal codes), they do not remain fully orthogonal in practice after the multipath channel and transmitter and receiver nonidealities. Thus, if a large dynamic range were used, power control would create a near-far situation for transmissions originating from the single transmitter. In the downlink, the dynamic range is on the order of 10 to 20 dB.

5.15.4 Power Control Command Rate

The second task for power control is to compensate for signal fluctuations caused by fast fading and to improve performance. It is very difficult to track fast fading at high mobile speeds due to measurement delay, signaling of the power control commands and processing delay caused by extracting the power control command in the receiver. In addition, coding and interleaving are most effective in fast mobile speeds. According to a field test trial with IS-95, the "grey" area, where neither power control nor the coding and interleaving are effective, is from 20 to 40 km/h. For wideband CDMA, this area is most likely smaller. Therefore, low mobile speeds set the requirements for fast power control.

In the 900-MHz band, the maximum Doppler frequency is 25 Hz; while in the 2-GHz band, 55 Hz is the maximum for a mobile speed 50 km/h. Coherence times are 4 and 1.8 ms, respectively. Thus, the power control update time needs to be around 1 ms or longer. As we can see, the optimal power control rate depends on the mobile speed and carrier frequency, as the Doppler frequency for higher carrier frequencies is higher. Thus, it would be beneficial to have a variable power control rate according to the mobile speed.

Due to delay requirements, power control commands are transmitted uncoded. Therefore, the error probability of power control commands is very high. However, since power control commands are transmitted very often and the step size is small, the power control loop converges to the right value despite the high probability of errors.

The net capacity gain of power control depends on the power control command rate consuming radio resources, the increase of other cell interference (see Chapter 7), and the performance increase due to compensation of fast fading and the reduction of the near-far effect. Especially for low bit rate services, the power control command rate of, for example, 2 Kbps is a high percentage of the overall service bit rate. Thus, with a smaller command rate, the overall capacity could be better. The impact of power control on system performance will be discussed further in Chapter 7.

5.15.5 Power Control and Multiuser Detection

Quite often, multiuser detection is advocated to relax power control requirements. However, in addition to tackling the near-far effect, power control improves performance in a fading channel. Thus, the basic requirements for the power control are the same with MUD as without MUD.

5.16 ADMISSION AND LOAD CONTROL

The purpose of admission control is to ensure that there are free radio resources for the intended call with required SIR and bit rate. The purpose of load control is to maintain in an objective manner the use of radio resources of the network within the given limits. Admission control is always performed when a mobile station initiates communications in a new cell, either through a new call or handover. Furthermore, admission control is performed when a new service is added during an active call. In general, the admission control procedure ensures that the interference created after adding a new call does not exceed a prespecified threshold.

The negotiation of a service contract is performed at the beginning of the call. Admission control can be involved since the user might accept, for example, a lower bit rate if it can be accommodated by the network.

The quantities used in admission and load control (e.g., SIR, total power) should be averaged to obtain stable values that are insensitive to power control errors. On the other hand, the averaging time cannot be too long since then outdated information would be used in the admission and load control processes.

5.16.1 Load Factor

Load factor η is used to measure the network congestion. Load factor for the reverse link can be defined as follows:

$$1 - \eta = \frac{\text{SIR}_{\text{loaded}}}{\text{SIR}_{\text{empty}}} = \frac{\dfrac{S}{I_{\text{tot}}}}{\dfrac{S}{I_0}} = \frac{N}{I_{\text{tot}}} \Leftrightarrow \eta = 1 - \frac{N_0}{I_{\text{tot}}} \tag{5.22}$$

where:
  N_0 is the thermal noise spectral density;
  I_{tot} is the total interference plus noise spectral density;
  S is the received power at the base station from each user;
  SIR is the signal-to-interference ratio.

From (5.5) we can see that the total received power at the base station can be used to measure the cell load. When the system is fully loaded then the load factor is one. Since a fully loaded system might get into an unstable state and drives the powers of all users to a maximum (see Chapter 7), a safety margin is required. Therefore, the load factor should be in the order of 0.4 to 0.8.

In the forward link, the load factor can be defined as the ratio of the maximum BTS transmission power to the predefined threshold value.

5.16.2 Admission Control Principles

Admission control needs to be done separately for uplink and downlink. This is especially important if the traffic is highly asymmetric. Typical criteria for admission control are call blocking and call dropping. Blocking occurs when a new user is denied access to the system. Call dropping means that a call of an existing user is terminated. Call dropping is considered to be more annoying than blocking. For detailed admission control algorithms, refer to [91-93].

5.16.3 Load Control Principles

The basic principle of load control is the same as admission control. While admission control is carried out as a single event, load control is a continuous process where the interference is monitored.

Load control measures the load factor, and, if the predefined load factor is exceeded, the network either reduces the bit rates of those users whose service contract allows it to be done, delays the transmission of those users without delay requirements, or drops low priority calls. If there is an underload, load control increases the bit rates of those users who can handle higher bit rates. The increase and decrease of the bit rates can be performed with priority order.

This kind of sophisticated control requires stability in all conditions. Thus, some time constants are required for the network not to react too fast and to prevent several users from simultaneously increasing their bit rates, resulting in too high a load.

REFERENCES

[1]   Recommendation ITU-R M.687, "Future Public Land Mobile Telecommunication Systems (FPLMTS)," 1998.

[2]   Recommendation ITU-R M.1034, "Requirements for the radio interface(s) for Future Public Land Mobile Telecommunication Systems (FPLMTS)," 1998.

[3]   ETSI ETR 291, "System Requirements for UMTS," *ETSI Technical Report*, May 1996.

[4]   ETSI UMTS 21.01, "Requirements for the UMTS Terrestrial Radio Access system (UTRA)," 1997.

[5]   Recommendation ITU-R M.1035, "Framework for the radio interface(s) and radio sub-system functionality for Future Public Land Mobile Telecommunication Systems (FPLMTS)," 1998.

[6]   Higashi, A., T. Taguchi and K. Ohno, "Performance of Coherent Detection and RAKE for DS-CDMA Uplink Channels," *Proceedings of PIMRC'95*, Toronto, Canada, September 1995, pp. 436–440.

[7]   Simon, M. K., J. K. Omura, R. A. Scholtz, and B. K. Levitt, *Spread Spectrum*

Communications Handbook, revised edition, New York: McGraw Hill, Inc., 1994.

[8] Peterson, R. L., R. E. Ziemer, and D. E. Borth, *Introduction to Spread Spectrum Communications*, Englewood Cliffs, NJ: Prentice Hall, 1995.

[9] Kärkkäinen, K., *Code Families and their Performance Measures for CDMA and Military Spread-Spectrum Systems*, Ph.D. Thesis, University of Oulu, Finland, 1996.

[10] TIA/EIA/IS-95-B, "Mobile Station-Base Station Compatibility Standard for Dual-Mode Wideband Spread Spectrum Cellular System," 1998.

[11] Koo, J. M., E. K. Hong and J. I. Lee, "Wideband CDMA technology for FPLMTS," *Proceedings of the 1st CDMA International Conference*, Seoul, Korea, November 1996.

[12] Ojanperä, T., K. Rikkinen, H. Hakkinen, K. Pehkonen, A. Hottinen and J. Lilleberg, "Design of a 3rd Generation Multirate CDMA System with Multiuser Detection, MUD-CDMA," *Proceedings of ISSSTA96*, Vol. 1, Mainz, Germany, Septemeber 1996, pp. 334–338.

[13] Yang, J., D. Bao, and M. Ali, "PN Offset Planning in IS-95 based CDMA system," *Proceedings of VTC'97*, Vol. 3, Phoenix, Arizona, USA, May 1997, pp. 1435–1439.

[14] Hammons, R. A., and V. J. Kumar, "On a recent 4-phase sequence design for CDMA," *IEICE Trans. Comm.*, Vol. E76-B, No. 8, August 1993, pp. 804–813.

[15] DaSilva, V. M., and E. S. Sousa, "Effect of Multipath Propagation on the forward Link of a CDMA Cellular System," *Wireless Personal Communications*, No. 1, Kluwer Academic Publishers, 1994, pp. 33–41.

[16] Adachi, F., M. Sawahashi, and K. Okawa, "Tree-Structured generation of orthogonal spreading codes with different lengths for forward link of DS-CDMA mobile radio," *IEE Electronics Letters*, Vol. 33, No. 1., January 1997, pp. 27–28.

[17] Hottinen, A., and K. Pehkonen, "A Flexible Multirate CDMA Concept with Multiuser Detection," *Proceedings of ISSSTA'96*, Vol. 1, Mainz, Germany, 1996, pp. 556–560.

[18] Ojanperä, T., J. Skold, J. Castro, L. Girard and A. Klein, "Comparison of Multiple Access Schemes for UMTS," *Proceedings of VTC'97*, Vol. 2, Phoenix, Arizona, USA, May 1997, pp. 490–494.

[19] TTA, "Global CDMA I: Multiband Direct-Sequence CDMA RTT System Description," June 1998.

[20] Dahlman, E., and K. Jamal, "Wide-band services in a DS-CDMA based FPLMTS system," *Proceedings of VTC'96*, Atlanta, Georgia, USA, May 1996, pp. 1656–1660.

[21] Hui, J. Y. N., "Throughput analysis for code division multiple accessing of the spread spectrum channel," *IEEE Trans. Vehicular Technology*, Vol. VT-33, No. 3, August 1984, pp. 98–102.

[22] Viterbi, A. J., "Very low rate convolutional code for maximum theoretical performance of spread-spectrum multiple-access channels," *IEEE Journal on Selected Areas in Communications*, Vol. 8, No. 4, May 1990, pp. 642–649.

[23] Rikkinen, K., "Comparison of very low rate coding methods for CDMA radio communications system," *Proceedings of IEEE ISSSTA'94*, Oulu, Finland, 1994, pp. 268–272.

[24] Torrieri, D. J., "Performance of Direct-Sequence Systems with Long Pseudo-noise Sequences," *IEEE Journal on Selected Areas in Communications*, Vol. 10, No. 4, May 1992, pp. 770–781.

[25] Lin, S., and J. Costello, Jr., *Error Control Coding: Fundamentals and Applications*, Englewood Cliffs, NJ: Prentice Hall, 1983.

[26] Cideciyan, R. D., E. Eleftheriou, and M. Rupf, "Concatenated Reed-Solomon/Convolutional Coding for Data Transmission in CDMA-Based Cellular Systems," *IEEE Transactions on Communications*, Vol. 45, No. 10, October 1997, pp. 1291–1303.

[27] Proakis, J. G., *Digital Communications*, 3rd ed., New York: McGraw-Hill, Inc., 1995.

[28] Wong, K. H. H., and L. Hanzo, "Channel Coding," *in Mobile Radio Communications*, R. Steele (ed.), New York: John Wiley & Sons Ltd, 1996, pp. 347–488.

[29] Berrou, C., A. Glavieux, and P. Thitimajshima, "Near Shannon Limit Error-Correcting Coding and Decoding: Turbo-Codes," *IEEE International Conference on Communications ICC'93*, Vol. 2, May 23-26, Geneva, Switzerland, pp. 1064–1070.

[30] Jung, P., J. Plechinger, M. Doetsch, and F. Berens, "Pragmatic approach to rate compatible punctured Turbo-Codes for mobile radio applications," *Proceedings of 6th International Conference on Advances in Communications and Control: Telecommunications/Signal Processing*, Corfu, Greece, June 23–27, 1997.

[31] Proceedings of International Symposium on Turbo-Codes and Related Topics, Brest, France, September 3-5, 1997.

[32] Benedetto, S., D. Divsalar, and J. Hagenauer, "Concatenated Coding Techniques and Iterative Decoding: Sailing Toward Channel Capacity," *guest editorial in Journal on Selected Areas in Communications*, Vol. 16, No. 2, Feb. 1998.

[33] Frenger, P., P. Orten, T. Ottosson and A. Svensson, "Rate Matching in Multichannel Systems using RCPC-Codes," *Proceedings of VTC'97*, Phoenix, Arizona, USA, May 1997, pp. 354–357.

[34] Okumura, Y., and F. Adachi, "Variable rate data transmission with blind rate detection for coherent DS-CDMA mobile radio," *IEE Electronics Letters*, Vol. 32, No. 20, September 1996, pp. 1865–1866.

[35] Parkvall, S., E. Ström, and B. Ottersen, "The Impact of Timing Errors on the Performance of Linear DS-CDMA Receivers," *IEEE Journal on Selected Areas in Communications*, Vol. 14, No. 8, October 1996, pp. 1660–1668.

[36] Sunay, O., and P. J. McLane, "Effects of Carrier Phase and Chip Timing Errors on the Capacity of Quadriphase Spread, BPSK Modulated DS CDMA System," *Proceedings of IEEE Globecom'95*, Singapore, November 1995, pp. 1114–1120.

[37] Glisic, S., and B. Vucetic, *Spread Spectrum CDMA Systems for Wireless Communications*, Boston: Artech House, 1996.

[38] Yang, S., *CDMA RF System Engineering*, Boston: Artech House, 1998.

[39] Hottinen, A., and R. Wichman, "Transmit Diversity by Antenna Selection in CDMA Downlink," *Proceedings of ISSSTA'98*, San City, South Africa, September 1998.

[40] Heikkinen, T., and A. Hottinen, "On Downlink Power Control and Capacity with

Multi-Antenna Transmission," *Proceedings of VTC'98*, Ottawa, Canada, May 1998, pp. 475–479.

[41] Gray, S. D., M. Kocic, and D. Brady, "Multiuser Detection in Mismatched Multiple-Access Channels," *IEEE Transactions on Communications*, Vol. 43, No. 12, December 1995.

[42] Holtzman, J. M., "DS/CDMA Successive Interference Cancellation," *in Code Division Multiple Access Communications*, S. G. Glisic and P. A. Leppänen (eds.), Boston: Kluwer, 1995, pp. 161–180.

[43] Wallace, M., and R. Walton, "CDMA Radio Network Planning," *Proceedings of IEEE VTC'94*, Stockholm, Sweden, June 1994, pp. 62–67.

[44] Westman, T., and H. Holma, "CDMA System for UMTS High Bit Rate Services," *Proceedings of VTC'97*, Phoenix, USA, May 1997, pp. 830–834.

[45] Schneider, K. S., "Optimum detection of code division multiplexed signals," *IEEE Trans. Aeros. Electron. Syst.*, Vol. AES-15, No. 1, January 1979, pp. 181–185.

[46] Kohno, R., H. Imai, and M. Hatori, "Cancellation technique of co-channel interference in asynchronous spread-spectrum multiple access systems," *IEICE Trans. Comm.*, Vol. 65-A, May 1983, pp. 416–423.

[47] Verdú, S., *Optimum Multiuser Signal Detection*, Ph.D. thesis, University of Illinois, Urbana-Champaign, 1984.

[48] Lupas, R., and S. Verdú, "Near-far resistance of multiuser detectors in asynchronous code-division multiple access communications," *IEEE Trans. Commun.*, COM-38, April 1990, pp. 496–508.

[49] Verdú, S, "Adaptive Multiuser Detection," in *Code Division Multiple Access Communications*, S. G. Glisic and P. A. Leppänen (eds.), Boston: Kluwer, 1994, pp. 97–116.

[50] Giallorenzi, T. R., and S. G. Wilson, "Suboptimum multiuser receivers for convolutionally coded asynchronous CDMA systems," *IEEE Trans. Commun.*, Vol. 44, No. 9, September 1996, pp. 1183–1196.

[51] Giallorenzi, T. R., and S. G. Wilson, "Multiuser ML sequence estimator for convolutionally coded asynchronous DS-CDMA systems," *IEEE Trans. Commun.*, Vol. 44, No. 8, August 1996, pp. 997–1007.

[52] Fawer, U., and B. Aazhang, "Multiuser receivers for code-division multiple access systems with trellis-coded modulation," *IEEE JSAC*, Vol. 14, No. 8, August 1996, pp. 1602–1609.

[53] Juntti, M, "Performance of Multiuser Detection in Multirate CDMA Systems," *Wireless Personal Communications*, submitted.

[54] Saquib, M., and R. Yates, " A Two Stage Decorrelator for a Dual Rate Synchronous DS/CDMA System," *Proceedings of ICC'97*, Montreal, Canada, June 1997, pp. 334–338.

[55] Hottinen, A., H. Holma, and A. Toskala, "Multiuser Detection for Multirate CDMA Communications," *Proceedings of ICC'96*, Dallas, Texas, US, June 1996, pp. 1819–1823.

[56] Hottinen, A., "Self-organizing multiuser detection," *Proceedings of ISSSTA'96*, Oulu, Finland, July 1996, pp. 152–156.

[57] Aazhang, B., B.-P. Paris, and G. Orsak, "Neural networks for multiuser detection in code-division multiple access systems," *IEEE Trans. Commun.*, Vol. 40, No. 7,

July 1992, pp. 1212–1222.

[58] Buehrer, R. M., *The Application of Multiuser Detection to Cellular CDMA*, Ph.D. Thesis, Virginia Polytechnic Institute and State University, Virginia, 1996.

[59] Juntti, M., *Multiuser Demodulation for CDMA Systems in Fading Channels*, Ph.D. thesis, University of Oulu, Finland, October 1997.

[60] Zvonar, Z., and D. Brady, "Multiuser Detection in Single-Path Fading Channels," *IEEE Transaction on Communications*, Vol. 42, No. 2/3/4, Feb./March/April 1994, pp. 1729–1739.

[61] Zheng, C., and S. K. Barton, "On the Performance of Near-Far Resistant CDMA Detectors in the Presence of Synchronization Errors," *IEEE Trans. on Commun.*, Vol. 43, No. 12, December 1995, pp. 3037–3045.

[62] Zheng, C., and M. Faulkner, "Power Control Requirements in Linear Decorrelating Detectors for CDMA," *Proceedings of VTC'97*, Phoenix, Arizona, USA, May 1997, pp. 213–217.

[63] Peng, M., Y. J. Guo, and S. K. Barton "One-shot Linear Decorrelating Detector for Asynchronous CDMA," *Proceedings of Globecom'96*, 1996, pp. 1301–1305.

[64] Duel-Hallen, A., J. Holtzman, and Z. Zvonar, "Multiuser Detection for CDMA Systems," *IEEE Personal Commun.*, April 1995, pp. 46–58.

[65] Madhow, U., and M. L. Honig, "MMSE Interference suppression for Direct-Sequence Spread-Spectrum CDMA," *IEEE Transactions on Commun.*, Vol. 42, No. 12, December 1994, pp. 3178–3188.

[66] Zecevic, N., *Techniques and Adaptation Algorithms for Direct-Sequence Spread-Spectrum CDMA Single-User Detection*, M.Sc. thesis, Virginia Tech, July 1996.

[67] Latva-aho, M. and M. Juntti, "Modified adaptive LMMSE receiver for DS-CDMA systems in fading channels", Proceedings of IEEE International Symposium on Personal, Indoor, and Mobile Radio Communications (PIMRC), Helsinki, Finland, September 1997, pp. 554–558.

[68] Monk, A. M., M. Davis, L. B. Milstein, and C. W. Helstrom, "A Noise-Whitening Approach to Multiple Access Noise Rejection-Part I: Theory and Background," *IEEE Journal on Selected Areas in Communications*, Vol. 12, No. 5, June 1994, pp. 817–827.

[69] Moshavi, S., "Multiuser Detection for DS-CDMA Communications," *IEEE Communications Magazine*, October 1996, pp. 124–136.

[70] Prasad, R., *CDMA for Wireless Personal Communications*, Boston-London: Artech House, 1996.

[71] Correal, N. S., R. M. Buehrer, and B. D. Woerner, "Improved CDMA Performance through Bias Reduction for Parallel Interference Cancellation," *Proceedings of PIMRC'97*, Helsinki, Finland, September 1997, pp. 565–569.

[72] Divsalar, D. and M. Simon, "A New Approach to Parallel Interference Cancellation," *Proceedings of Globecom'96*, London, UK, November 1996, pp. 1452–1457.

[73] Haifeng, W., J. Lilleberg, and K. Rikkinen, "A new sub-optimal multiuser detection approach for CDMA systems in Rayleigh fading channel", *Proceedings of Conference on Information Sciences and Systems (CISS)*, The Johns Hopkins University, Baltimore, MD, USA, March 1997, pp. 276–280.

[74] I, C.-L., C. A. Webb III, H. C. Huang, S. ten Brink, S. Nada, and R. D. Gitlin, "IS-

95 Enhancements for Multimedia Services," *Bell Labs Technical Journal*, Autumn 1996, pp. 60–87.

[75] Kailath, K., "Correlation detection of signals perturbed by a random channel," *IRE Trans. Information Theory*, Vol. 6, No. 3, June 1960, pp. 361–366.

[76] Lilleberg, J., E. Nieminen, and M. Latva-aho, "Blind Iterative Multiuser Delay Estimator for CDMA," *Proceedings of PIMRC'96*, Taipei, Taiwan, October 1996, pp. 565–568.

[77] P. Luukkanen and J. Joutsensalo, "Comparison of MUSIC and Matched Filter Delay Estimators in DS-CDMA," *Proceedings of PIMRC'97*, Helsinki, Finland, September 1997, pp. 830–834.

[78] Joutsensalo, J., J. Lilleberg, A. Hottinen, and J. Karhunen, "A Hierarchic Maximum Likelihood Method for Delay Estimation in CDMA," *Proceedings of VTC'96*, Atlanta, Georgia, USA, April 1996, pp. 188–192.

[79] Latva-aho, M., and J. Lilleberg, "Parallel Interference Cancellation in Multiuser CDMA Channel Estimation," *Wireless Personal Communications*, Boston: Kluwer, to appear 1998.

[80] Juntti, M. J., B. Aazhang, and J. O. Lilleberg, "Iterative Implementation of Linear Multiuser Detection for Dynamic Asynchronous CDMA Systems", *IEEE Transactions on Communications*, Vol. 46, No. 4, April 1994, pp. 503–508.

[81] Ojanperä, T., "Overview of Multiuser Detection/Interference Cancellation for DS-CDMA," *Proceedings of ICPWC'97*, Mumbay, India, Dec. 1997, pp. 115–119.

[82] Mandayan, N., and S. Verdú, "Analysis of Approximate Decorrelating Detector," *Journal on Wireless Personal Communications*, submitted.

[83] Pedersen, K. I., T. E. Kolding, I. Seskar, and J. Holtzman, "Practical Implementation of Successive Interference Cancellation in DS/CDMA Systems," *Technical Report*, WINLAB, Rutgers University, 1996.

[84] Orten, P., and T. Ottoson, "Robustness of DS-CDMA Multiuser Detectors, *Proceedings of CTMC Globecom'97*, Phoenix, Arizona, USA, November 1997, pp. 144-148.

[85] Keurulainen, J., *Delay-Locked Loop Performance in CDMA Receivers*, M.Sc. thesis, Helsinki University of Technology, 1994.

[86] Wichman, R., and A. Hottinen, "Multiuser Detection for Downlink CDMA Communications in Multipath Fading Channels," *Proceedings of VTC'97*, Vol. 2, Phoenix, Arizona, USA, May 1997, pp. 572–576.

[87] Hatrack, P., and J. M. Holzman, "Reduction of Other-Cell Interference with Integrated Interference Cancellation," *Proceedings of VTC'97*, Phoenix, Arizona, USA, May 1997, pp. 1842–1846.

[88] Andersson, T., "Tuning the Macro Diversity Performance in a DS-CDMA System," *Proceedings of VTC´94*, Stockholm, Sweden, June 8-10, 1994, pp. 41–45.

[89] Gustafsson, M., K. Jamal, and E. Dahlman, "Compressed Mode Techniques for Inter-Frequency Measurements in a Wide-band DS-CDMA System," *Proceedings of PIMRC´97*, Helsinki, Finland, September 1997, pp. 231–235.

[90] Fukasawa, A., T. Sato, Y. Takizawa, T. Kato, M. Kawabe, and R. E. Fisher, "Wideband CDMA System for Personal Radio Communications," *IEEE Communications Magazine*, October 1996, pp. 116–123.

[91] Knutsson, J., P. Butovitsch, M. Persson, and R. D. Yates, "Evaluation of Admission Control Algorithms for CDMA System in a Manhattan Environment," *Proceedings of The 2nd CDMA International Conference*, Seoul, Korea, October 1997, pp.414–418.

[92] Huang, C. Y., and R. D. Yates, "Call Admission in Power Controlled CDMA Systems," *Proceedings of IEEE VTC'96*, Atlanta, Georgia, USA, May 1996, pp. 1665–1669.

[93] Liu, Z., and M. E. Zarki, "SIR Based Call Admission Control for DS-CDMA Cellular System," *IEEE Journal on Selected Areas in Communications*, Vol. 12, 1994, pp. 638–644.

Chapter 6

WIDEBAND CDMA SCHEMES

6.1   INTRODUCTION

This chapter presents the wideband CDMA air interfaces being developed by the standardization organizations in Europe, Japan, the United States, and Korea for third generation communication systems. Figure 6.1 illustrates the different schemes and their relations to standards bodies and to each other.

The third generation air interface standardization for the schemes based on CDMA seems to focus on two main types of wideband CDMA: network asynchronous and network synchronous. In network asynchronous schemes the base stations are not synchronized, while in network synchronous schemes the base stations are synchronized to each other within a few microseconds. There are three network asynchronous CDMA proposals: WCDMA[1] in ETSI and ARIB, and TTA II[2] wideband CDMA in Korea have similar parameters [1]. In addition, T1P1 in the United States has joined the development of WCDMA. TR46.1 in the United States is also developing a wideband CDMA scheme (Wireless Multimedia & Messaging Services, WIMS), which has been recently harmonized with WCDMA. A network synchronous wideband CDMA scheme has been proposed by TR45.5 (cdma2000) and is being considered by Korea (TTA I[3]) [1]. All schemes are geared towards the IMT-2000 radio transmission technology selection process in ITU-R TG8/1. In addition to the above main wideband CDMA schemes, we introduce two more that are interesting from a wideband CDMA development perspective. These are CODIT and IS-665 W-CDMA.

Several attempts have been made to harmonize the different wideband CDMA approaches in search of a unified global air interface. However, due to the evolution of current systems and the strong commercial interests of their supporters, at the moment it seems that there will be at least two wideband CDMA standards for third generation. It should be noted that several changes of parameters have occurred during the development of these proposals and the detailed concepts and standards will be

---

[1] WCDMA is written without a dash when used for the ARIB/ETSI system. For other wideband CDMA proposals it can be written as W-CDMA.

[2] For the ITU RTT submission TTA II was renamed Global CDMA II

[3] For the ITU RTT submission TTA I was renamed Global CDMA I

developed during 1998 and 1999. In this chapter we try to reflect the latest information available in the literature. The ITU radio transmission technology descriptions of different wideband CDMA schemes can be found in [2–8] (available from the ITU web site http://www.itu.int/imt/).

We first describe the technical approaches for the network asynchronous and synchronous schemes and discuss the reasoning for 5-MHz bandwidth for third generation wideband CDMA. WCDMA and cdma2000 are described in detail covering carrier spacing and deployment scenarios, physical channels, spreading, multirate schemes (variable data rates), packet data, and handover. The main technical parameters for the TTA I and TTA II schemes are presented and their differences compared to WCDMA and cdma2000 are highlighted. Finally, the CODIT and IS-665 schemes are briefly described.

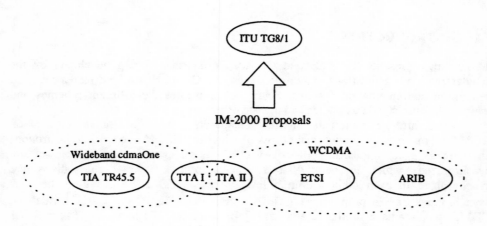

Figure 6.1   Relationships between wideband CDMA schemes and standards bodies.

6.2   TECHNICAL APPROACHES

In the following, we discuss the main technical approaches of WCDMA and cdma2000. These differences apply to the TTA I and TTA II as well. The main differences between WCDMA and cdma2000 systems are chip rate, downlink channel structure, and network synchronization. cdma2000 uses a chip rate of 3.6864 Mcps for the 5-MHz band allocation with the direct spread downlink and a 1.2288-Mcps chip rate for the multicarrier downlink [1]. WCDMA uses direct spread with a chip rate of 4.096 Mcps. The multicarrier approach is motivated by a spectrum overlay of cdma2000 with existing IS-95 carriers. Similar to IS-95, the spreading codes of cdma2000 are generated using different phase shifts of the same M-sequence. This is possible because of the synchronous network operation. Since WCDMA has an asynchronous network, different long codes rather than different phase shifts of the same code are used for the cell and user separation. The code structure further impacts how code synchronization, cell acquisition, and handover synchronization are performed.

The nominal bandwidth for all third generation proposals is 5-MHz. There are several reasons for choosing this bandwidth. First, data rates of 144 and 384 Kbps are achievable within 5-MHz bandwidth for third generation systems and can be provided with reasonable capacity. Even 2-Mbps peak rate can be provided under limited conditions. Second, lack of spectrum calls for reasonably small minimum spectrum allocation, especially if the system has to be deployed within the existing frequency bands already occupied by the second generation systems. Third, the large 5-MHz bandwidth can resolve more multipaths than a narrower bandwidth, thus increasing diversity and improving performance. Larger bandwidths of 10, 15, and 20 MHz have been proposed to support highest data rates more effectively.

6.3   WCDMA

The WCDMA scheme has been developed as a joint effort between ETSI and ARIB during the second half of 1997 [9]. The ETSI WCDMA scheme has been developed from the FMA2 scheme in Europe [10-16] and the ARIB WCDMA from the Core-A scheme in Japan [17-22]. The uplink of the WCDMA scheme is based mainly on the FMA2 scheme, and the downlink on the Core-A scheme. In this section, we present the main technical features of the ARIB/ETSI WCDMA scheme. Table 6.1 lists the main parameters of WCDMA.

6.3.1   Carrier Spacing and Deployment Scenarios

The carrier spacing has a raster of 200 kHz and can vary from 4.2 to 5.4 MHz. The different carrier spacings can be used to obtain suitable adjacent channel protections depending on the interference scenario. Figure 6.2 shows an example for the operator bandwidth of 15 MHz with three cell layers. Larger carrier spacing can be applied between operators than within one operator's band in order to avoid inter-operator interference. Interfrequency measurements and handovers are supported by WCDMA to utilize several cell layers and carriers.

6.3.2   Logical Channels

WCDMA basically follows the ITU Recommendation M.1035 in the definition of logical channels [23]. The following logical channels are defined for WCDMA. The three available common control channels are:

- Broadcast control channel (BCCH) carries system and cell specific information;
- Paging channel (PCH) for messages to the mobiles in the paging area;
- Forward access channel (FACH) for messages from the base station to the mobile in one cell.

Table 6.1

Parameters of WCDMA

Channel bandwidth	(1.25), 5, 10, 20 MHz
Downlink RF channel structure	Direct spread
Chip rate	(1.024)[a]/4.096/8.192/16.384 Mcps
Roll-off factor for chip shaping	0.22
Frame length	10 ms / 20 ms (optional)
Spreading modulation	Balanced QPSK (downlink) Dual channel QPSK(uplink) Complex spreading circuit
Data modulation	QPSK (downlink) BPSK (uplink)
Coherent detection	User dedicated time multiplexed pilot (downlink and uplink), no common pilot in downlink
Channel multiplexing in uplink	Control and pilot channel time multiplexed I&Q multiplexing for data and control channel
Multirate	Variable spreading and multicode
Spreading factors	4 − 256
Power control	Open and fast closed loop (1.6 kHz)
Spreading (downlink)	Variable length orthogonal sequences for channel separation Gold sequences 2^{18} for cell and user separation (truncated cycle 10 ms)
Spreading (uplink)	Variable length orthogonal sequences for channel separation, Gold sequence 2^{41} for user separation (different time shifts in I and Q channel, truncated cycle 10 ms)
Handover	Soft handover Interfrequency handover

[a] In the ARIB WCDMA proposal chip rate of 1.024 Mcps has been specified, while in the ETSI WCDMA it has not.

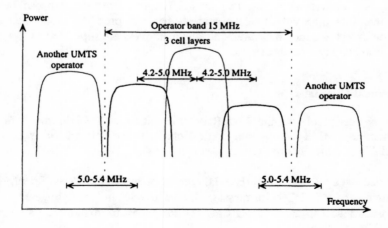

Figure 6.2   Frequency utilization with WCDMA.

In addition, there are two dedicated channels:

- Dedicated control channel (DCCH) covers two channels: stand-alone dedicated control channel (SDCCH) and associated control channel (ACCH);
- Dedicated traffic channel (DTCH) for point-to-point data transmission in the uplink and downlink.

6.3.3 Physical Channels

6.3.3.1 Uplink Physical Channels

There are two dedicated channels and one common channel on the uplink. User data is transmitted on the dedicated physical data channel (DPDCH), and control information is transmitted on the dedicated physical data channel (DPDCH). The random access channel is a common access channel.

Figure 6.3 shows the principle frame structure of the uplink DPDCH. Each DPDCH frame on a single code carries 160×2^k bits (16×2^k Kbps), where $k = 0, 1, ..., 6$, corresponding to a spreading factor of $256/2^k$ with the 4.096-Mcps chip rate. Multiple parallel variable rate services (= dedicated logical traffic and control channels) can be time multiplexed within each DPDCH frame. The overall DPDCH bit rate is variable on a frame-by-frame basis.

In most cases, only one DPDCH is allocated per connection, and services are jointly interleaved sharing the same DPDCH. However, multiple DPDCHs can also be allocated (e.g., to avoid a too low spreading factor at high data rates).

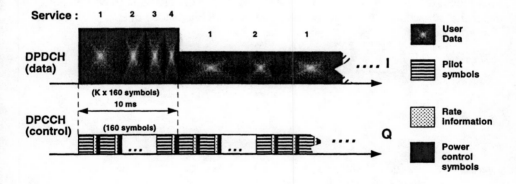

Figure 6.3   WCDMA uplink multirate transmission.

The dedicated physical control channel (DPCCH) is needed to transmit pilot symbols for coherent reception, power control signaling bits, and rate information for rate detection. Two basic solutions for multiplexing physical control and data channels are time multiplexing and code multiplexing. A combined IQ and code multiplexing solution (dual-channel QPSK) is used in WCDMA uplink to avoid electromagnetic compatibility (EMC) problems with discontinuous transmission (DTX).

The major drawback of the time multiplexed control channel are the EMC problems that arise when DTX is used for user data. One example of a DTX service is speech. During silent periods no information bits need to be transmitted, which results in pulsed transmission as control data must be transmitted in any case. This is illustrated in Figure 6.4. Because the rate of transmission of pilot and power control symbols is on the order of 1 to 2 kHz, they cause severe EMC problems to both external equipment and terminal interiors. This EMC problem is more difficult in the uplink direction since mobile stations can be close to other electrical equipment, like hearing aids.

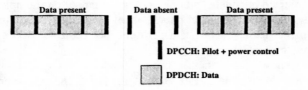

Figure 6.4   Illustration of pulsed transmission with time multiplexed control channel.

The IQ/code multiplexed control channel is shown in Figure 6.5. Now, since pilot and power control are on a separate channel, no pulse-like transmission takes place. Interference to other users and cellular capacity remains the same as in the time multiplexed solution. In addition, link level performance is the same in both schemes if the energy allocated to the pilot and the power control bits is the same.

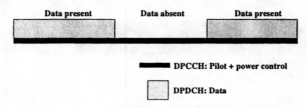

Figure 6.5   Illustration of parallel transmission of DPDCH and DPCCH channel when data is present/absent (DTX).

The structure of the random access burst is shown in Figure 6.6. The random access burst consists of two parts, a preamble part of length 16×256 chips (1 ms) and a data part of variable length.

The WCDMA random access scheme is based on a slotted ALOHA technique with the random access burst structure shown in Figure 6.6. Before the transmission of a random access request, the mobile terminal should carry out the following tasks:

- Achieve chip, slot, and frame synchronization to the target base station from the synchronization channel (SCH) and obtain information about the downlink scrambling code also from the SCH.
- Retrieve information from BCCH about the random access code(s) used in the target cell/sector.
- Estimate the downlink path loss, which is used together with a signal strength target to calculate the required transmit power of the random access request.

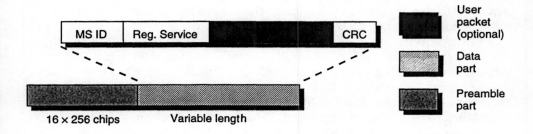

Figure 6.6   Structure of WCDMA random access burst.

It is possible to transmit a short packet together with a random access burst without settting up a scheduled packet channel. No separate access channel is used for packet traffic related random access, but all traffic shares the same random access channel. More than one random access channel can be used if the random access capacity requires such an arrangement. The performance of the selected solution is presented in [24]. The use of the random access burst for packet access is described in Section 6.3.6.

6.3.3.2   Downlink physical channels

In the downlink, there are three common physical channels. The primary and secondary common control physical channels (CCPCH) carry the downlink common control logical channels (BCCH, PCH, and FACH); the SCH provides timing information and is used for handover measurements by the mobile station.

The dedicated channels (DPDCH and DPCCH) are time multiplexed. The EMC problem caused by discontinuous transmission is not considered significant in downlink since (1) there are signals to several users transmitted in parallel and at the same time and (2) base stations are not so close to other electrical equipment, like hearing aids.

In the downlink, time multiplexed pilot symbols are used for coherent detection. Since the pilot symbols are connection dedicated, they can be used for channel estimation with adaptive antennas as well. Furthermore, the connection dedicated pilot symbols can be used to support downlink fast power control. In addition, a common pilot time multiplexed in the BCCH channel can be used for coherent detection.

The primary CCPCH carries the BCCH channel and a time multiplexed common pilot channel. It is of fixed rate and is mapped to the DPDCH in the same way as dedicated traffic channels. The primary CCPCH is allocated the same channelization code in all cells. A mobile terminal can thus always find the BCCH, once the base station's unique scrambling code has been detected during the initial cell search.

The secondary physical channel for common control carries the PCH and FACH in time multiplex within the super frame structure. The rate of the secondary CCPCH may be different for different cells and is set to provide the required capacity for PCH and FACH in each specific environment. The channelization code of the secondary CCPCH is transmitted on the primary CCPCH.

The SCH consists of two subchannels, the primary and secondary SCHs. Figure 6.7 illustrates the structure of the SCH. The SCH applies short code masking to minimize the acquisition time of the long code [29]. The SCH is masked with two short codes (primary and secondary SCH). The unmodulated primary SCH is used to acquire the timing for the secondary SCH. The modulated secondary SCH code carries information about the long code group to which the long code of the BS belongs. In this way, the search of long codes can be limited to a subset of all the codes.

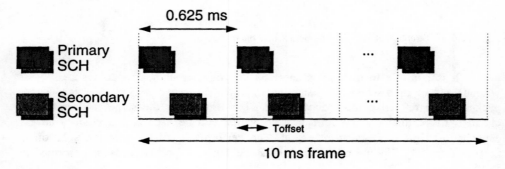

Figure 6.7   Structure of the synchronization channel (SCH).

The primary SCH consists of an unmodulated code of length 256 chips, which is transmitted once every slot. The primary synchronization code is the same for every base station in the system and is transmitted time aligned with the slot boundary, as illustrated in Figure 6.7.

The secondary SCH consists of one modulated code of length 256 chips, which is transmitted in parallel with the primary SCH. The secondary synchronization code is chosen from a set of 16 different codes depending on to which of the 32 different code groups the base station downlink scrambling code c_{sc} belongs.

The secondary SCH is modulated with a binary sequence length of 16 bits, which is repeated for each frame. The modulation sequence, which is the same for all base stations, has good cyclic autocorrelation properties.

The multiplexing of the SCH with the other downlink physical channels (DPDCH/DPCCH and CCPCH) is illustrated in Figure 6.8. The SCH is transmitted only intermittently (one codeword per slot), and it is multiplexed with the DPDCH/DPCCH and CCPCH after long code scrambling is applied on DPDCH/DPCCH and CCPCH. Consequently, the SCH is nonorthogonal to the other downlink physical channels.

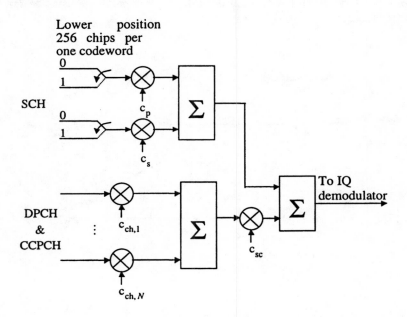

Figure 6.8 Multiplexing of the SCH (s_p = primary spreading code, s_c = secondary spreading code, c_{ch} = orthogonal code, and c_{sc} = long scrambling code).

6.3.4 Spreading

The WCDMA scheme employs long spreading codes. Different spreading codes are used for cell separation in the downlink and user separation in the uplink. In the downlink, Gold codes of length 2^{18} are used, but they are truncated to form a cycle of a 10-ms frame. The total number of available scrambling codes is 512, divided into 32 code groups with 16 codes in each group to facilitate a fast cell search procedure. In the uplink, either short or long spreading (scrambling codes) are used. The short codes are used to ease the implementation of advanced multiuser receiver techniques; otherwise long spreading codes can be used. Short codes are VL-Kasami codes of length 256 and lond codes are Gold sequences of length 2^{41}, but the latter are truncated to form a cycle of a 10-ms frame.

For channelization, orthogonal codes are used. Orthogonality between the different spreading factors can be achieved by the tree-structured orthogonal codes whose construction was described in Chapter 5.

IQ/code multiplexing leads to parallel transmission of two channels, and therefore, attention must be paid to modulated signal constellation and related peak-to-average power ratio (crest factor). By using the complex spreading circuit shown in Figure 6.9, the transmitter power amplifier efficiency remains the same as for QPSK transmission in general.

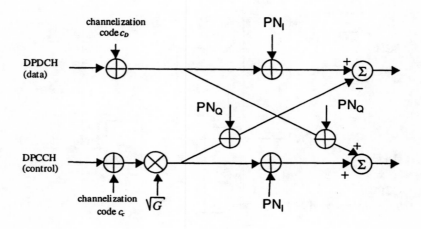

Figure 6.9   IQ/code multiplexing with complex spreading circuit.

Moreover, the efficiency remains constant irrespective of the power difference G between DPDCH and DPCCH. This can be explained with Figure 6.10, which shows the signal constellation for IQ/code multiplexed control channel with complex spreading. In the middle constellation with $G = 0.5$ all eight constellation points are at the same distance from the origin. The same is true for all values of G. Thus, signal envelope variations are very similar to the QPSK transmission for all values of G. The IQ/code multiplexing solution with complex scrambling results in power amplifier output backoff requirements that remain constant as a function of power difference. Furthermore, the achieved output backoff is the same as for one QPSK signal.

6.3.5 Multirate

Multiple services of the same connection are multiplexed on one DPDCH. Multiplexing may take place either before or after the inner or outer coding, as illustrated in Figure 6.11. After service multiplexing and channel coding, the multiservice data stream is mapped to one DPDCH. If the total rate exceeds the upper limit for single code transmission, several DPDCHs can be allocated.

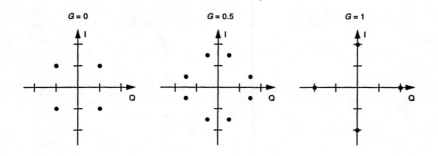

Figure 6.10 Signal constellation for IQ/code multiplexed control channel with complex spreading. G is the power difference between DPCCH and DPDCH.

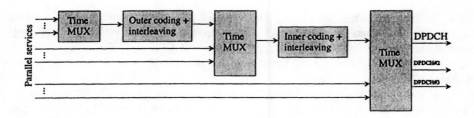

Figure 6.11 Service multiplexing in WCDMA.

A second alternative for service multiplexing would be to map parallel services to different DPDCHs in a multicode fashion with separate channel coding/interleaving. With this alternative scheme, the power, and consequently, the quality of each service, can be separately and independently controlled. The disadvantage is the need for multicode transmission, which will have an impact on mobile station complexity. Multicode transmission sets higher requirements for the power amplifier linearity in transmission, and more correlators are needed in reception.

For BER = 10^{-3} services, convolutional coding of 1/3 is used. For high bit rates a code rate of 1/2 can be applied. For higher quality service classes outer Reed-Solomon coding is used to reach the 10^{-6} BER level. Retransmissions can be utilized to guarantee service quality for non real-time packet data services.

After channel coding and service multiplexing, the total bit rate can be almost arbitrary. The rate matching adapts this rate to the limited set of possible bit rates of a DPDCH. Repetition or puncturing is used to match the coded bit stream to the channel gross rate. The rate matching for uplink and downlink are introduced below.

For the uplink, rate matching to the closest uplink DPDCH bit rate is always based on unequal repetition (a subset of the bits repeated) or code puncturing. In general, code puncturing is chosen for bit rates less than ≈20% above the closest lower DPDCH bit rate. For all other cases, unequal repetition is performed to the closest higher DPDCH bit rate. The repetition/puncturing patterns follow a regular predefined rule (i.e., only the amount of repetition/puncturing needs to be agreed on). The correct repetition/puncturing pattern can then be directly derived by both the transmitter and receiver side.

For the downlink, rate matching to the closest DPDCH bit rate, using either unequal repetition or code puncturing, is only made for the highest rate (after channel coding and service multiplexing) of a variable rate connection and for fixed-rate connections. For lower rates of a variable rate connection, the same repetition/puncturing pattern as for the highest rate is used, and the remaining rate matching is based on discontinuous transmission where only a part of each slot is used for transmission. This approach is used in order to simplify the implementation of blind rate detection in the mobile station.

6.3.6 Packet Data

WCDMA has two different types of packet data transmission possibilities. Short data packets can be appended directly to a random access burst. This method, called *common channel packet transmission*, is used for short infrequent packets, where the link maintenance needed for a dedicated channel would lead to an unacceptable overhead.

When using the uplink common channel, a packet is appended directly to a random access burst. Common channel packet transmission is typically used for short, infrequent packets, where the link maintenance needed for a dedicated channel would lead to unacceptable overhead. Also, the delay associated with a transfer to a dedicated channel is avoided. Note that for common channel packet transmission only open loop power control is in operation. Common channel packet transmission should therefore be limited to short packets that only use a limited capacity. Figure 6.12 illustrates packet transmission on a common channel.

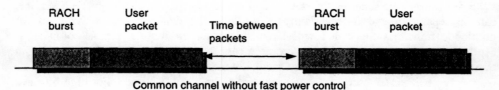

Common channel without fast power control

Figure 6.12   Packet transmission on the common channel.

Larger or more frequent packets are transmitted on a dedicated channel. A large single packet is transmitted using a *single-packet scheme* where the dedicated channel is released immediately after the packet has been transmitted. In a *multipacket scheme* the dedicated channel is maintained by transmitting power control and synchronization information between subsequent packets.

6.3.7 Handover

Base stations in WCDMA need not be synchronized, and therefore, no external source of synchronization, like GPS, is needed for the base stations. Asynchronous base stations must be considered when designing soft handover algorithms and when implementing position location services.

Before entering soft handover, the mobile station measures observed timing differences of the downlink SCHs from the two base stations. The structure of SCH is presented in Section 6.3.3. The mobile station reports the timing differences back to the serving base station. The timing of a new downlink soft handover connection is adjusted with a resolution of one symbol (i.e., the dedicated downlink signals from the two base stations are synchronized with an accuracy of one symbol). That enables the mobile RAKE receiver to collect the macro diversity energy from the two base stations. Timing adjustments of dedicated downlink channels can be carried out with a resolution of one symbol without losing orthogonality of downlink codes.

6.3.7.1 Interfrequency Handovers

Interfrequency handovers are needed for utilization of hierarchical cell structures; macro, micro, and indoor cells (see Chapter 8). Several carriers and interfrequency handovers may also be used for taking care of high capacity needs in hot spots. Interfrequency handovers will be needed also for handovers to second generation systems, like GSM or IS-95. In order to complete interfrequency handovers, an efficient method is needed for making measurements on other frequencies while still having the connection running on the current frequency. Two methods are considered for interfrequency measurements in WCDMA

- Dual receiver;
- Slotted mode.

The dual receiver approach is considered suitable especially if the mobile terminal employs antenna diversity. During the interfrequency measurements, one receiver branch is switched to another frequency for measurements, while the other keeps receiving from the current frequency. The loss of diversity gain during measurements needs to be compensated for with higher downlink transmission power. The advantage of the dual receiver approach is that there is no break in the current frequency connection. Fast closed loop power control is running all the time.

The slotted mode approach depicted in Figure 6.13 is considered attractive for the mobile station without antenna diversity. The information normally transmitted during a 10-ms frame is compressed time either by code puncturing or by changing the FEC rate. For the performance of different interfrequency handover options, see Section 5.14.5.

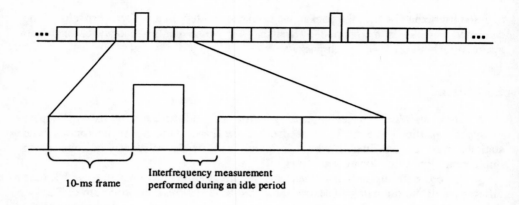

Figure 6.13 Slotted mode structure.

6.3.8 Inter-operability Between GSM and WCDMA

The handover between the WCDMA system and the GSM system, offering world-wide coverage already today, has been one of the main design criteria taken into account in the WCDMA frame timing definition. The GSM compatible multiframe structure, with the superframe being multiple of 120 ms, allows similar timing for inter-system measurements as in the GSM system itself. Apparently the needed measurement interval does not need to be as frequent as for GSM terminal operating in a GSM system, as inter-system handover is less critical from intra-system interference point of view. Rather the compatibility in timing is important that when operating in WCDMA mode, a multimode terminal is able to catch the desired information from the synchronization bursts in the synchronisation frame on a GSM carrier with the aid of frequency correction burst. This way the relative timing between a GSM and WCDMA carriers is maintained similar to the timing between two asynchronous GSM carriers. The timing relation between WCDMA channels and GSM channels is indicated in Figure 6.14, where the GSM traffic channel and WCDMA channels use similar 120 ms multiframe structure. The GSM frequency correction channel (FCCH) and GSM synchronization channel (SCH) use one slot out of the eight GSM slots in the indicated frames with the FCCH frame with one time slot for FCCH always preceding the SCH frame with one time slot for SCH as indicated in the Figure 6.14. Further details on GSM common channel structures can be found in [30].

A WCDMA terminal can do the measurements either by requesting the measurement intervals in a form of slotted mode where there are breaks in the downlink transmission or then it can perform the measurements independently with a suitable measurement pattern. With independent measurements the dual receiver approach is used instead of the slotted mode since then GSM receiver branch can operate independently of the WCDMA receiver branch.

For smooth inter-operation between the systems, information needs to be exchanged between the systems, in order to allow the WCDMA base station to notify the terminal of the existing GSM frequencies in the area. In addition, more integrated operation is needed for the actual handover where the current service is maintained, taking naturally into account the lower data rate capabilities in GSM when compared to UMTS maximum data rates reaching all the way to 2 Mbps.

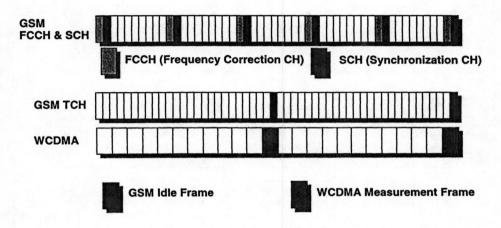

Figure 6.14   Measurements timing relation between WCDMA and GSM frame structures.

The GSM system is likewise expected to be able to also indicate the WCDMA spreading codes in the area to make the cell identification simpler. After that the existing measurement practices in GSM can be used for measuring the WCDMA when operating in GSM mode.

As the WCDMA does not rely on any superframe structure as with GSM to find out synchronization, the terminal operating in GSM mode is able to obtain the WCDMA frame synchronization once the WCDMA base station scrambling code timing is acquired. The base station scrambling code has 10-ms period and its frame timing is synchronized to WCDMA common channels.

6.4   cdma2000

Within standardization committee TIA TR45.5, the subcommittee TR45.5.4 was responsible for the selection of the basic cdma2000 concept. Like for all the other wideband CDMA schemes, the goal has been to provide data rates that meet the IMT-2000 performance requirements of at least 144 Kbps in a vehicular environment, 384 Kbps in a pedestrian environment, and 2048 Kbps in an indoor office environment. The main focus of standardization has been providing 144 Kbps and 384 Kbps with approximately 5-MHz bandwidth. The main parameters of cdma2000 are listed in Table 6.2.

Table 6.2
cdma2000 Parameter Summary

Channel bandwidth	1.25, 5, 10, 15, 20 MHz
Downlink RF channel structure	Direct spread or multicarrier
Chip rate	1.2288/3.6864/7.3728/11.0593/14.7456 Mcps for direct spread n × 1.2288 Mcps (n = 1,3,6,9,12) for multicarrier
Roll-off factor	Similar to IS-95 (see [31])
Frame length	20 ms for data and control / 5ms for control information on the fundamental and dedicated control channel
Spreading modulation	Balanced QPSK (downlink) Dual-channel QPSK (uplink) Complex spreading circuit
Data modulation	QPSK (downlink) BPSK (uplink)
Coherent detection	Pilot time multiplexed with PC and EIB (uplink) Common continuous pilot channel and auxiliary pilot (downlink)
Channel multiplexing in uplink	Control, pilot, fundamental and supplemental code multiplexed I&Q multiplexing for data and control channels
Multirate	Variable spreading and multicode
Spreading factors	4 – 256
Power control	Open loop and fast closed loop (800 Hz, higher rates under study)
Spreading (downlink)	Variable length Walsh sequences for channel separation, M-sequence 2^{15} (same sequence with time shift utilized in different cells, different sequence in I&Q channel)
Spreading (uplink)	Variable length orthogonal sequences for channel separation, M-sequence 2^{15} (same for all users different sequences in I&Q channels), M-sequence $2^{41}-1$ for user separation (different time shifts for different users)
Handover	Soft handover Interfrequency handover

6.4.1 Carrier Spacing and Deployment Scenarios

In the following we highlight the channel structures of cdma2000. Currently there exist two main alternatives for the downlink: multicarrier and direct spread options. The multicarrier approach maintains orthogonality between the cdma2000 and IS-95 carriers [32]. In the downlink this is more important because the power control cannot balance the interfering powers between different layers, as it can in the uplink. As illustrated in Figure 6.15, transmission on the multicarrier downlink (nominal 5-MHz band) is achieved by using three consecutive IS-95B carriers[4] where each carrier has a chip rate of 1.2288 Mcps. For the direct spread option, transmission on the downlink is achieved by using a nominal chip rate of 3.6864 Mcps. The multicarrier approach has been proposed since it might provide an easier overlay with the existing IS-95 systems. This is because without multipath it retains orthogonality with existing IS-95 carriers. However, in certain conditions the spectrum efficiency of multicarrier is 5% to 10%

---

[4] Assuming a three-times increase in bandwidth compared to IS-95B. Options using 6 times, 9 times, and 12 times IS-95 bandwidths are possible.

worse than direct spread since it can resolve a smaller number of multipath components [32]. Regardless of the downlink solution, if an operator has a 5-MHz allocation and if at least 1.25 MHz is already in use, the implementation of either the multicarrier or the direct spread overlay could be challenging [32].

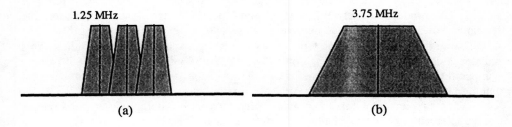

Figure 6.15   Illustration of (a) multicarrier and (b) direct spread downlink.

The starting point for bandwidth design of cdma2000 has been the PCS spectrum allocation in the United States. The PCS spectrum is allocated in 5-MHz blocks (D, E, and F blocks) and 15-MHz blocks (A, B, and C blocks). One 3.6864-Mcps carrier can be deployed within 5-MHz spectrum allocation including guardbands. For the 15-MHz block, three 3.6864 Mcps carriers plus two 1.2288-Mcps carriers can be deployed. For a 10-MHz block, two 3.6864-Mcps carriers plus one 1.2288-Mcps carrier can be deployed [33].

6.4.2   Logical Channels

At the time of writing, the logical channels for cdma2000 were still under development. The reader is referred to the latest standards documents in TIA [6].

6.4.3   Physical Channels

6.4.3.1   Uplink Physical Channels

In the uplink, there are four different dedicated channels. The fundamental and supplemental channels carry user data. A dedicated control channel, with a frame length 5 or 20 ms, carries control information such as measurement data, and a pilot channel is used as a reference signal for coherent detection. The pilot channel also carries time multiplexed power control symbols. Figure 6.16 illustrates the different uplink dedicated channels separated by Walsh codes.

The reverse access channel (R-ACH) and the reverse common control channel (R-CCCH) are common channels used for communication of layer 3 and MAC layer messages. The R-ACH is used for initial access, while the R-CCCH is used for fast packet access.

The fundamental channel conveys voice, signaling, and low rate data. Basically, it will operate at low FER (around 1%). The fundamental channel supports basic rates of 9.6 Kbps and 14.4 Kbps and their corresponding subrates (i.e., Rate Set 1 and 2 of IS-95). The fundamental channel will always operate in soft handover mode. The fundamental channel does not operate in a scheduled manner; thus permitting the mobile station to transmit acknowledgments or short packets without scheduling. This reduces delay and the processing load due to scheduling [33]. Its main difference compared to the IS-95 voice channel is that discontinuous transmission is implemented using repetition coding rather than gated transmission.

The supplemental channel provides high data rates. The uplink supports one or two supplemental channels. If only one supplemental channel is transmitted, then the Walsh code (+-) is used on the first supplemental channel, and if two supplemental channels are transmitted then the Walsh code (+-+-) is used. A repetition scheme is used for variable data rates on the supplemental channel.

6.4.3.2 Downlink Physical Channels

Downlink has three different dedicated channels and three common control channels. Similar to the uplink, the fundamental and supplemental channels carry user data and the dedicated control channel control messages. The dedicated control channel contains

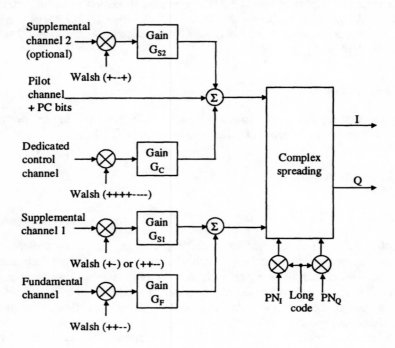

Figure 6.16 The uplink dedicated channel structure.

power control bits and rate information. The synchronization channel is used by the mobile stations to acquire initial time synchronization. One or more paging channels are used for paging the mobiles. The pilot channel provides a reference signal for coherent detection, cell acquisition, and handover.

In the downlink, cdma2000 has a common pilot channel, which is used as a reference signal for coherent detection when adaptive antennas are not employed. The pilot channel is similar to IS-95 (i.e., it is comprised of a long PN-code and Walsh sequence number 0). When adaptive antennas are used, auxiliary pilot is used as a reference signal for coherent detection. Code multiplexed auxiliary pilots are generated by assigning a different orthogonal code to each auxiliary pilot. This approach reduces the number of orthogonal codes available for the traffic channels. This limitation is alleviated by expanding the size of the orthogonal code set used for the auxiliary pilots. Since a pilot signal is not modulated by data, the pilot orthogonal code length can be extended, thereby yielding an increased number of available codes, which can be used as additional pilots

As mentioned, two alternatives for downlink modulation still exist: direct spread and multicarrier. The multicarrier transmission principle is illustrated in Figure 6.17. A performance comparison of direct spread and multicarrier can be found in [34].

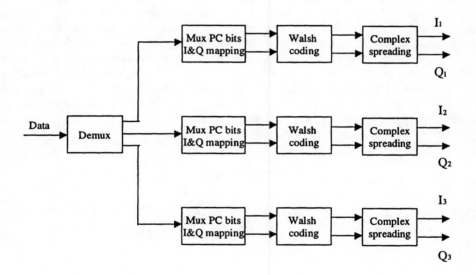

Figure 6.17   Multicarrier downlink.

6.4.4 Spreading

On the downlink, the cell separation for cdma2000 is performed by two M-sequences of length 2^{15}, one for the I channel and one for the Q channel, which are phase shifted by PN-offset for different cells. Thus, during the cell search process only these sequences need to be searched. Since there is only a limited number of PN-offsets, they need to be planned in order to avoid PN-confusion [35]. In the uplink, user separation is performed by different phase shifts of M-sequence of length 2^{41}. The channel separation is performed using variable spreading factor Walsh sequences, which are orthogonal to each other. Fundamental and supplemental channels are transmitted with the multicode principle. The variable spreading factor scheme is used for higher data rates in the supplemental channel.

Similar to WCDMA, complex spreading is used. In the uplink, it is used with dual-channel modulation.

6.4.5 Multirate

The fundamental and supplemental channels can have different coding and interleaving schemes. In the downlink, high bit rate services with different QoS requirements are code multiplexed into supplemental channels, as illustrated in Figure 6.18. In the uplink, one or two supplemental channels can be transmitted. The user data frame length of cdma2000 is 20 ms. For the transmission of control information, 5- and 20-ms frames can be used on the fundamental channel. Also on the fundamental channel, a convolutional code with constraint length of 9 is used. On supplemental channels, a convolutional code is used up to 14.4 Kbps. For higher rates, Turbo codes with constraint length 4 and rate 1/4 are preferred. Rate matching is performed by puncturing, symbol repetition, and sequence repetition.

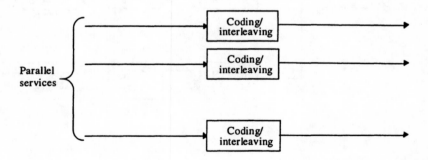

Figure 6.18   Principle of code multiplexing.

6.4.6 Packet Data

cdma2000 also uses the slotted Aloha principle for packet data transmission. However, instead of fixed transmission power, it increases the transmission power for the random access burst after an unsuccessful access attempt. When the mobile station has been allocated a traffic channel, it can transmit without scheduling up to a predefined bit rate. If the transmission rate exceeds the defined rate, a new access request has to be made. When the mobile station stops transmitting, it releases the traffic channel but not the dedicated control channel. After a while, it also releases the dedicated control channel but maintains the link layer and network layer connections in order to shorten the channel setup time when new data need to be transmitted. Short data bursts can be transmitted over a common traffic channel in which a simple ARQ is used to improve the error rate performance.

6.4.7 Handover

It is expected that soft handover of the fundamental channel will operate similarly to the soft handover in IS-95. In IS-95, the active set is the set of base stations transmitting to the mobile station. For the supplemental channel, the active set can be a subset of the Active Set for the fundamental channel. This has two advantages. First, when diversity is not needed to counter fading, it is preferable to transmit from fewer base stations. This increases the overall downlink capacity. For stationary conditions, an optimal policy is to transmit only from one base station — the base station that would radiate the smallest amount of downlink power. Second, for packet operation, the control processes can also be substantially simplified if the supplemental channel is not in soft handover. However, maintaining the fundamental channel in soft handover provides the ability to reliably signal the preferred base station to transmit the supplemental channel when channel conditions change [33].

6.4.8 Transmit Diversity

The downlink performance can be improved by transmit diversity. For direct spread CDMA schemes, this can be performed by splitting the data stream and spreading the two streams using orthogonal sequences. For multicarrier CDMA, the different carriers can be mapped into different antennas.

6.5 KOREAN AIR INTERFACES

In Korea, two wideband CDMA air interfaces are being considered: TTA I and TTA II. The Electronics and Telecommunications Research Institute (ETRI) has established an R&D consortium to define the Korean proposal for IMT-2000 during 1997 and 1999. A wideband CDMA proposal has been developed within ETRI [36–38]. SK Telecom has also developed a wideband CDMA air interface [39–43]. This has been combined with a number of other proposals to form the basis for the TTA II scheme [43]. The main

Table 6.3
Parameters of Korean Wideband CDMA Schemes

	TTA I	TTA II
Channel spacing	1.25 MHz / 5 MHz / 20 MHz	1.25 MHz / 5 MHz / 10 MHz / 20 MHz
Chip rate	0.9216 Mcps / 3.6864 Mcps / 14.7456 Mcps	1.024 Mcps / 4.096 Mcps / 8.192 Mcps / 16.384 Mcps
Frame length	20 ms	10 ms
Spreading	Walsh + long codes	Walsh + long codes
Pilot for coherent detection	UL: pilot symbol based DL: common pilot	UL: pilot channel based (multiplexed with power control symbols) DL: common pilot
Base station synchronization	Synchronous	Asynchronous

parameters of these two air interfaces are listed in Table 6.3. The TTA II concept is closer to cdma2000, and TTA I resembles WCDMA. For a more details of these schemes, refer to [7–8].

6.5.1 Differences Between TTA I and cdma2000

The differences between TTA I and cdma2000 include

- A 1.6-kHz power control rate instead of 800 Hz;
- A 10-ms frame length instead of 20 ms;
- Orthogonal complex QPSK (OCQPSK) modulation in the uplink;
- Selectable forward error correction code;
- Time division transmit diversity (TDTD) instead of orthogonal transmit diversity in the downlink;
- Quasi orthogonal code spreading to increase the number of orthogonal codes for packet operation;
- Intercell asynchronous mode;
- The lowest chip rate of 0.9216 Mcps instead of 1.2288 Mcps.

OCQPSK constraints phase transitions within certain period to be $\pi/2$. The possible advantages of this scheme are [7]
- Reduced linearity requirements for power amplifier;
- Small complexity since only one PN code is used.

6.5.2 Differences Between TTA II and WCDMA

The differences between TTA II and WCDMA include

- Continuous pilot in the uplink;
- QPSK spreading in the downlink;
- OCQPSK modulation in the uplink;
- Selectable forward error correction code;
- Quasi orthogonal code spreading to reduce the intracell interference;
- The downlink pilot structure;
- Optional synchronization in the uplink.

The original chip rate of the SK Telecom scheme was 4.068 Mchip [41]. This was changed to 4.096 Mcps as a result of harmonization with the Japanese Core-A proposal.

In TTA II, the different cells in the downlink and the users in the uplink are distinguished by long spreading codes. Since TTA II has long spreading codes, it uses two pilot channels in the downlink, a cluster pilot and a cell pilot, to reduce long synchronization time. A cluster consists of several cells, and under each cluster the same long spreading code pilots are reused. Each cluster has a cluster pilot that is also a long spreading sequence. There are 16 cluster pilots, and each cluster can have 32 cell sequences. Thus, a maximum of 48 pilot codes need to be searched (i.e., 16 cluster pilot codes and 32 cell pilot codes). A cluster pilot can be transmitted by the center cell of a cluster or by each cell. The former technique is suited for a hierarchical cell system.

To reduce the intracell interference, the TTA II wideband CDMA scheme time synchronizes all users in the uplink with an accuracy of 1/8 chip. This is done by measuring the timing in the base station and signaling the timing adjustment commands with a rate of 2 Kbps to the mobile station. However, multipath results in intracell interference, and the gain from the orthogonal uplink depends on the channel profile. In addition, the signaling traffic reduces the downlink capacity for each user by 2 Kbps.

6.6 CODIT

The CODIT air interface was developed between 1992 and 1995 as part of the RACE II program [44–46]. CODIT was supposed to be a candidate for the UMTS air interface. However, its development was stopped after the project ended in 1995. Some of the concepts, such as seamless interfrequency handover, that were developed in the CODIT project have been adopted by other wideband CDMA candidates. Table 6.4 lists the main parameters of the CODIT air interface. Some of the CODIT features are described in the following sections.

6.6.1 Modulation

QPSK data modulation is used in both uplink and downlink. Spreading modulation is QPSK in the downlink and O-QPSK in the uplink. Coherent detection is employed in both uplink and downlink. In the uplink, CODIT performs coherent detection using the demodulated control channel as a pilot [46]. The control channel itself uses differentially coherent detection. Since the CODIT data and control channels are transmitted in parallel, the samples of the 10-ms data frame have to be buffered before despreading and demodulation.

Table 6.4
Main Parameters of CODIT

Channel bandwidth	1, 5, 20 MHz
Downlink RF channel structure	Direct spread
Chip rate	1.023/5.115/20.46 Mcps
Frame length	10 ms
Spreading modulation	QPSK (downlink)
	Offset QPSK(uplink)
Data modulation	QPSK (downlink)
	BPSK (uplink)
Coherent detection	Common pilot channel (downlink)
	With diff. coherent pilot channel (uplink)
Multirate	Variable spreading factor
Spreading factors	Arbitrary
Power Control	Open and fast closed loop (uplink) (1.6 kHz)
	Slow quality based (downlink)
Base station synchronization	Asynchronous

6.6.2 Spreading

CODIT has long spreading codes of length $2^{41}-1$ for data channels and short Gold codes of length 1023 for control channels. The reasons for selecting long codes include the large number of codes available and flexibility to multiple bit rates, variable spreading factors, and relaxed requirements for interchannel and intercell synchronization [45]. Furthermore, long codes were selected instead of short ones because of expected problems with short codes in code planning, restrictions in applicable spreading factors, and timing control. It should be noted that these problems have been solved in the current wideband CDMA schemes (WCDMA).

6.6.3 Multirate

CODIT uses variable spreading both in the uplink and downlink. Long codes facilitate arbitrary spreading ratios. The different services of the same user are time multiplexed into the frame. Control information is transmitted on a parallel code channel. CODIT can control the data rate on a frame-by-frame basis. Since CODIT can have arbitrary spreading ratios, the sampled signal needs to be buffered before despreading until the rate information has been decoded.

The frame length is 10 ms, which is also the basic interleaving period. For data services a longer interleaving period is applied. CODIT employs a convolutional code for services with BER 10^{-3} and a concatenated convolutional code and RS code for services with BER 10^{-6}. For nontransparent services, the type-I hybrid ARQ scheme that exploits the error detection capabilities of the RS-code is used.

6.6.4 Handover

Similar to other wideband CDMA proposals CODIT uses soft handover. Since it employs long codes, the mobile station can measure the timing uncertainty between the base stations involved in the soft handover and perform handover in an asynchronous network.

6.7 IS-665 W-CDMA

The origins of IS-655 W-CDMA standard dates back to the broadband CDMA (B-CDMA) proposal by Interdigital. The B-CDMA concept was introduced in 1989 [47]. The original bandwidth was as large as 48 MHz, and the chip rate 24 Mcps. This system was field-tested with microwave signals in the PCS band [47]. In 1994 an AMPS overlay was tested in the cellular band. Microwave and AMPS carriers were supposed to be filtered out by notch filters. As a result of the U.S. PCS frequency plan, where 5 and 15 MHz band allocations were adopted, the original B-CDMA proposal was modified to be 5 and 15 MHz and the overlay idea was given up.

In 1993, a wideband CDMA scheme was proposed by OKI as a PCS standard. Lockheed Sanders, Interdigital, and AT&T co-operated with OKI during the standardization, and the proposed W-CDMA system became the TIA/EIA (Electronics Industry Association) IS-665 standard, the T1P1 Trial Use Standard J-STD-015, and ITU-R Recommendation M.1073 [48]. It was also used as a basis for the Core-B proposal in Japan [17]. However, IS-665 W-CDMA has not been commercially successful, and except for trial systems it has not been deployed.

The main parameters of the system are listed in Table 6.5 [49]. The system has three bandwidths of 5, 10, and 15 MHz to fit into the U.S. PCS spectrum allocations [49]. Basic data rates are up to 16, 32, and 64 Kbps. The 64 Kbps data rate is transmitted using rate 1/2 convolutional coding, and lower rates use symbol repetition. The Core-B proposal in Japan is an enhanced version of IS-665 W-CDMA offering 128 Kbps within a single code and 2 Mbps using multicode transmission. The frame length of the system is 5 ms and interleaving can be over 5, 10, or 20 ms. IS-665 W-CDMA has open and closed loop control. The closed loop power control uses variable step sizes of 0.5, 1, 2, and 4 dB, which are adaptively controlled [49]. IS-665 W-CDMA supports the use of interference cancellation.

Table 6.5
IS-665 W-CDMA System Parameters

Basic chip rate		4.096, 8.192, and 12.288 Mcps
Base station synchronization		Synchronous
Frame length		5 ms
Multirate / variable rate		Symbol repetition / multicode
Coherent detection	UL	Pilot channel
	DL	Pilot channel
Power control		2 Kbps time multiplexed

REFERENCES

[1] Ojanperä, T., and R. Prasad, "Overview of air interface multiple access for IMT-2000/UMTS," *IEEE Communications Magazine*, to appear in September 1998 issue.

[2] ETSI, "The ETSI UMTS Terrestrial Radio Access (UTRA) ITU-R RTT Candidate Submission," June 1998.

[3] ARIB, "Japan's Proposal for Candidate Radio Transmission Technology on IMT-2000:W-CDMA," June 1998.

[4] T1P1, "IMT-2000 Radio Transmission Technology Candidate," June 1998.

[5] TR46, "TR46 RTT Candidate based on the WIMS W-CDMA Proposal," June 1998.

[6] TIA, "The cdma2000 ITU-R RTT Candidate Submission," June 1998.

[7] TTA, Global CDMA I: Multiband Direct-Sequence CDMA System RTT System Description," June 1998.

[8] TTA, "Global CDMA II for IMT-2000 RTT System Description," June 1998.

[9] Ishida, Y., "Recent Study on Candidate Radio Transmission Technology for IMT-2000," *First Annual CDMA European Congress*, London, UK, October 1997.

[10] Ojanperä, T., M. Gudmundson, P. Jung, J. Sköld, R. Pirhonen, G. Kramer, and A. Toskala, "FRAMES – Hybrid Multiple Access Technology", *Proceedings of ISSSTA'96*, Vol. 1, Mainz, Germany, September 1996, pp. 320–324.

[11] Ojanperä, T., P. O. Anderson, J. Castro, L. Girard, A. Klein, and R. Prasad, "A Comparative Study of Hybrid Multiple Access Schemes for UMTS," *Proceedings of ACTS Mobile Summit Conference*, Vol. 1, Granada, Spain, November 1996, pp. 124–130.

[12] Ojanperä, T., J. Sköld, J. Castro, L. Girard, and A. Klein, "Comparison of Multiple Access Schemes for UMTS," *Proceedings of VTC'97*, Vol. 2, Phoenix, Arizona, USA, May 1997, pp. 490–494.

[13] Ojanperä, T., A. Klein, and P.-O. Anderson, "FRAMES Multiple Access for UMTS," *IEE Colloquium on CDMA Techniques and Applications for Third Generation Mobile Systems*, London, May 1997.

[14] Ovesjö, F., E. Dahlman, T. Ojanperä, A. Toskala, and A. Klein, "FRAMES Multiple Access Mode 2 – Wideband CDMA," *Proceedings of PIMRC97*, Helsinki, Finland, September 1997, pp. 42–46.

[15] Nikula, E., A. Toskala, E. Dahlman, L. Girard, and A. Klein, "FRAMES Multiple Access for UMTS and IMT-2000," *IEEE Personal Communications Magazine*, April 1998, pp. 16–24.

[16] CSEM/Pro Telecom, Ericsson, France Télécom – CNET, Nokia, Siemens, "FMA - FRAMES Multiple Access A Harmonized Concept for UMTS / IMT-2000," *ITU Workshop on Radio Transmission Technologies for IMT-2000*, Toronto, Canada, September 10 – 11, 1997.

[17] ARIB FPLMTS Study Committee, "Report on FPLMTS Radio Transmission Technology SPECIAL GROUP, (Round 2 Activity Report)," Draft v.E1.1, January 1997.

[18] Adachi, F., K. Ohno, M. Sawahashi, and A. Higashi, "Multimedia mobile radio access based on coherent DS-CDMA," *Proceedings of 2nd International workshop on Mobile Multimedia Commun.*, A2.3, Bristol University, UK April 1995.

[19] Ohno, K., M. Sawahashi, and F. Adachi, "Wideband coherent DS-CDMA," *Proceedings of VTC'95*, Chicago, Illinois, USA, July 1995, pp. 779–783.

[20] Dohi, T., Y. Okumura, A. Higashi, K. Ohno, and F. Adachi, "Experiments on Coherent Multicode DS-CDMA," *Proceedings of VTC'96*, Atlanta, Georgia, USA, April 1996, pp. 889–893.

[21] Adachi, F., M. Sawahashi, and K. Ohno, "Coherent DS-CDMA: Promising Multiple Access for Wireless Multimedia Mobile Communications," *Proceedings of ISSSTA'96*, Mainz, Germany, September 1996, pp. 351–358.

[22] Onoe, S., K. Ohno, K. Yamagata and T. Nakamura, "Wideband-CDMA Radio Control Techniques for Third Generation Mobile Communication Systems," *Proceedings of VTC97*, Vol. 2, Phoenix, Arizona, USA, May 1997, pp. 835–839.

[23] Recommendation ITU-R M.1035, "Framework for the Radio Interface(s) and Radio Sub-System Functionality for Future Public Land Mobile Telecommunication Systems (FPLMTS)," 1994.

[24] Toskala, A., E. Dahlman, M. Gustafsson, M. Latva-Aho, and M. J. Rinne, "Frames Multiple Access Mode 2 Physical Transport Control Functions," *Proceedings of ACTS Summit*, Aalborg, Denmark, October 1997, pp. 697–702.

[25] Esmailzadeh, R., and M. Gustafson, "A New Slotted ALOHA Based Random Access Method for CDMA Systems," *Proceedings of ICUPC'97*, San Diego, California, USA, October 1997, pp. 43–47.

[26] Ojanperä, T., K. Rikkinen, H. Hakkinen, K. Pehkonen, A. Hottinen, and J. Lilleberg, "Design of a 3rd Generation Multirate CDMA System with Multiuser Detection, MUD-CDMA", *Proceedings of ISSSTA'96*, Vol. 1, Mainz, Germany, September 1996, pp. 334–338.

[27] Westman, T., and H. Holma, "CDMA System for UMTS High Bit Rate Services," *Proceedings of VTC97*, Phoenix, Arizona, USA, May 1997, pp. 824–829.

[28] Pehkonen, K., H. Holma, I. Keskitalo, E. Nikula, and T. Westman, "A Performance Analysis of TDMA and CDMA Based Air Interface Solutions for UMTS High Bit Rate Services," *Proceedings of PIMRC'97*, Helsinki, Finland, September 1997, pp. 22-26.

[29] Higuchi, K., M. Sawahashi and F. Adachi, "Fast Cell Search Algorithm in DS-CDMA Mobile Radio Using Long Spreading Codes," *Proceedings of VTC97*,

Vol.3, Phoenix, Arizona, USA, May 1997, pp. 1430-1434.

[30] Mouly, M., and M.-B. Pautet, *The GSM System for Mobile Communications*, published by the authors, 1992

[31] TIA/EIA/IS-95-A, "Mobile Station-Base Station Compatibility Standard for Dual-Mode Wideband Spread Spectrum Cellular System," Telecommunications Industry Association, Washington, D.C., May 1995.

[32] Chia, S., "Will cdmaOne be the third choice," *CDMA Spectrum*, Issue 2, pp. 30-34, September 1997.

[33] Tiedemann, Jr., E. G., Y.-C. Jou, and J. P. Odenwalder, "The Evolution of IS-95 to a Third Generation System and to the IMT-2000 Era," *Proceedings of ACTS Summit 1997*, Aalborg, Denmark, October 1997, pp. 924–929.

[34] Jalali, A., and A. Gutierrez, "Performance Comparison of Alternative Wideband CDMA Systems", *Proceedings of ICC'98*, Atlanta, Georgia, USA, June 1998, pp. 38–42.

[35] Yang, J., D. Bao and M. Ali, "PN Offset Planning in IS-95 based CDMA system" *Proceedings of VTC'97*, Vol. 3, Phoenix, Arizona, USA, May 1997, pp. 1435–1439.

[36] Bang, S. C., H-R. Park and Y. Han, "Performance Analysis of Wideband CDMA System for FPLMTS," *Proceedings of VTC'97*, Phoenix, Arizona, USA, May 1997, pp. 830–834

[37] Park, H.-R., "A Third Generation CDMA System for FPLMTS Application," *The 1st CDMA International Conference*, Seoul Korea, November 1996.

[38] Han, Y., S. C. Banh, H.-R. Park and B.-J. Kang, "Performance of Wideband CDMA System for IMT-2000," *2nd CDMA international Conference (CIC)*, Seoul, Korea, October 1997, pp. 583–587.

[39] Koo, J. M., Y. I. Kim, J. H. Ryu and J. I. Lee, "Implementation of prototype wideband CDMA system," *Proceedings of ICUPC'96*, Cambridge, Massachusetts, USA, September/October 1996, pp. 797–800.

[40] Koo, J. M., E. K. Hong and J. I. Lee, "Wideband CDMA Technology for FPLMTS," *The 1st CDMA International Conference*, Seoul, Korea, November 1996.

[41] Y.-W. Park, E. K. Hong, T.-Y. Lee, Y.-D. Yang and S.-M. Ryu, "Radio Characteristics of PCS using CDMA," *Proceedings of IEEE VTC'96*, Atlanta, Georgia, USA, April 1996, pp. 1661–1664.

[42] Hong, E. K., T.-Y. Lee, Y.-D. Yang, B.-C. Ahn and Y.-W. Park, "Radio Interface Design for CDMA-Based PCS," *Proceedings of ICUPC'96*, Cambridge, Massachusetts, USA, September/October 1996, pp. 365–368.

[43] Hong, E. K., "Intercell Asynchronous W-CDMA System for IMT-2000," *ITU-R TG 8/1 Workshop*, Toronto, Canada, September 1997.

[44] Baier, A., U.-C. Fiebig, W. Granzow, W. Koch, P. Teder and J. Thielecke, "Design Study for a CDMA-Based Third Generation Mobile Radio System," *IEEE Journal on Selected Areas in Communications*, Vol. 12, No. 4, May 1994, pp. 733–743.

[45] Andermo, P.-G., (ed.), "UMTS Code Division Testbed (CODIT)," *CODIT Final Review Report*, September 1995.

[46] Andermo, P.-G., and L-M. Ewerbring, "A CDMA-Based Radio Access Design for

UMTS," *IEEE Personal Communications*, Vol. 2, No. 1, February 1995, pp. 48–53.

[47] Milstein, L. B., "The CDMA Overlay Concept," *Proceedings of IEEE ISSSTA '96*, Mainz, Germany, September 1996, pp. 476–480.

[48] Fisher, R. E., A. Fukasawa, T. Sato, Y. Takizawa, T. Kato, and M. Kawabe "Wideband CDMA System for Personal Communications Services," *Proceedings of VTC'96*, Atlanta, Georgia, USA, April 1996, pp. 1625–1655.

[49] Fukasawa, A., T. Sato, Y. Takizawa, T. Kato, M. Kawabe, and R. E. Fisher, "Wideband CDMA System for Personal Radio Communications," *IEEE Communications Magazine*, October 1996, pp. 116–123.

Chapter 7

PERFORMANCE ANALYSIS

7.1   INTRODUCTION

In this chapter, the performance of a wideband CDMA cellular system in different radio environments is investigated. Since absolute performance results are valid only for the used assumptions and models, the intention of this chapter is to study relative performance figures and to analyze the impact of different techniques on wideband CDMA performance.

In Section 7.2, the effect of diversity, including multipath diversity, antenna and polarization diversity, macro diversity, and time diversity on wideband CDMA performance is investigated. In Section 7.3, the principles of the wideband CDMA link and system simulations to obtain spectrum efficiency are explained. Section 7.4 investigates the impact of fast power control on the system's performance. Especially, the effect of fast power control on the average transmission powers is studied. In addition, the impact of imperfect power control on CDMA cellular capacity is discussed.

Section 7.5 investigates spectrum efficiency for different performance enhancement techniques and radio environments. In the uplink, the impact of MUD with different efficiencies compared to a conventional receiver is simulated. Furthermore, analytical results are presented. As multipath partially destroys the orthogonality of orthogonal Walsh codes used in the downlink, the impact of multipath delay spread of different radio environments on the orthogonality factor is studied. The uplink and downlink spectrum efficiencies for service bit rates of 12 and 144 Kbps and 2 Mbps are simulated. Finally, the spectrum efficiency differences between the uplink and the downlink are analyzed.

Section 7.6 presents the principles of cell range calculation. Furthermore, the impact of system load on CDMA range is explained. Multiuser detection can increase range by canceling intracell interference. Range extension from MUD in the uplink is calculated analytically, and the impact of different MUD efficiencies is studied. If the gain from MUD is not used for range extension, it can reduce the mobile station transmission power. The decrease in the transmission power of a mobile station as a function of multiuser detection efficiency is calculated analytically.

7.2   DIVERSITY

Diversity is essential in achieving good performance in a fading channel. With wideband CDMA, the means for achieving diversity are multipath, antenna, polarization, macro, and time diversity. The use of frequency hopping typical of TDMA systems, like GSM, is not employed with wideband CDMA to gain frequency diversity.

7.2.1   Gain From Multipath Diversity

The uncoded bit error rate as a function of SNR and the number of diversity branches is given as [1]

$$P_b = \left[\frac{1}{2}(1-\mu)^L \sum_{k=0}^{L-1} \left(\frac{L-1+k}{k} \right) \left[\frac{1}{2}(1+\mu) \right]^k \right] \tag{7.1}$$

where L is the number of diversity branches and

$$\mu = \sqrt{\frac{\bar{\gamma}_c}{1+\bar{\gamma}_c}} \tag{7.2}$$

where $\bar{\gamma}_c$ is the average SNR per diversity branch. It is assumed that all diversity branches have equal SNR on average.

The importance of diversity can be seen in Figure 7.1, where the average uncoded BER is plotted as a function of the total SNR in a Rayleigh fading channel for PSK modulation with Lth order diversity. Ideal coherent maximal ratio combining of uncorrelated diversity branches is assumed. Maximal ratio combining requires estimation of the amplitudes and phases of the diversity components. When the number of multipath components increases, the SNR per each multipath gets lower, and it becomes more difficult to perform coherent combining in a RAKE receiver. Also, when the diversity branches are correlated, the diversity gain gets lower.

7.2.2   Multipath Diversity in Different Radio Environments

One of the key factors that differentiate the third generation wideband CDMA from the second generation narrowband CDMA is the wider bandwidth. In addition to the ability to provide wideband services, the increased bandwidth makes it possible to resolve more multipath components in a mobile radio channel. If the transmission bandwidth is wider than the coherence bandwidth of the channel, the receiver is able to separate

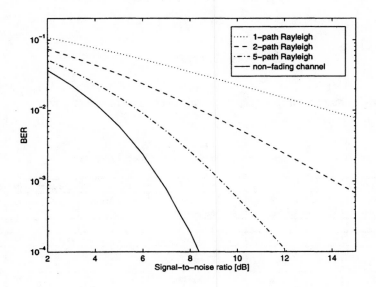

Figure 7.1 Uncoded BER with diversity assuming ideal coherent combining of diversity branches.

multipath components. This brings more diversity and higher capacity, along with power control.

We already discussed the effect of bandwidth on the diversity order in Chapter 4, where ATDMA and CODIT channel models were used. In this section, we use the ITU channel models presented in [2-3]. ITU channel models are based on the JTC (Joint Technical Committee for the U.S. PCS standardization) and ATDMA models [2]. Table 7.1 presents the coherence bandwidths for different environments. Also, the theoretical number of separable paths with chip rates of 1.2288, 4.096, and 8.192 Mcps is calculated.

The coherence bandwidth is calculated as in [4]:

$$(\Delta f)_c = \frac{1}{2\pi S} \tag{7.3}$$

where S is the rms delay spread defined as in [5]:

$$S = \sqrt{\frac{\int_0^\infty (\tau - D)^2 P(\tau) d\tau}{\int_0^\infty P(\tau) d\tau}} \tag{7.4}$$

and $P(\tau)$ is the power delay profile, and D is the average delay defined as

$$D = \frac{\int_0^\infty \tau P(\tau)d\tau}{\int_0^\infty P(\tau)d\tau} \tag{7.5}$$

The maximum number of rays (paths) in the ITU channel models is 6.

Table 7.1
Coherence Bandwidths of ITU Channel Models

ITU Channel model	Number of paths in channel model	Coherence bandwidth	Maximum number of separable paths		
			1.25 Mcps	4.0 Mcps	8.0 Mcps
Indoor A	6	4.3 MHz	1	1	2
Indoor B	6	1.6 MHz	1	3	6
Outdoor to indoor A	4	3.5 MHz	1	2	3
Outdoor to indoor B	6	250 kHz	5	17	33
Vehicular A	6	430 kHz	3	10	20
Vehicular B	6	53 kHz	24	78	155

Source: [6].

With a chip rate of 1.2288 Mcps, only very limited multipath diversity is available in indoor and in outdoor-to-indoor environments, for ITU channel models.

In Table 7.2, the number of useful multipath components is presented with three different environments measured in Australia [7]. The measured results indicate clearly the superiority of the wideband CDMA compared to narrowband CDMA in terms of achieving diversity gain with the RAKE receiver. With the 1.25-MHz bandwidth only one or two multipath components are available, where a single finger collects most of the energy. Very limited multipath diversity can be achieved with 1.25-MHz. With the 4-MHz bandwidth the achieved number of multipath components is in all measurements either two or four, which means a clearly achievable diversity gain.

If the autocorrelation properties of the spreading codes are not ideal, multipath components interfere with each other degrading the multipath diversity gain. This effect will be noticeable with very low spreading factors (i.e., with high bit rates).

Table 7.2
Number of Useful Multipath Components

	1.25 MHz	4.0 MHz	8.0 MHz
Melbourne	1	2	2
Adelaide	1	2	3
Sydney	2	4	6

Source:[7].

7.2.3 Antenna Diversity

If antenna diversity branches are close to each other, they are correlated and diversity gain is smaller. Therefore, antenna diversity is best suited for base stations and for large sized mobile stations.

Receiver antenna diversity can be used to average out receiver noise in addition to providing diversity against fading and interference. At base stations, receiver antenna diversity can be used to increase cell coverage by increasing the noise limited uplink range.

Transmission antenna diversity can be used to generate multipath diversity in such environments where only limited multipath diversity is available (i.e., in indoor and macrocell). Transmission diversity can be obtained in several ways (See Section 5.11.2.1). The same signal can be transmitted from more than one antenna with a delay that the RAKE receiver can separate. Alternatively, the data can be divided between transmission antennas without any delays, thus maintaining the orthogonality of the diversity signals. Transmission antenna diversity is best suited for the downlink transmission in the base station together with the uplink diversity reception.

7.2.4 Polarization Diversity

Different polarization directions may experience different fading. Especially in indoor environments, polarization directions have been shown to be nearly uncorrelated, thus providing diversity [8]. The advantage of polarization diversity over antenna diversity is that polarization diversity does not require separation between antennas, and thus, it could be applied to small sized equipment as well.

7.2.5 Macro Diversity

With CDMA systems, the use of macro diversity (i.e., soft handover) is essential for achieving reasonable system performance because of frequency reuse 1 and fast power control. If the mobile station is not connected to the base station to which the attenuation is the lowest, unnecessary interference is generated in adjacent cells. In the uplink, the macro diversity effects are only positive, since the more base stations try to detect the signals, the higher the probability is for at least one to succeed. In the uplink direction, the detection process itself does not utilize the information from the other base stations receiving the same signal, but the diversity is selection diversity, where the best frame is selected in the network based on the frame error indication from a CRC check. The selection can be done in the base station controller (BSC) or in some other network element, depending on the implementation.

In the downlink, macro diversity is different as the transmission now originates from several sources and diversity reception is handled by one receiver unit in the mobile station. All extra transmissions contribute to the interference. Capacity improvement is based on a similar principle as with a RAKE receiver in a multipath channel, where the received power level fluctuations tend to decrease as the number of separable paths increases. With downlink macro diversity, the RAKE receiver capability to gain from the extra diversity depends also on the number of available RAKE fingers. If a RAKE receiver is not able to collect enough energy from the transmissions from two or, in some cases, three base stations due to a limited number of RAKE fingers, the extra transmissions to the mobile can have a negative effect on the total system capacity due to increased interference. This is most likely in the macro cellular environments because the typical number of RAKE fingers considered adequate for capturing the channel energy in most cases is four. If all connections offered that amount of diversity, then the receiver has only one or two branches to allocate for each connection.

Table 7.3 illustrates the gain of soft handover over hard handover in the downlink. The results are generated with the system simulator as presented later in Section 7.3.2. The CODIT macro- and microcell environments, described in Chapter 4, were used in simulations [9].

From the macrocell results it can be seen that system downlink capacity decreases by 10% if macro diversity is not utilized. Capacity loss is not remarkable even if soft handover gain and macro diversity gain is lost. A CODIT macrocell channel gives sufficient diversity even if the mobile station was in a hard handover state. On the other hand, the limited number of RAKE fingers cannot fully exploit the attained diversity from soft handover in the macrocellular environment. On the contrary, a remarkable part of the energy generated by soft handover base stations is lost, which contributes to the interference. The situation is different for microcell, since a CODIT microcell channel provides only little diversity. Thus, macro diversity is essential for high capacity in environments with a low diversity order.

Table 7.3
Downlink Capacity with Hard and Soft Handover

[Kbps/cell/MHz]	Hard handover	Soft handover
Macro	155	169
Micro	139	222

In addition to providing gain against multipath fading, macro diversity also gives gain against shadowing. The shadow fade margin for soft handover is analyzed to be 2 to 3 dB smaller than the fade margin for hard handover [10,11]. Macro diversity can thus be used to increase the cellular range.

Soft handovers should be used with circuit switched services since macro diversity is important for low delay services to guarantee high quality with short delays. For packet data there are no strict delay requirements, and therefore, soft handover is not that important for nonreal-time packet data services that can take advantage of time diversity through retransmissions.

7.2.6 Time Diversity

Time diversity is achieved by coding, interleaving, and by retransmissions. Channel coding is applied to achieve lower power levels and required signal quality in terms of

BER/FER. Interleaving and channel coding processes are used to correct the errors due to channel fades and interference peaks. The longer the interleaving, the better an error burst is spread over the interleaving period. Consequently, the individual errors can be much better handled with error correction coding than when errors appeared to the decoder in bursts. In practice, it is the delay requirements that limit the allowed interleaving depth. For nonreal-time services the requirements are clearly looser, and retransmissions can be employed to ensure near error-free transmission of data. The total delay requirement for real-time services is typically on the order of 20 to 100 ms.

As the interleaving period gets shorter, the requirements for fast power control get tighter. Generally with CDMA signal transmission there is a peak in the BER curve when plotting the BER as a function of mobile speed; until certain velocities, the power control eliminates the fades and, at some speed, the power control is not able to follow the fades anymore [12–13]. But as the fades occur more often and get shorter in the time domain, the interleaving is able to spread the information over several fades and thus uncorrelate the signal errors to allow them to be corrected with forward error correction such as convolutional coding. However, when antenna diversity is applied, the performance is almost flat over different mobile speeds [12].

7.3 W-CDMA SIMULATORS

In this section the principles of the evaluation of the WCDMA radio network performance are analyzed. There are two approaches to simulate the overall performance of a DS-CDMA system. One is a combined approach where the link level and cellular network level simulations are combined into one simulator. Another approach is to separate the link and system level simulations to reduce the complexity of the simulators. For the performance evaluation of cellular networks a single simulator approach would be preferred. The complexity of such simulator — including everything from transmitted waveforms to a cellular network with tens of base stations — is far too high. Therefore, separate link and system level simulators are needed.

In the link level, one communication link between a mobile station and a base station is modeled. The time resolution is typically 1 to 4 samples per chip. For 4.0 Mcps with one sample per chip, the time resolution is 0.25 µs. In the system level there are tens of base stations and all the mobiles connected to those base stations. The time resolution is coarser than in the link level, typically one power control period (0.5 ms) or longer.

The interface between the link level simulator and the system level simulator is presented in Figure 7.2 and in [14]. The link level simulator provides the system level simulator with the required input parameters. Those parameters are E_b/N_0 both for uplink and downlink, MUD efficiency for uplink, orthogonality factor for downlink, and multipath channel model including antenna diversity. Here, E_b is energy per received bit and N_0 is the interference power per bit. The orthogonality factor could be also calculated theoretically based on the multipath profile.

MUD efficiency states the percentage of intracell interference that can be cancelled with base station multiuser detection. In system level simulations it is assumed that MUD efficiency is the same for different system loads.

202

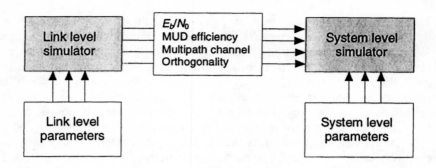

Figure 7.2 The interface between the link level and the system level simulator.

7.3.1 Link Level Simulation

The simulation program used in the link level performance evaluations was the Communications Simulation and System Analysis Program (COSSAP) by Synopsys Inc. The COSSAP simulation program is a stream driven simulation program, with support for asynchronous operation since no timing control between simulation elements is required. All the receiver blocks were generated with the C language for efficiency and flexibility in the configuration.

Link level simulations included a multipath channel model that was modeled as a tapped delay line with one sample per chip. The CODIT macro-, micro-, and picocell channel models described in Chapter 4 were used. Amplitudes and phases of the multipath components were estimated in the receiver, but the multipath delays were known a priori. Fast power control in the link level simulations was based on the estimated received SIR values.

7.3.2 System Level Simulation

7.3.2.1 Simulator Principle

The system level capacity is studied in an interference-limited case. One capacity simulation typically contains 10,000 local mean SIR values. One local mean SIR value is simulated over a 500-ms period for each user, consisting of random mobile placement, virtual handover, and fast power control. Figure 7.3 presents the block diagram of the system level simulator. The simulations give the outage probability for the system, where outage probability represents the number of users having a worse average SIR than required. An outage probability of 5% was selected as the target value for capacity.

The mobile stations in this simulator are not actually moving, such as in dynamic simulators with mobility. The effect of mobility is modeled by adding a fast fading process to the signal in the system level. In this semistatic approach, signal strength varies due to fading as a function of the mobile speed given as a parameter. In the system level it is important to model the effect fast fading and fast power control as discussed in Section 7.4. The simulator can also be used for multioperator simulations.

The propagation model of the system simulator consists of attenuation, shadowing, and statistically generated fast fading. Both macro- and microcell models are adopted from [15].

7.3.2.2 Path Loss Models

The distance-dependent attenuation for macro- and microcell environments is used as presented in Section 4.5 for macro- and microcell environments. The simulation environment parameters are given in Table 7.4.

The path loss between a mobile station and a macrocell base station is modeled as

$$L = 29 + 36\log(R) + 31\log(f) \tag{7.6}$$

where f is the carrier frequency in megahertz and d the distance between the transmitter and the receiver in kilometers. Resulting path loss L is given in decibels.

Path loss between a mobile station and a microcell base station is calculated with a multislope model. The microcell base stations are located on every second street intersection. The slopes are non-line-of-sight slope L_{NLOS}, line-of-sight slope L_{LOS} for small distance and line-of-sight slope for long distances. If the connection between transmitter and receiver is line of sight, attenuation is calculated as

$$L_{LOS} = \begin{cases} 82 + 20\log(\dfrac{d}{300}), & if \quad d \le 300\text{m} \\[2mm] 82 + 40\log(\dfrac{d}{300}), & if \quad d > 300\text{m} \end{cases} \tag{7.7}$$

At a distance of 300m a breakpoint marks the separation between two line-of-sight segments.

Turning a corner causes an additional loss in (7.7). Attenuation between a transmitter and a receiver that have non-line-of-sight connections constitutes a line-of-sight segment, a non-line-of-sight segment, and an additional corner attenuation.

$$L_{NLOS} = L_{LOS}(d_{corner}) + 17 + 0.05d_{corner} + (25 + 0.2d_{corner})\log\left(\frac{d}{d_{corner}}\right) \tag{7.8}$$

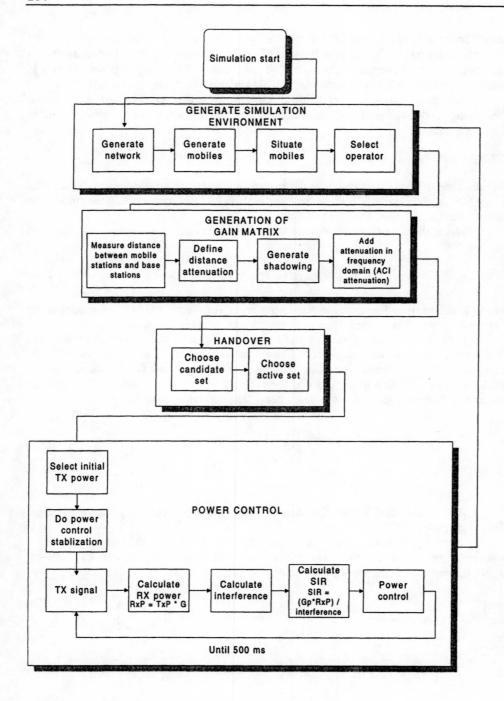

Figure 7.3 The block diagram of the system level simulator.

Line-of-sight attenuation is calculated between a corner and receiver d_{corner}, and non-line-of-sight connection between a corner and transmitter.

Table 7.4
Simulation Environment

	Micro	Macro
Shadowing	mean 0 dB, std dev 4dB	mean 0 dB, std dev 10dB
Number of base stations	128	19
Base station layout	Manhattan	hexagonal
Base station spacing	200m	800m
Building width	100m	-
Street width	30m	-
Mobile speed	3 km/h 36 km/h	5 km/h 50 km/h

Figures 7.4 and 7.5 depict, respectively, the macrocellular base station location and the base station deployment in microcellular system level simulations.

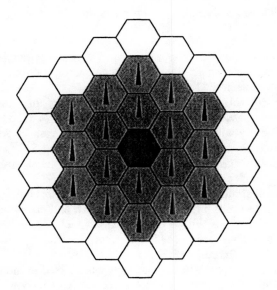

Figure 7.4   Macrocellular base station location.

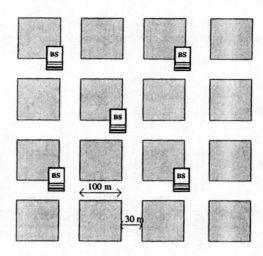

Figure 7.5   Base station deployment in a microcellular system level simulator.

7.3.2.3   Fast Fading and Fast Power Control Modeling

In order to correctly model CDMA effects such as fast power control, a fast fading functionality was included in the system simulator, with the principle presented in [4]. The same kind of multipath channel with Rayleigh fading multipath components was used in the system level as in the link level simulations.

The power control algorithm is implemented as in [16]. The signal-to-interference ratio (SIR)-based power control algorithm aims to keep SIR levels of users at an appropriate level by adjusting the transmission power up or down. The received SIR level is compared to the SIR threshold. If the received SIR was lower than the threshold, a "power up" command is sent. If the SIR level was higher than the threshold, a "power down" command is sent.

In the link level simulations, realistic fast power control has been modeled. Real SIR estimation has been used in the WCDMA receiver, and errors have been added to the power control commands in the feedback channel. Also, real power control frequency and step size have been used. The effect of nonideal power control can be seen in the obtained E_b/N_0 figures.

In the system level, ideal SIR estimation has been used in the fast power control since the effect of nonideal power control is taken into account in the link level simulations. In the system level, fast fading and fast power control are still needed because we can observe how power control effects the interference in the network. Therefore, real power control frequency and step sizes are used also in the system level.

7.3.2.4   Handover Modeling

In the handover modeling, a maximum of 3 BTSs are selected in the active set. The BTSs are selected randomly with a uniform distribution from a pool of BTSs that fits in the handover margin. Such a BTS selection procedure takes the effects of nonideal handovers into account.

In the uplink, macro diversity (i.e., soft handover) is taken into account by selecting the best source (the frame with highest average SIR) on a frame-by-frame basis. In the downlink, macro diversity can be modeled in two ways at the system level. The first approach is to simulate macro diversity in the link level so that separate E_b/N_0 levels are obtained for users in the soft handover state. There is a difference in E_b/N_0 levels because a different amount of diversity is available during soft handover and because part of the transmitted energy may not be captured by the limited number of RAKE fingers in the mobile station. Then, when we have separate E_b/N_0 values for the soft handover state in the system level, the mobile can be assumed to be able to receive all the paths directed to it in the system level.

Another way to model downlink macro diversity is to transmit the signal from several base stations to mobile stations so that the mobile station cannot receive all the paths directed to it. Paths not captured by RAKE processing contribute to the interference. Now the same E_b/N_0 threshold is used for all the mobile stations. The second method has been used in the system simulator due to straightforward modeling. The first method could be more accurate because it can also take into account the effect of increased multipath diversity on the E_b/N_0 performance.

7.3.2.5   Micro Diversity Modeling

Micro diversity (i.e., multipath diversity and antenna diversity in the BTS) is modeled by uncorrelated Rayleigh fading multipath components. The diversity combining is modeled as coherent maximal ratio combining. In this system, level micro diversity modeling, it is assumed that the number of RAKE fingers at the base station is sufficient and all modeled multipath components are received. The performance loss due to the limited number of RAKE fingers is modeled in the link level, and can be seen in the E_b/N_0 values.

Micro diversity modeling is needed in the system level because the amount of diversity affects the intercell interference, as shown in Section 7.4.2.

7.3.2.6   Modeling Downlink Pilot and RAKE

In the downlink, the pilot signal is essential for system operation, and it is generally transmitted with a higher power level than any traffic channel. Modeling of the common pilot is not included in the link level performance figures. For this reason, its effect on downlink interference must be included in the system level modeling. In the simulator, each base station transmits user-specific signals and a common pilot signal. Powers of all the transmitted signals are summed together at the base station transmitter. Coherent maximal ratio combining at the RAKE receiver is assumed, and it is modeled in the system level by taking the sum of SNR values of paths [4].

7.3.2.7 Interference Cancellation

Intracell interference cancellation is modeled in the system simulator by removing β percentage of intracell interference in the interference calculation. The factor β is called MUD efficiency, and it is obtained from the link level simulations shown in Section 7.5.1.

In the uplink, when measuring SIR for a single user, the cancellation process includes all the users connected to the same BTS. Interference cancellation is nonideal, with the efficiency derived from the link level simulations. Thus, the total interference that the BTS experiences consists of the interference from all the mobiles in the network not having a connection with the BTS and from the partly canceled interference from the mobiles having connection to the BTS.

7.3.2.8 Sectorization and Adaptive Antennas

The effects of sectorization are not included in this study. The capacity gain from sectorization is $\xi \cdot N$; where ξ is the sectorization efficiency, and N is the number of sectors. Coverage overlap between antenna beams causes interference spillover, which causes sectorization efficiency to be less than 1 and reduces capacity. For 3-dB overlap between antennas, sin(x)/x antenna pattern gives a sectorization efficiency of 0.88. When the beam overlap increases, the efficiency decreases. The efficiency also depends on whether softer handoff is used between overlapping sectors [17]. The use of adaptive antennas for improving capacity or coverage is not covered here, but evaluation results can be found in [18].

7.3.3 Simulation Parameters

The wideband CDMA system parameters used in the simulation are listed in Table 7.5. These system parameters are from the MUD-CDMA system concept [19], a wideband CDMA system that has been used as a basis for the FRAMES FMA2 wideband CDMA air interface [20]. Since the main system parameters are similar to the other wideband CDMA system presented in Chapter 6, the results are representative for third generation wideband CDMA systems.

7.4 FAST POWER CONTROL

As was discussed in Chapter 5, fast power control impacts the performance of a wideband CDMA system in three different ways. First, it compensates for the fast fading and, in a perfect case, turns the fading channel into a nonfading channel in the receiver, reducing the required E_b/N_0. The second effect is a consequence of the first: the compensation of the fading channel by power control leads to peaks in mobile station transmission power, which affect the intercell interference in the cellular network. And thirdly, fast power control equalizes the mobile station powers in the base station and thus prevents the near-far effect. The effect of fast power control on a CDMA system can be summarized as follows:

- Lowers the E_b/N_0 requirement in the receiver;
- Introduces interference peaks in the transmitter;
- Avoids near-far problems in the uplink.

Table 7.5
Simulation Parameters

CDMA System Parameters	
Carrier spacing	6 MHz
Chip rate	Basic chip rate 5.1 Mcps
Carrier frequency	2 GHz
Spreading codes	Uplink: Extended VL-Kasami Downlink: Augmented Walsh-Hadamard
Bit rates	12 Kbps 144 Kbps 2 Mbps
Interleaving	20 ms (12 Kbps) 40 ms (144 Kbps) 100 ms (2 Mbps)
PC dynamics	Uplink 80 dB Downlink 20 dB
Link Level Parameters	
Voice activity	100%
Number of stages in base station MUD	2 (1 interference canceling)
Antenna diversity (uplink)	2 antennas, with fading correlation 0.7 and 0.0
Path combining method	Maximal ratio combining
Target BER	10^{-3}
Coding	Uplink: 1/2 rate convolutional with K = 9 Downlink: 1/3 rate convolutional with K = 9
Fast power control	2 kHz, step size 1.0 dB, 5% errors in feedback signaling
RAKE fingers	4 or 9
System Level Parameters	
Outage requirement	95% satisfied users
Active set size in soft handover	3
Handover margin in nonideal handover	3 dB
PC step size	1 dB
Pilot strength	6 dB higher than maximum traffic channel power

In practice there are imperfections in the power control because of the transmission and processing delay, command errors, limited dynamic range, and step size; the system performance is degraded because of the increased residual variation in the signal-to-noise ratio.

7.4.1 Impact of Fast Power Control in E_b/N_0

The gain in E_b/N_0 depends on the channel characteristics, the bandwidth of the system, and the mobile speed. Since a wide bandwidth gives better diversity, power control is not as critical for the wideband CDMA as for the narrowband CDMA. In a small delay spread channel, the improvement of performance from fast power control is largest. Furthermore, as discussed in Section 7.2.6, the improvement of E_b/N_0 is highest at slow mobile speeds. This is due to the ability of fast power control to follow the fast fading. As shown in Section 7.5.1 for a 2-Mbps service, fast power control is beneficial especially in micro- and picocell channels due to a limited multipath diversity. In a macrocell channel, the gain of fast power control in E_b/N_0 is not very large because of the high degree of multipath diversity. Fast power control is compared to slow power control in the downlink in [13,21].

7.4.2   Intercell Interference With Uplink Fast Power Control

When uplink fast power control is able to follow the fast fading, the intercell interference will increase due to peaks in the transmission (TX) power of the mobile MS in Figure 7.6. The received (RX) interference level in the neighboring cells experiences peaks. In the downlink, this effect is smaller due to limited power control dynamics. Downlink power control is not considered in this subsection. The less diversity available, the higher the average transmission power is and the more intercell interference is generated. This phenomenon of increased interference is imminent with such systems with fast power control that cannot exploit the multipath diversity of the channel and do not employ antenna diversity. On the other hand, soft handovers provide more diversity and reduce intercell interference. As explained below, if a system employs fast power control and soft handovers, they should be modeled together with a relevant channel model in the system simulations to give valid capacity results.

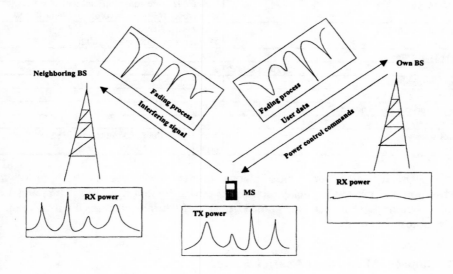

Figure 7.6   Interference to neighboring base stations in fading channel with fast power control without macro diversity.

In this section, we fix the SIR target to be the same for all channels in order to study only the impact of increased intercell interference. It should be noted that in practice this is not the case, but different channels require different SIR to achieve the required QoS.

In the following treatment, we first provide an analytical model and then the corresponding simulation results. Due to modeling difficulties, handovers are not considered in the analysis corresponding to the assumption of slow hard handovers, where a mobile is connected to the base station with the smallest long-term attenuation. This hard handover analysis leads to a worst case analysis in terms of transmission power and interference. In reality, soft handover increases diversity for mobile stations at the cell border, and thus, the interference is reduced. This is visible in the system simulations where both hard handover and soft handover are modeled.

7.4.2.1   Average Uplink Transmission Power With Fast Power Control

In this section an analysis on the average uplink transmission power with ideal fast power control is presented. The terminal transmission power level depends on the multipath diversity of the channel model. This result is then used in the next subsection to analyze system capacity with fast power control.

The average transmission powers with different degrees of diversity are calculated as follows. The probability density function (PDF) of the total channel power with L statistically independent Rayleigh fading multipath components, given in [1], is

$$p(\gamma) = \sum_{k=1}^{L} \frac{\pi_k}{\overline{\gamma_k}} e^{-\gamma/\overline{\gamma_k}} \quad \gamma \geq 0 \tag{7.9}$$

where $\overline{\gamma_k}$ is the average power of the kth multipath component and

$$\pi_k = \prod_{\substack{i=1 \\ i \neq k}}^{L} \frac{\overline{\gamma_k}}{\overline{\gamma_k} - \overline{\gamma_i}} \tag{7.10}$$

If the multipath components are equally strong, the PDF is written as

$$p(\gamma) = \frac{L^L \gamma^{L-1}}{(L-1)!} e^{-L\gamma_b} \quad \gamma \geq 0 \tag{7.11}$$

Now we want to find out the PDF of the transmission power ρ of a given mobile station. In this section, we assume that the number of interfering users is so high that the interference at the base station can be approximated by Gaussian noise. The more users there are in the system, the better this approximation holds. With this assumption, the interference level at the base station is constant. Therefore, SIR-based power control can be replaced by power control based on received signal power only. Since we assume ideal power control, mobile transmission power $\rho = 1/\gamma$ and the PDF of ρ is [22]

$$p_\rho(\rho) = \frac{1}{\rho^2} p\left(\frac{1}{\rho}\right) \qquad \rho \geq 0 \tag{7.12}$$

The average transmission power can be obtained by

$$\overline{\rho} = \int_0^\infty \rho p_\rho(\rho) d\rho \tag{7.13}$$

In case of equally strong Rayleigh fading components and unlimited power control dynamics, the increase in transmission power can be shown to be

$$\frac{L}{L-1} \tag{7.14}$$

where L is the number of multipath components. In a 1-path Rayleigh fading channel ($L = 1$) the result would be infinity with unlimited power control dynamics. Typical total dynamics of uplink CDMA power control are 80 dB, which takes care not only of fast fading but also distance attenuation and slow fading. In these calculations only fast fading is taken into account, and therefore, the power control dynamics allowed for the fast power control is set to be 50 dB. The minimum and maximum transmission powers are chosen to be $\rho_{min} = 0.2$ and $\rho_{max} = 2.0 \times 10^4$. Now, the average transmission power can be calculated as

$$\overline{\rho} = \int_0^{\rho_{min}} \rho_{min} P_\rho(\rho) d\rho + \int_{\rho_{min}}^{\rho_{max}} \rho p_\rho(\rho) d\rho + \int_{\rho_{max}}^\infty \rho_{max} P_\rho(\rho) d\rho \tag{7.15}$$

The average transmission powers are calculated in the AWGN channel, in the 1-tap Rayleigh fading channel, in the 2-tap Rayleigh fading channel (2 equally strong taps on average), in the 4-tap Rayleigh fading channel (4 equally strong taps on average), and in the ATDMA macrocell channel with and without base station antenna diversity. The 2-tap and 4-tap Rayleigh fading channels, with all taps equally strong, can be considered either as multipath channels or as antenna diversity. The impulse response of the ATDMA macrocell channel model used is shown in Figure 7.7. The used bandwidth is chosen to be so high that all the multipath components in the ATDMA macrocell channel model can be resolved. The minimum bandwidth needed for separating all the paths in the ATDMA macrocell channel is approximately 3.5 MHz. Maximal ratio combining of multipath components is assumed in this analysis.

The calculated increase in average transmission power compared to nonfading channel ΔP_{tx} is shown in Table 7.6.

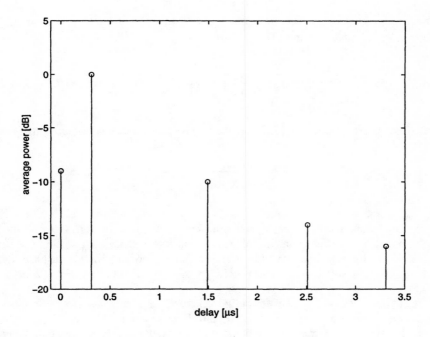

Figure 7.7 Impulse response of ATDMA macrocell channel.

Table 7.6
Average Increase of Transmission Power in a Fading Channel Compared to a Nonfading Channel With Hard Handovers

Channel	ΔP_{tx}(dB)
Nonfading channel	0
1-tap Rayleigh fading channel	10.1
2-tap Rayleigh fading channel	3.0
4-tap Rayleigh fading channel	1.2
ATDMA macro channel	2.5
ATDMA macro channel with antenna diversity	1.3

The average transmission power in a fading channel with fast power and with hard handovers is higher than in a nonfading channel. Thus, the average intercell interference from the mobile stations to the base stations in the surrounding cells is also higher.

7.4.2.2 Analysis of Impact of Fast Power Control Into Capacity

The maximum number of simultaneous users N on one carrier in a cellular CDMA system can be calculated as in [23]:

$$N = F \left[G_p \left(\frac{E_b}{N_0} \right)^{-1} + 1 \right] \tag{7.16}$$

where G_p is the processing gain, E_b/N_0 is the received bit energy per noise density, and

$$F = \frac{I_{intra}}{I_{intra} + I_{inter}} \tag{7.17}$$

where I_{intra} is the intracell interference, I_{inter} is the intercell interference in a nonfading channel, and F denotes the percentage of interference originating from intracell mobiles at the base station. In this definition of F, intracell interference I_{intra} at the base station includes all the interference from the mobiles that are connected to that base station. Therefore, users in soft handover are included in intracell interference in 2 to 3 base stations.

Let this F be valid in a nonfading channel. In a fading channel the average transmission power increases, and thus, the interference from other cells I'_{inter} is higher than the intercell interference in a nonfading channel I_{inter}. Now the percentage of interference from the own cell F' is

$$F' = \frac{I_{intra}}{I'_{intra} + I_{inter}} \tag{7.18}$$

where I'_{inter} is the intercell interference in a fading channel.

The increase in interference is equal to the increase in the average mobile transmission power ΔP_{tx} shown in Table 7.6, where ΔP_{tx} is given by

$$\frac{I'_{inter}}{I_{inter}} = \Delta P_{tx} \tag{7.19}$$

Now, we can obtain F'

$$F' = \frac{1}{1 + \Delta P_{tx} \left(\frac{1}{F} - 1 \right)} \tag{7.20}$$

F is here assumed to be 55% in a nonfading channel with hexagonal base stations. This value is obtained from the system simulator for the macrocell environment and is a typical value for this environment. The values for F' in different channels are shown in

Table 7.7. If we assume in (7.16) that the processing gain G_p and the E_b/N_0 requirement do not change between different channels, then $N = F \cdot$constant, (i.e., the relative capacity):

$$\frac{N'}{N} = \frac{F'}{F} \tag{7.21}$$

The relative capacities N'/N are shown in Table 7.7.

Table 7.7
Relative Capacities Compared to a Nonfading AWGN Channel With Hard Handovers

Channel model	Percentage of interference from intracell mobiles F'	Relative capacities N'/N compared to AWGN (%)
AWGN channel	55	100
1-tap Rayleigh	11	19
2-tap Rayleigh	38	69
4-tap Rayleigh	48	87
ATDMA macro	41	74
ATDMA macro with antenna diversity	48	86

7.4.2.3 Simulated Cellular Capacity in Fading Channel

In this subsection the analytical capacity estimates are compared to the simulated values. The power control dynamic range is 80 dB, which is larger than the value used in the analytical calculations. This is due to the additional distance-dependent attenuation as well as to the slow fading that must be compensated for in the simulator by the available power control dynamic range. The parameters of the power control scheme are shown in Table 7.8.

The propagation model of the system simulator consists of attenuation, shadowing, and statistically generated fast fading. The attenuation model used is presented in Section 7.3.2.

Even if the power control algorithm in the simulations has a perfect knowledge of the received SIR, the fast fading process cannot be followed perfectly because of a finite step size.

Table 7.8
Simulation Parameters for Fast Power Control Simulations

Power control frequency	1 kHz
Power control step size	1.0 dB
Power control dynamics	80 dB
Mobile speed	1 km/h
Active set size in soft handover	3
Handover margin in non-ideal handover	5 dB
Log normal shadowing	mean 0 dB, std dev 6 dB
User bit rate	10 Kbps

Both hard and soft handovers are supported by the system simulation. When soft handover is used, the mobile station is able to communicate simultaneously with one or more base stations. The uplink transmission power in the mobile station is increased only if all base stations in the active set request more transmission power through fast power control signaling. Uplink macro diversity in soft handover is considered by selecting the best source (the frame with the highest average SIR) on a frame-by-frame basis. Each active base station receives the frame transmitted by the mobile station. The quality of the frame is measured by calculating the average SIR over the frame. The frame with the highest quality is utilized, while other frames are discarded.

Hard handover is modeled in simulations by connecting the mobile station to the base station to which the distance dependent attenuation and the slow fading attenuation is the lowest. This can be done by setting the maximum active set size to one.

Simulated and calculated capacities for hard handovers can be compared in Table 7.9. The simulated capacities are lower than the calculated capacities except in the 1-tap Rayleigh fading channel. These differences are due to nonidealities in the power control in the network simulation. The power control step size was fixed and the same in different fading channels. It should be noted that the SIR target was set to be the same for all capacity simulations. In practice, however, the 1-tap Rayleigh type channel with power control imperfections would require the highest SIR target for power control to achieve comparable link quality with other cases. In other channels, the power control step size was too large, and thus, unnecessary power fluctuations increased the interference and resulted in lower capacities.

Soft handover (macro diversity) shows a considerable gain in the 1-tap Rayleigh fading channel over hard handover.

The differences in capacities between different channel models point out the importance of including fast fading, fast power control, and realistic handovers in CDMA system simulations if realistic capacity results are needed for network planning. The simulated capacity with 1-tap Rayleigh fading channel and soft handover is only 43% of that calculated and simulated for nonfading channel even if the same SIR target was used in the system simulations.

The results of Table 7.9 and Figure 7.8 show capacity differences between different amounts of diversity. Therefore, if the channel does not have enough diversity, other means such as receiver antenna diversity or transmit diversity need to be used. Furthermore, in order to obtain reliable capacity results the multipath channel needs to be modeled in the same way both in link level simulations as well as in system level simulations.

Table 7.9
Simulated Relative Capacities With Different Channel Models

Channel model	Hard handover (slow) (%)	Soft handover (fast) (%)
Non-fading AWGN channel	100	100
1-tap Rayleigh	25	43
2-tap Rayleigh	57	Not simulated
4-tap Rayleigh	78	Not simulated
ATDMA macro	60	Not simulated
ATDMA macro with antenna diversity	79	Not simulated

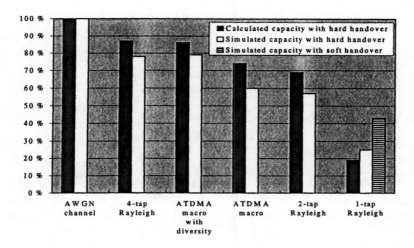

Figure 7.8   Calculated and simulated relative capacities with different channel models.

7.4.3   Capacity Degradation due to Imperfect Power Control

One purpose of power control is to equalize the received powers in the base station to eliminate the near-far effect. However, in practice it cannot eliminate the near-far effect completely due to imperfections in the power control loop. Thus, the residual variation in SIR causes degradation of the system capacity [24]. The amount of residual variation, and thus capacity degradations, depends on channel response, the system chip rate, SIR estimation accuracy, power control loop delay, command rate, and power control step size. The effect of imperfect power control is included in the E_b/N_0 results of the link level simulation since a realistic power control algorithm has been used.

The effect of imperfect power control has been studied widely in connection with IS-95 uplink performance analysis. As indicated in several studies, the performance in terms of system capacity decreases by the percentages listed in Table 7.10, as the power control error increases. The power control error causes variations in the received E_b/N_0. Empirical evidence suggests that E_b/N_0 is log-normally distributed and that the standard deviation for the narrowband CDMA employing both open loop and fast closed loop power control is between 1.5 and 2.5 dB [25].

Wideband CDMA reduces the impact of imperfect power control. This is because better diversity results in less variation in residual SIR. In [24] power control feedback delay was shown to be less critical for wideband systems than narrowband systems.

Table 7.10
Capacity Reduction With Power Control Error

Power control error, standard deviation (dB)	Capacity reduction (%)
0	0
1	3
1.5	8
2.0	13
2.5	20

Source: [26]

MUD alleviates the near-far effect, and thus the degradation of performance due to imperfect power control is smaller. In practice the gain of MUD against power control errors depends on the performance of MUD (i.e., how much of the generated intracell interference a particular receiver solution is able to remove).

7.5   SPECTRUM EFFICIENCY

In this section the evaluation of W-CDMA spectrum efficiency is analyzed. First, the link level performance of the uplink is simulated, taking into account multiuser detection efficiency. Data rates of 12 Kbps, 144 Kbps and 2 Mbps are considered. Second, spectrum efficiency results are presented based on the system level simulation using the link level results as depicted in Figure 7.3. Third, to verify the simulation results, analytical spectrum efficiency calculations are presented.

7.5.1   Link Level Performance of W-CDMA Uplink

As an input for the uplink capacity studies, link level performance, E_b/N_0 for the desired BER, and MUD efficiency are needed. The performance was studied in both CODIT macro and microcell environments. The performance figures are given in Figures 7.9 and 7.10.

The BER performances in COSSAP Monte Carlo simulations are used to assess the efficiency of multiuser detection β. This efficiency denotes the percentage of intracell interference being removed by multiuser detection at the base station receiver. The efficiency of MUD is estimated from the load that can be accommodated with a specific E_b/N_0 value with a conventional RAKE receiver and with a multiuser receiver. The same efficiency of MUD is assumed for different loads in the system simulator. The additive white Gaussian noise N_0 is used to represent both thermal noise and intercell interference, while intracell interference is represented by real transmitters. The target BER is 10^{-3}. In the analysis, the number of users with a RAKE receiver is denoted by K_{RAKE} and that with a MUD receiver by K_{MUD}. The efficiency of MUD β at a given E_b/N_0 value is given by

$$K_{RAKE} = (1 - \beta)K_{MUD} \tag{7.22}$$

This applies to the macrocell where multipath interference is significant. In a microcell, the channel model is close to a single path channel and thus, self-interference from multipath components is negligible. Therefore, we can subtract the desired user from the total number of users and calculate the MUD efficiency for the microcell environment by

$$K_{RAKE}-1 = (1-\beta)(K_{MUD}-1) \tag{7.23}$$

To keep the complexity of the receiver to a moderate level, only four RAKE fingers were used in the macrocell with which 63% of the total energy could be collected on average. In the microcell, two RAKE fingers were used with which 95% of the energy could be captured. If only part of the energy is captured by the receiver, a higher E_b/N_0 is required, resulting into lower capacity figures. Therefore, the required E_b/N_0 in the macrocell is higher than in microcell. In the CODIT macrocell channel, the capacity was increased by about 30% by using nine RAKE fingers instead of four fingers [27].

We performed the link level simulations for the MUD efficiency study with some differences to the assumptions mentioned earlier. The service studied in this case was 74 Kbps in the CODIT macro- and microcell environments with BPSK data and chip modulation, and antenna diversity with a 0.7 correlation factor. Pilot symbols for channel estimation were inserted in the data stream. It should be noted that since delay estimation was assumed to be perfect, these results are only indicative. Furthermore, the MUD efficiency varies according to system load, which has not been considered in these simulations. The multiuser detection algorithm that has been used in the link level simulations has been presented in [28]. The impact of imperfect delay estimation on the multiuser detection has been considered, for example, in [29,30].

Simulated BER curves for the microcell are shown in Figure 7.9 and for the macrocell in Figure 7.10. All the users have equal average power. The bit error probabilities shown are the mean values of all the users. The single user case without intracell interference is shown for comparison.

In the microcell we have used simulations to estimate the required E_b/N_0 when 15 users are transmitting in the multiuser detection system. The capacity of the conventional RAKE receiver with the same E_b/N_0 is found to be between five and six users yielding a MUD efficiency of 64% to 71%. In the macrocell, the corresponding figures are 10 users with MUD and three to four users without MUD. The efficiency of MUD in macrocell is then 60% to 70%. The link level simulation results and the resulting MUD efficiencies are summarized in Table 7.11. Since the calculated MUD efficiency is valid for the intracell interference only, it does not directly convert to the gain in the cellular capacity. The interference from all the cells in the network must be included in the study to obtain the capacity for the whole network (see (7.27)).

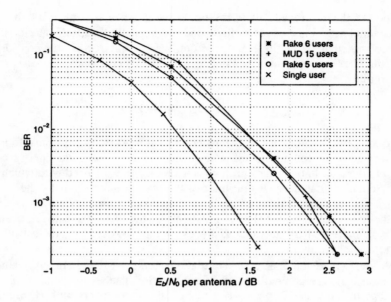

Figure 7.9   BER as a function of E_b/N_0 for the CODIT micro-cell with 36 km/h.

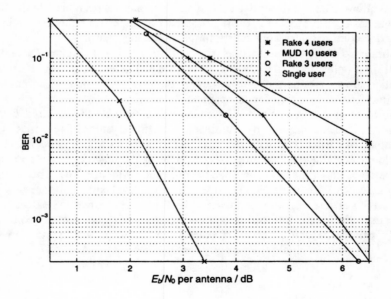

Figure 7.10   BER as a function of E_b/N_0 for the CODIT macro-cell with 50 km/h.

Table 7.11
Link Level Results With and Without Base Station Multiuser Detection

Channel model, E_b/N_0 value	Number of users, no MUD, K_{RAKE}	Number of users with MUD, K_{MUD}	Efficiency of MUD β
micro, 2.2 dB	5 – 6	15	64%-71%
macro, 5.9 dB	3 – 4	10	60%-70%

For the spectrum efficiency comparison for the uplink, results were produced for 12- and 144-Kbps services in the macrocell environment with the channel model presented in Figure 7.7. The main difference with the downlink results is the use of uplink antenna diversity. In the simulations for macrocell, the fading on the different antenna branches has been assumed to be uncorrelated. For the studied 12- and 144-Kbps services, the performance results are given in Table 7.12 for CODIT microcell multipath profile.

Table 7.12
Link Level Results of 12- and 144-Kbps in CODIT Microcell

Studied service in the uplink	E_b/N_0
12 Kbps	5.7
144 Kbps	1.7

The difference between the uplink 12- and 144-Kbps services is because of the larger interleaving period for 144 Kbps and because more energy is available for the decision directed channel estimator, thus giving better channel estimates for 144 Kbps than for 12 Kbps.

7.5.1.1   Transmission of 2 Mbps

Link level performance simulations for the 2-Mbps user bit rate and BER of 10^{-3} were carried out using similar wideband channel models as in the CODIT project (described in Chapter 4), for urban macrocell and urban street microcell at speeds of 3 and 36 km/h and for indoor environments [9] at a speed of 3 km/h using three different chip rates of 5, 10, and 20 Mcps. The actual target BER of 10^{-6} was not simulated because of the very high computational efforts required.

The performance of uplink with 2-Mbps service is good (4 to 6 dB without antenna diversity) with all simulated bandwidths. Receiver antenna diversity was not used in the link level simulations, but diversity gain is taken into account in the range and capacity simulations in the following sections. Fast power control offered considerable gains in the microcell and indoor environments. The E_b/N_0 performance was in the best cases within 1 dB of the simulation results in nonfading channels for the coding scheme given in [1]. Uplink simulation results for BER of 10^{-3} with fast power control and without antenna diversity are shown in Table 7.13. The corresponding results without power control are presented in Table 7.14.

Table 7.13
Uplink Performance With Fast Power Control but Without Antenna Diversity (BER 10^{-3})

Bandwidth	5.12 Mcps	10.24 Mcps	20.48 Mcps
Channel	E_b/N_0	E_b/N_0	E_b/N_0
Macro 3 km/h	4.6	5.5	6.2
Micro 3 km/h	3.9	3.6	4.1
Indoor 3 km/h	N/A	3.6	3.3

Table 7.14
Uplink Performance Without Power Control and Without Antenna Diversity (BER 10^{-3})

Bandwidth	5.12 Mcps	10.24 Mcps	20.48 Mcps
Channel	E_b/N_0	E_b/N_0	E_b/N_0
Macro 3 km/h	7.0	7.0	6.5
Micro 3 km/h	10.4 (with antenna diversity)	9.9	7.5
Indoor 3 km/h	N/A	6.0	5.3

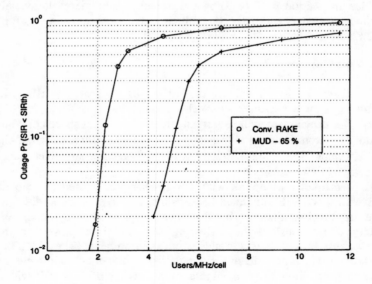

Figure 7.11 Outage probability curves for microcells with conventional receiver and with MUD receiver with $\beta = 65\%$.

7.5.2 Spectrum Efficiency of W-CDMA Uplink

Figure 7.11 shows the outage figures as a function of the load in the microcell. The capacity with a conventional RAKE receiver is 2.0 users/MHz/cell, and for the system with MUD it is 4.7 users/MHz/cell with the 74-Kbps service used in the analysis for MUD performance. The capacity is obtained with 5% outage.

In Figure 7.12 the simulated outage figures for the macrocell are presented as a function of load for different MUD efficiencies. Lines indicate performances with different MUD efficiency. It can be seen that the capacity of MUD with 65% efficiency is almost doubled compared to the conventional RAKE receiver.

The high capacity of CDMA with MUD results from the fact that most of the interference is intracell interference. Because of soft handover, users in a handover state can be included in the MUD process, which allows an extra 20% of interference to be included in the interference cancellation in the macrocellular environment. In the microcellular environment, cell separation is better, fewer users are in the soft handover state, and the amount of interference from other cells is lower than that in the macrocellular environment.

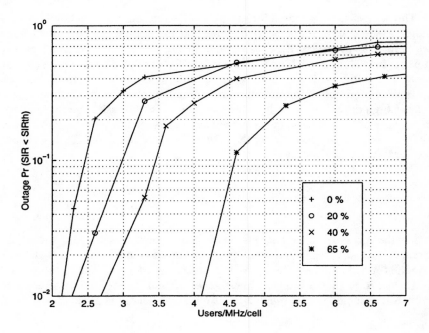

Figure 7.12 Outage probability curves as a function of MUD efficiency in macrocell. The line with 0% efficiency corresponds to the conventional RAKE receiver.

7.5.2.1 Analytical Spectrum Efficiency Calculations

To validate the simulation results, analytical spectrum efficiency calculations for the macrocell environment were carried out. The simulated performance compared with analytical results is given in Table 7.15. The reason for the conventional RAKE receiver having differences between simulated and analytical results is in the power control modeling. The analytical approach assumes perfect power control, whereas the simulator has the actual power control algorithm running with limited dynamics and accuracy. Furthermore, the analytical results assume a nonfading channel, while the simulated capacities are obtained in a fading channel.

Table 7.15
Analytical and Simulated (5% Outage) Capacities in Macrocell (users/MHz/cell)

	Conventional RAKE	MUD 20%	MUD 40%	MUD 65%
Simulated	2.3	2.7	3.3	4.5
Analytical	2.8	3.3	4.0	5.3

The Gaussian approximation is generally the basis for analytical capacity calculations. The central limit theorem states that the normalized sum of independent random variables approaches a Gaussian random variable as the number of terms (here users) increases [31]. This is highly applicable in the case of a large number of low bit rate users; but as the data rates get higher and the interference originates from only a few users, Gaussian approximation underestimates the interference. It should be noted again that a high degree of diversity allows the use of the Gaussian approximation a bit further because of the randomizing effect on the interference.

First, the capacity formula for the network without MUD in BTS is defined. The attained E_b/N_0 value is given by

$$\frac{E_b}{N_0} = \frac{S \cdot G_p}{I_{\text{intra}} + I_{\text{inter}} + N_0} \tag{7.24}$$

where S is the received signal strength, G_p is the processing gain, I_{intra} is the intracell interference, I_{inter} is the intercell interference from other cells and N_0 is the thermal noise. In the following, thermal noise N_0 is neglected and I_{intra} is equal to $(N-1) S$.

The fraction of the intracell interference caused by the users operating in the same cell as the studied user compared to the total interference is given as

$$F = \frac{I_{\text{intra}}}{I_{\text{intra}} + I_{\text{inter}}} \tag{7.25}$$

It should be noted that in (7.25) the intracell interference I_{intra} is equal to $N \cdot S$, instead of $(N-1) \cdot S$. The simulated value for F in macrocells is 0.73. For a single cell F is equal to 1. From the two equations above we get

$$\frac{E_b}{N_0} = \frac{G_p}{\frac{N}{F} - 1} \Leftrightarrow N = F \cdot (G_p (\frac{E_b}{N_0})^{-1} + 1) \tag{7.26}$$

The value N is the number of users that are associated with the BTS. N also includes users that are connected to more than one BTS while in a soft handover state. The number of users being connected to two or three BTSs is obtained from the simulator and used to adjust the analysis correspondingly. Simulations show that typically in the simulated macrocellular environment, 80% of the users are connected to only one BTS, while 15% of the users are connected to two BTSs, and 5% are connected to three BTSs. The calculated capacity must be then scaled by 1.25, as the effect of soft handover is seen in F and the total number of connections in the system is higher than the number of mobiles in the system.

The corresponding analysis for the MUD case can be performed by

$$\frac{E_b}{N_0} = \frac{S \cdot G_p}{(1 - \beta)I_{\text{intra}} + I_{\text{inter}} + N_0} \tag{7.27}$$

Intercell interference can be calculated by

$$I_{\text{inter}} = \frac{1 - F}{F} I_{\text{intra}} \tag{7.28}$$

From (7.23) and (7.24) we have

$$\frac{E_b}{N_0} = \frac{S \cdot G_p}{(1 - \beta)(I_{\text{intra}} - S) + \frac{1 - F}{F} I_{\text{intra}}} \tag{7.29}$$

Since I_{intra} is equal to NS, we can write

$$\frac{E_b}{N_0} = \frac{S \cdot G_p}{(1 - \beta)(NS - S) + \frac{1 - F}{F} NS} \tag{7.30}$$

The capacity of the system is now given by

$$N = F \left(\frac{G_p (\frac{E_b}{N_0})^{-1} - (\beta - 1)}{1 - F\beta} \right) \tag{7.31}$$

If β is set to 0, then the equation above becomes the same as (7.16). The β with value 0 represents the capacity of the conventional RAKE receiver-based system.

Capacity as a function of MUD efficiency is shown in Table 7.15. Both analytical and simulated results are shown. The simulated results compare well to the analytical results. The offset between analytical and simulated results is due to the non-ideal power control and due to real interferers in the simulator instead of Gaussian approximation.

Capacity strongly depends on the radio environment, which is defined by the pathloss attenuation factor, shadowing, and the wideband channel model. The proportion of the interference coming from the own cell from the total interference F in (7.31) is greatly dependent on the environment. It ranges from the single isolated cell case with no intercell interference to the macro cell environment with up to 40% of interference having intercell origin.

Multiuser detection is shown to have the potential to increase capacity. It is more suited for the uplink, where all users need to be detected in any case, and the base station has more processing power available than the mobile station. A suboptimal multiuser receiver has been shown to offer a clearly improved link performance over the RAKE receiver by removing 60% to 70% of the intracell interference in urban micro- and macrocell environments. This interference reduction achieved in the link level with a multiuser detector was used as an input to the system level simulator, which led to a considerable increase in cellular capacity. The capacity gain depends on the ratio of intracell interference to intercell interference, and therefore, the microcellular environment offering high cell isolation gained even more from the use of MUD. Alternative solutions for capacity enhancements are, for example, the use of adaptive antennas or cell splitting, both of which require extra hardware such as antennas. With the use of MUD for capacity enhancement, only the baseband hardware needs to be modified.

7.5.3 Link Level Performance of W-CDMA Downlink

As an input for the downlink system level simulations, we need link level performance results and the orthogonality factor. We first derive the orthogonality factor and then present the link level simulation results.

7.5.3.1 Derivation of the Orthogonality Factor

The interference in the downlink from the same BS originates from a single point and the parallel code channels can be synchronized. When using orthogonal spreading codes in the ideal case, the intracell interference could be completely avoided; but after the multipath channel, part of the orthogonality is lost and intracell interference exists also in the downlink.

For system level studies in the downlink, an estimate of the achievable degree of orthogonality needs to be drawn from the link level studies. An analytical capacity calculation of the orthogonality factor assuming Gaussian approximation is presented below. The attained SIR can be written as

$$\text{SIR} = \frac{\sum_{i=0}^{N} g_i L_{p,l} P_{tx}}{\sum_{i=0}^{N} \overline{g}_i} \frac{1}{I_{\text{total}}} \tag{7.32}$$

where N is the number of perceived paths in the channel model with the selected bandwidth, $L_{p,l}$ is the pathloss between mobile stations and base station l, P_{tx} is the transmission power for the selected user, I_{total} is the total interference experienced by that user, g_i is the instantaneous path gain, and $\overline{g}_i$ the average gain for path i.

Instantaneous SIR is calculated by dividing the received signal by the interference, and multiplying by the processing gain. The downlink intracell interference is multiplied by $1-\alpha$ where α is the orthogonality factor. An orthogonality factor of 1 corresponds to perfectly orthogonal intracell users, while with the value of 0 the intracell interference is completely asynchronous. Signal-to-interference ratio is then given by

$$\text{SIR} = \frac{\sum_{i=0}^{N} g_i L_{p,l} P_{tx}}{\sum_{i=0}^{N} \overline{g}_i} \frac{G_p}{(1-\alpha)I_{\text{intra}} + I_{\text{inter}} + N_0} \tag{7.33}$$

where N_0 is the thermal noise, G_p is the processing gain, I_{intra} is the intracell interference, and I_{inter} is the intercell interference. Since the system is interference limited, thermal noise N_0 is assumed small and therefore neglected. I_{intra} and I_{inter} are equal to

$$I = \sum_{l=0}^{M} L_{p,l} \frac{\sum_{i=0}^{N} g_{i,l}}{\sum_{i=0}^{N} \overline{g}_i} \left(\sum_{k=0}^{R} P_{tx,k} + P_{\text{pilot},l} \right) \tag{7.34}$$

where R is the number of interferers within one cell, and $P_{tx,k}$ is the transmission power to the kth user. Each path of the desired user experiences the same interference on average. M is the number of base stations causing inter- or intracell interference and $P_{\text{pilot},l}$ is the pilot power of base station l.

Orthogonality factor α is given by

$$\alpha = 1 - \frac{E_b}{I_0} \left(\frac{E_b}{N_0} \right)^{-1} \tag{7.35}$$

where N_0 is intercell interference, I_0 is intracell interference, and E_b/N_0 and E_b/I_0 are given in absolute figures, not in dB. E_b/N_0 is the performance figure with Gaussian noise, and E_b/I_0 is the corresponding figure dominating intracell interference. The values of E_b/N_0 and E_b/I_0 are obtained from link level simulations. The block diagram of downlink simulation with intracell and intercell interference is shown in Figure 7.13.

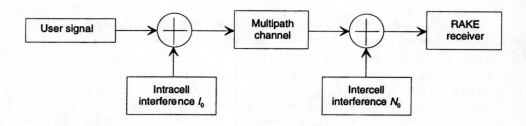

Figure 7.13 The block diagram of downlink simulation with intracell and intercell interference.

In the uplink, the signals are always asynchronous and do not have orthogonality unless the mobiles are synchronized with very high accuracy and operated in an environment where only one multipath component existed.

Figure 7.14 and Tables 7.16 and 7.17 show the results for downlink 12- and 144-Kbps link level E_b/N_0 and E_b/I_0 (in dB) in the CODIT microcell multipath channel. For 12 and for 144 Kbps results, the main differences to the uplink were the lack of antenna diversity and the use of a pilot channel for channel estimation. The pilot channel is not included in the E_b/N_0, and thus must be included in the system level modeling.

E_b/I_0 values have been calculated using the following equation:

$$E_b / I_0 \equiv \frac{WP_{\text{user}}}{R(NP + P_{\text{pilot}})} \tag{7.36}$$

where, W is the chip rate of the downlink, R is the user data rate, P_{user} is the power of the wanted signal, N is the number of interfering channels, P is the power of interfering channel, and P_{pilot} the power of the pilot signal.

Table 7.16
Downlink 12 Kbps Link Level E_b/N_0 and E_b/I_0 Results in CODIT Microcell Multipath Channel

12 Kbps E_b/I_0	12 Kbps E_b/N_0	Orthogonality factor α
2.3	6.4 dB	0.61

Table 7.17

Downlink 144 Kbps Link Level E_b/N_0 and E_b/I_0 Results in CODIT Microcell Multipath Channel

144 Kbps E_b/I_0	144 Kbps E_b/N_0	Orthogonality factor α
1.9 dB	6.1 dB	0.62

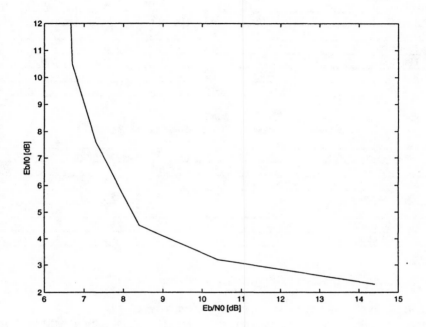

Figure 7.14 Downlink 12 Kbps with intracell interference (I_0) and with intercell interference (N_0).

7.5.3.2   Transmission of 2 Mbps

Transmission of 2-Mbps service was simulated in the CODIT macro-, micro-, and picocell channels. In each channel, enough RAKE fingers were allocated to gather all the signal energy. The maximum number of RAKE fingers was nine. The most important parameter used in the comparison of different configurations in noise limited cases is E_b/N_0, which represents the received energy per bit versus noise power density. In the intracell interference case, energy per bit versus intracell interference power E_b/I_0 was compared.

The corresponding link performance results are presented in Tables 7.18 and 7.19. According to the downlink simulation results, the requirements of providing 2-Mbps service are achieved using reasonable E_b/N_0 values (4 to 10 dB), even with the narrowest 5-Mcps bandwidth. Only the performance in the macro channel seems to be somewhat poorer when the 5-Mcps bandwidth is used. This is due to severe interpath interference and low processing gain.

Table 7.18
Downlink Performance Without Power Control (BER 10^{-3})

Bandwidth	5.12 Mcps		10.24 Mcps		20.48 Mcps	
Channel	E_b/N_0	E_b/I_0	E_b/N_0	E_b/I_0	E_b/N_0	E_b/I_0
Macro 3 km/h	10.3	9.8	6.8	5.7	5.3	4.2
Macro 36 km/h	9.5	8.8	6.3	5.5	4.8	4.2
Micro 3 km/h	14.3	3.5	12.3	2.2	9.0	2.5
Micro 36 km/h	9.8	2.2	8.8	1.9	6.8	2.5
Indoor 3 km/h	9.8	0.4	9.3	-0.2	6.8	1.7

Table 7.19
Downlink Performance With Fast Power Control (BER 10^{-3})

Bandwidth	5.12 Mcps		10.24 Mcps		20.48 Mcps	
Channel	E_b/N_0	E_b/I_0	E_b/N_0	E_b/I_0	E_b/N_0	E_b/I_0
Macro 3 km/h	10.3	9.8	6.0	5.7	5.3	4.2
Macro 36 km/h	9.5	8.8	6.3	5.5	4.8	4.2
Micro 3 km/h	8.3	3.5	6.4	2.2	4.0	2.5
Micro 36 km/h	6.0	2.2	4.5	1.9	4.0	2.5
Indoor 3 km/h	4.7	0.4	4.2	-0.2	3.6	1.7

7.5.4 Spectrum Efficiency of W-CDMA Downlink

Figure 7.15 shows the downlink spectrum efficiency of 144 Kbps for micro- and macrocells as a function of the orthogonality factor. The actual orthogonality factor was calculated to be 0.7 and 0.2 in micro- and macrocells, respectively. Therefore, multiuser detection used to remove intracell interference would still improve the spectrum efficiency, since the orthogonality is not perfect (=1). This contrast often prompted results that downlink multiuser detection was not beneficial because of orthogonality.

As shown in Figure 7.16, canceling the intercell interference clearly has a smaller effect on the total system capacity. Curves indicate the capacity improvement with a different number of intercell interferers that were included in the interference cancellation process. The curves are shown as a function of the efficiency of the interference cancellation.

Link level E_b/N_0 and E_b/I_0 values used in 12- and 144-Kbps services for spectrum efficiency comparison were shown in Tables 7.16 and 7.17.

It was noted that in the downlink macrocell environment, intracell interference is most important due to multipath propagation. In the macrocell environment, significant capacity improvements could be expected if intracell interference could be canceled in the mobile. Also, in a microcell environment, intracell interference cancellation could improve the capacity but not as much as in the macrocell environment. The difference is because there is more intracell interference in macrocell due to multipath propagation.

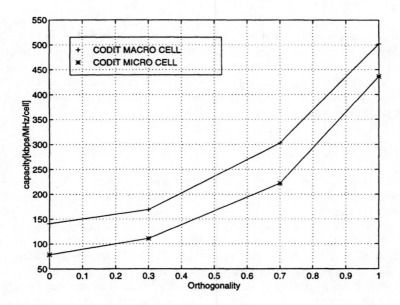

Figure 7.15 Downlink capacity as a function of intracell interference.

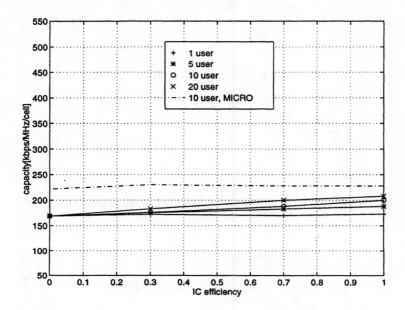

Figure 7.16 Downlink capacity as a function of intercell interference.

Additional diversity obtained from macro diversity was important in the microcell environment as there were limited number of resolvable multipath components due to large coherence bandwidth; similarly the mobile antenna diversity would be beneficial if allowed by complexity.

The downlink spectrum efficiency results were almost the same for the 12- and 144-Kbps services. This is reasonable since the downlink link level results were also very near to each other.

7.5.5 Comparison of Spectrum Efficiency for Uplink and Downlink

Table 7.20 shows capacity results for the uplink and the downlink. Downlink capacity is approximately 2 to 2.5 times lower than the corresponding uplink capacity, if both services are compared.

In addition to the base station antenna diversity, the major reason for the higher capacity of the uplink was the use of MUD, providing almost two-fold capacity if compared to a conventional receiver. In the downlink, the corresponding factor for MUD gain is the orthogonality factor. There is a difference between uplink MUD gain and downlink orthogonality in the case of soft handover. In soft handover one uplink transmission is received at two base stations, and that interference can be canceled at both base stations and MUD gain can be obtained. In the downlink there are transmissions from two base stations to one mobile station, and those two signals are not orthogonal. It should also be noticed that the orthogonality factor gives only a modest capacity gain in a macrocell channel because of multipath propagation.

Table 7.20
Cellular Capacity in Kbps/MHz/cell

	Uplink	Downlink
Macro 12 Kbps	192	108
Macro 144 Kbps	388	108

7.5.5.1   Spectrum Efficiency With Circuit Switched 2 Mbps

Table 7.21 shows the spectral efficiency in Kbps/MHz/cell for 2-Mbps communication for the downlink and for the uplink.

Table 7.21
Spectrum Efficiency in Kbps/MHz/cell.

	Uplink	Downlink
Micro 5 Mcps	1040	80
Micro 10 Mcps	1080	160
Macro 5 Mcps	360	40
Macro 10 Mcps	260	100

The cellular capacity of networks offering 2-Mbps circuit switched services seems to be quite low due to the poor downlink capacity. According to the results in Table 7.21, 15- to 20-MHz bandwidth is required to support one active 2-Mbps user in every cell. However, since all the users seldom transmit simultaneously, packet reservation multiple access techniques can be exploited to alleviate the capacity problem.

The difference in spectrum efficiency between uplink and downlink is due to the following features. As in the uplink, BTS antenna diversity can be used with less complexity considerations. This easily brings a 3- or 4-dB gain to the uplink direction. This difference becomes even larger if multiuser detection is applied in the base station receiver. Even in the case of a single user, multiuser detection can be used to remove the interpath interference in multipath channels. Interpath interference is significant with low processing gain (i.e., with high bit rates). The large dynamics allowed for uplink power control also make it possible to largely compensate for the fading, especially in the microchannel case with 2 Mbps, which allows the link level results to approach the AWGN performance. In CODIT microcell, delay spread is rather narrow. As a result, the downlink receiver cannot obtain enough diversity, although the code channels remain more orthogonal. In the uplink, antenna diversity offers the required diversity.

7.6   COVERAGE

In this section, we first explain range calculation principles. Second, we consider the coverage of a DS-CDMA cellular system in both an unloaded and a loaded case. In addition, we show the impact of multiuser detection on the uplink range. If the gain from the multiuser detection is not used for coverage improvement, it can be used to reduce the mobile station transmission power. Finally, we discuss the effect of the user bit rate on the range.

7.6.1   Range Calculation

This section lists and explains the range calculation principles. The absolute range values given in this section depend on the used propagation model and cannot be directly used for network dimensioning. The values given here are relative and are to be used for comparison purposes between different services and receiver solutions. Table 7.22, which follows these principles, shows an example calculation. The parameters used in the range calculations are shown in Table 7.23 and are listed below. The link budget model is based on [3,6]. Link budget assessment is also discussed in Chapter 11.

(a)   **Average transmitter power per traffic channel, $P_{TX,avg}$ (dBm).** The average transmitter power per traffic channel is defined as the mean of the total transmitted power over an entire transmission cycle with maximum transmitted power when transmitting.

(b) **Cable, connector, and combiner losses at the transmitter, L_{TX} (dB).** These are the combined losses of all transmission system components between the transmitter output and the antenna input (all losses in positive dB values).

(c) **Transmitter antenna gain, G_{TX} (dBi).** Transmitter antenna gain is the maximum gain of the transmitter antenna in the horizontal plane (specified as dB relative to an isotropic radiator).

(d) **Transmitter E.I.R.P. per traffic channel, P_{RX} (dBm).** This is the summation of transmitter power output per traffic channel (dBm), transmission system losses (dB), and the transmitter antenna gain (dBi), in the direction of maximum radiation:

$$P_{RX} = P_{TX,avg} - L_{TX} + G_{TX} \tag{7.37}$$

(e) **Receiver antenna gain, G_{RX} (dBi).** Receiver antenna gain is the maximum gain of the receiver antenna in the horizontal plane (specified as dB relative to an isotropic radiator). The received power P_{RX} at the base station can be written as

$$P_{RX} = P_{TX,avg} - L_{TX} + G_{TX} + G_{RX} \tag{7.38}$$

(f) **Cable, connector, and splitter losses at the receiver, L_{RX} (dB).** These are the combined losses of all transmission system components between the receiving antenna output and the receiver input (all losses in positive dB values).

(g) **Receiver noise figure, NF (dB).** Receiver noise figure is the noise figure of the receiving system referenced to the receiver input.

(h) **Thermal noise density, N_0 (dBm/Hz).** Thermal noise density is defined as the noise power per hertz at the receiver input. This can be calculated from kT, where k is the Bolzman constant and T is the temperature.

(i) **Receiver interference density, I_0 (dBm/Hz).** Receiver interference density is the interference power per hertz at the receiver front end. This is the in-band interference power divided by the system bandwidth. The in-band interference power consists of both co-channel interference as well as adjacent channel interference. Thus, the receiver and transmitter spectrum masks must be taken into account. Receiver interference density for the forward link is the interference power per hertz at the mobile station receiver located at the edge of coverage, in an interior cell. In this example, an unloaded cell is assumed. The effect of interference can also be modeled under other losses calculated as presented in Chapter 11 (Figure 11.4.)

(j) **Total effective noise plus interference density (dBm/Hz).** Total effective noise plus interference density is the logarithmic sum of the receiver noise density and the receiver noise figure and the arithmetic sum of the receiver interference density.

$$(\text{Noise} + \text{Interference}) = 10\log\left(10^{((NF + N_0)/10)} + I_0\right) \tag{7.39}$$

(k)   **Information rate, $10\log(R_b)$ (dB Hz).** Information rate is the channel bit rate in (dB Hz); the choice of R_b must be consistent with the E_b assumptions.

(l)   **Required $E_b/(N_0+I_0)$ (dB).** The ratio between the received energy per information bit to the total effective noise and interference power density needed to satisfy the quality objectives.

The translation of the threshold error performance to $E_b/(N_0+I_0)$ performance depends on the particular multipath conditions assumed. The CDMA uplink range can be calculated based on the link level E_b/N_0 value where E_b is energy per user bit and N_0 represents thermal and receiver noise. When reaching the maximum range, the mobile is transmitting at constant full power. Therefore, this E_b/N_0 value is obtained from link level simulations without fast power control, and it is higher than the value used in capacity simulations.

(m)   **Receiver sensitivity, $P_{RX,min}$ (dBm).** This is the signal level needed at the receiver input that just satisfies the required $E_b/(N_0+I_0)$. The minimum reception power $P_{RX,min}$ represents a long time average power. If real receiver sensitivity is required, an activity factor must be added to this value. This is the case if voice activity is employed. In this case, continuous reception is assumed. The sensitivity can be calculated as

$$P_{RX,min} = kT + NF + 10log(R_b) + E_b/(N_0+I_0) \qquad (7.40)$$

(n)   **Handoff gain, G_{HO} (dB).** This is the gain brought by handoff to maintain specified reliability at the boundary. In case of a single base station, there is probability P_{OUT} to have outage at distance r from the base station. In case of multiple base stations, at a point at distance r from, say, two base stations, the probability for outage relative to one base station is the same, but the probability to have outage from both the base stations is smaller and the resulting P_{OUT} is thus smaller than in the single cell case. In network planning, this reduces the shadowing margin needed in multiple cell case. The difference in shadowing margins is called handover gain. Depending on how the handover is done, the gain may be different. Here the handover gain G_{HO} is assumed to be 5 dB [10].

(o)   **Other gain, Gother (dB).** An additional gain may be achieved as a result of future technologies. For instance, space diversity multiple access (SDMA) may provide an excess antenna gain.

(p)   **Log-normal fade margin, ξ (dB).** The log-normal fade margin is defined at the cell boundary for isolated cells. This is the margin required to provide a specified coverage availability over the individual cells. In practice the propagation conditions vary considerably, and the pathloss attenuation factor ranges from 3 to 4 [5]. The coverage requirement of 95% sets the actual shadowing margin. If 95% coverage is to be achieved over the whole cell area and we know that we have shadowing with a standard deviation σ of 6.0 dB and pathloss model with $n = 3.6$ we get from [4] that the coverage probability at the area boundary is 84%. To have this, we need a shadowing margin of approximately 6.0 dB.

(q)   **Maximum path loss, L_{max} (dB).** This is the maximum loss that permits the required performance at the cell boundary.

$$L_{max} = P_{TX} - P_{RX,min} + (G_{RX} - L_{RX}) + G_{HO} + G_{other} - \xi \qquad (7.41)$$

(r) **Maximum range, d_{max} (km).** The maximum range is computed for each deployment scenario. Maximum range is given by the range associated with the maximum path loss. The propagation loss in decibels in a macrocell environment is calculated as in [32]:

$$PL = 123 + \alpha \times 10\log(d) + \sigma \qquad (7.42)$$

where α is the pathloss attenuation factor, d is distance in kilometers, and σ is a Gaussian stochastic variable that models the shadow fading. It is assumed that the average of the shadow fading equals zero and the standard deviation of the shadow fading is 6 dB. However, this might vary between 6 and 10 dB [5]. The maximum range d_{MAX} is given by

$$d_{max} = 10^{\frac{L_{max} - 123 - \sigma}{36}} \qquad (7.43)$$

Table 7.22
Link Budget Calculation Template

		Downlink	Uplink	Unit
(a)	Average TX power/TCH	30	24	dBm
(b)	Cable, connector, and combiner losses	2	0	dB
(c)	Transmitter antenna gain	13	0	dBi
(d)	TX EIRP/TCH = a – b + c	41.00	24.00	dBm
(e)	Receiver antenna gain	0	13	dB
(f)	Cable, connector, and splitter losses at the receiver	0	2	dB
(g)	Receiver noise figure	5	5	dB
(h)	Thermal noise density	−174	−174	dBm/Hz
(i)	Receiver interference density	−1000	−1000	dBm/Hz
(j)	Total noise + interference density (g + h + i)	−169	−169	dBm
	Information rate R_b	8	8	kHz
(k)	$10\log(R_b)$	39.03	39.03	dBHz
(l)	$E_b/(N_0 + I_0)$ (link level sim. result)	8	6.6	dB
(m)	Receiver sensitivity (j + j + l)	−122.0	−123.4	dBm
(n)	Handoff gain	5	5	dB
(o)	Other gains	0	0	dB
(p)	Log normal fade margin	11.3	11.3	dB
(q)	Maximum path loss (d – m + (e - f) + n + o)	153.67	149.07	dB
	Range	4.84	3.61	km

Table 7.23
Parameters for Range Calculations

Service	Medium bit rate data
User data rate	144 Kbps
BER	10^{-3}
Delay	100 ms
Environment	Macrocellular
Cell layout	Hexagonal
Pathloss exponent α	3.6
Log-normal shadowing σ	6 dB
Handover gain G_{HO}	5 dB
Fractional cell loading F	70%
Transceiver parameters	
Base station antenna gain G_{BS}	6 dBi
Mobile station antenna gain G_{MS}	0 dB
Maximum mobile transmission power $P_{TX,MS}$	1W = 30 dBm
Thermal noise kT	−174 dBm
Receiver noise figure NF	7 dB
Capacity with 5% outage [33]	
Downlink capacity	169 Kbps/MHz/cell = 7.0 users at 6.0 MHz
Uplink E_b/N_0 at constant (full) transmission power	4.0 dB

7.6.2   Range in Unloaded and Loaded Networks

In cellular CDMA, the achievable range is very much dependent on cell load: If the number of users increases, the range decreases. This agrees with all cellular solutions where system capacity is considered to be mainly limited by the interference generated by the system itself. With CDMA, the effect is also seen in a single cell case, unlike in systems where intracell users are completely orthogonal, like in pure TDMA-based systems.

In this section, the decrease in cell range is calculated when network load is increased, and the effect of base station multiuser detection in increasing the range in a loaded network is also analyzed. Base station multiuser detection is shown to lower the mobile station transmission power in a loaded network, thus making uplink range and average mobile transmission power more insensitive to network load. The impact of intercell interference at the base station in different propagation environments is also studied. For actual network planning, multiuser detection and propagation environment must be taken into account.

In the case of an unloaded network, the uplink direction limits the achievable range and coverage, as the maximum transmission power of the mobile station is low compared to the maximum transmission power of the base station in the downlink. In a loaded network the downlink may limit the range if there is more load and thus more interference in the downlink than in the uplink.

The received SIR at the base station is obtained by [23]

$$\frac{E_b}{I_0 + N_0} = \frac{E_b}{I_{intra} + I_{inter} + N_0}$$ (7.44)

where E_b is the received energy per bit, I_{intra} is the intracell interference from own cell mobiles, I_{inter} is the intercell interference from those mobiles not connected to this particular base station, and N_0 is the thermal noise.

In the case of an unloaded network, $I_{intra} = 0$, $I_{inter} = 0$, and the required E_b/N_0 for range calculations is simply equal to $E_b/(I_0 + N_0)$. In a loaded network, the percentage of own cell interference from the total interference (fractional cell loading) is defined as

$$F = \frac{I_{intra} + S}{I_{intra} + S + I_{inter}}$$ (7.45)

where $S = E_b/G_p$ is the received signal power from one user and G_p is the processing gain.

The value for F depends on the propagation environment. The higher the pathloss attenuation factor, the higher the F. Intracell interference I_{intra} can be given in terms of intercell interference I_{inter} as

$$I_{inter} = I_{intra}\left(\frac{1}{F} - 1\right) + \frac{E_b}{G_p}\left(\frac{1}{F} - 1\right)$$ (7.46)

Intracell interference is in terms of number of users N given by

$$I_{intra} = (N - 1)\frac{E_b}{G_p}$$ (7.47)

and the total interference can be written as

$$I_{intra} + I_{inter} = \left(\frac{N}{F} - 1\right)\frac{E_b}{G_p}$$ (7.48)

Now, setting the required $E_b/(I_0 + N_0)$ in the loaded network equal to the required E_b/N_0 in the unloaded network gives

$$\frac{E_{b,\text{loaded}}}{N_0 + I_0} = \frac{E_{b,\text{loaded}}}{\left(\dfrac{N}{F}-1\right)\dfrac{E_b}{G_p} + N_0} = \left(\frac{E_b}{N_0}\right)_{\text{unloaded}} \tag{7.49}$$

and solving the needed E_b/N_0 in the loaded case gives

$$\left(\frac{E_b}{N_0}\right)_{\text{loaded}} = \frac{1}{\left(\dfrac{E_b}{N_0}\right)_{\text{unloaded}}^{-1} - \left(\dfrac{N}{F}-1\right)\dfrac{1}{G_p}} \tag{7.50}$$

In Figure 7.17, range is presented as a function of uplink load for fractional cell load, $F = 70\%$. Load is shown as a percentage of maximum downlink capacity.

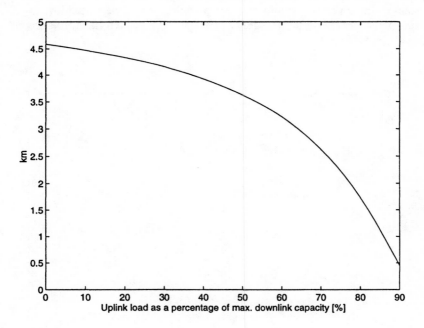

Figure 7.17 Range as a function of load. Load is shown as a percentage of maximum downlink capacity.

The maximum cell range decreases significantly when the load increases beyond about 50% of maximum downlink load. In an unloaded network the maximum cell range is 4.6 km, at 50% load range is 3.6 km, and at 90% load range is only 0.5 km. In

Figure 7.18 the E_b/N_0 requirements as a function of number of users is shown for $F = 50\%, 70\%$, and 90% corresponding to different propagation environments.

For $F = 50\%$ the increase in E_b/N_0 requirements is much higher than for $F = 70\%$ since there is, in addition to intracell interference, high intercell interference.

7.6.3 Range Extension With Base Station MUD in Loaded Networks

The range decrease as a function of load can be partially avoided by using interference cancellation or multiuser detection at the base station. Base station MUD can be used to increase the uplink capacity, but it can also be used to extend the range in a loaded network. It is assumed that MUD is able to cancel part of the intracell interference at the base station.

Base station MUD can be used to make range and cellular coverage more insensitive to uplink load. In practice, MUD efficiency β will depend on the channel estimation algorithm, interference cancellation algorithm, and mobile speed. This introduces one new parameter into network planning. Also, the actual impact of MUD into cell range depends on the fractional cell loading F, which has to be predicted for each cell individually.

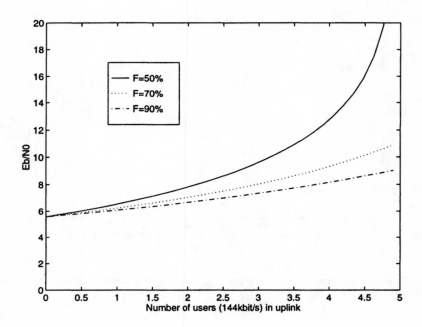

Figure 7.18 E_b/N_0 requirements in dB as a function of load in different propagation environments.

The effect of MUD can be taken into account by using the efficiency of MUD β as a measure of performance of MUD. That gives the percentage of intracell interference that is removed by MUD. With MUD the intercell interference $I_{\text{intra,MUD}}$ can be written as

$$I_{\text{intra,MUD}} = (1 - \beta)I_{\text{intra}} \tag{7.51}$$

The total interference is now

$$I_{\text{intra,MUD}} + I_{\text{inter}} = \left(\frac{N(1-\beta)+\beta}{F} - 1 \right) \frac{E_b}{G_p} \tag{7.52}$$

The required E_b/N_0 in the loaded network with MUD becomes

$$\left(\frac{E_b}{N_0} \right)_{\text{loaded,MUD}} = \frac{1}{\left(\dfrac{E_b}{N_0} \right)_{\text{unloaded}}^{-1} - \left(\dfrac{N(1-\beta)+\beta}{F} - 1 \right) \dfrac{1}{G_p}} \tag{7.53}$$

Range as a function of efficiency of base station MUD is shown in Figure 7.19. An efficiency of 0% corresponds to the base station without MUD. We can see that the range varies considerably depending on MUD efficiency.

From Section 7.5 we obtain a simulated figure for efficiency of MUD β: 70%. Now, the ranges with and without MUD are presented in Table 7.24 and in Figure 7.20.

Table 7.24
Range With and Without Base Station MUD.

Uplink load (%)	Range without MUD (km)	Range with MUD (km) (β=70%)
0	4.6	4.6
50	3.6	4.4
70	2.6	4.2
90	0.5	4.0

7.6.4 Mobile Transmission Power Savings with Base Station MUD

Even if the required coverage in the network can be achieved without base station MUD, advanced receiver algorithms at the base station can be used to decrease the transmission power of the mobile stations. The transmission power at maximum range $P_{\text{TX,MS}}$ can be written from (7.36)

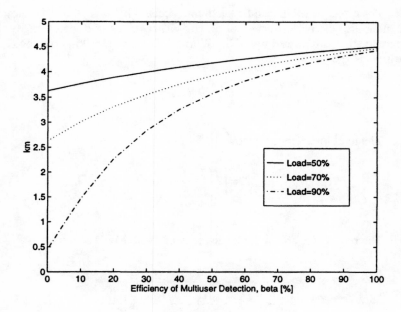

Figure 7.19 Range extension with base station MUD.

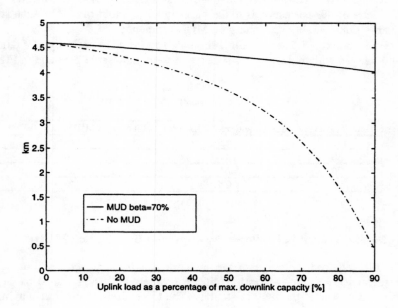

Figure 7.20 Range as a function of load with and without MUD.

$$P_{TX,MS} = P_{RX,min} + L_{HO} - G_{MS} - G_{BS} \qquad (7.54)$$

and by replacing $P_{RX,min}$ with (7.35)

$$P_{TX;MS} = \frac{E_b}{N_0} + R_b + NF + kT + L_{HO} - G_{MS} - G_{BS} \qquad (7.55)$$

Since all the terms in (7.50) are the same except for E_b/N_0, regardless of the base station receiver algorithm, $P_{TX,MS}$ is determined only by the E_b/N_0 requirement. The decrease in required transmission power with MUD is therefore

$$\frac{P_{TX,MS}}{P_{TX,MS,MUD}} = \frac{\left(E_b/N_0\right)_{loaded}}{\left(E_b/N_0\right)_{loaded,MUD}} \qquad (7.56)$$

and the gain in E_b/N_0 can be written as

$$\frac{\left(\dfrac{E_b}{N_0}\right)_{loaded}}{\left(\dfrac{E_b}{N_0}\right)_{loaded,MUD}} = \frac{\left(\dfrac{E_b}{N_0}\right)_{unloaded}^{-1} - \left(\dfrac{N(1-\beta)+\beta}{F} - 1\right)\dfrac{1}{G_p}}{\left(\dfrac{E_b}{N_0}\right)_{unloaded}^{-1} - \left(\dfrac{N}{F} - 1\right)\dfrac{1}{G_p}} \qquad (7.57)$$

In the above equations, it must be assumed that the propagation loss is lower than the maximum loss allowed for a system without MUD. If the mobile is not transmitting at full power but is able to use fast power control, then (7.52) is not exact but an approximation. In Figure 7.21 the estimated decrease in average transmission power of mobile station is shown as a function of MUD efficiency.

Assuming MUD efficiency $\beta = 70\%$ and fractional cell loading $F = 70\%$, the decrease of transmission power is 3 dB at 50% load and 7 dB at 70 % load.

7.6.5 Effect of User Bit Rate on the Range

The maximum range depends on the user bit rate. The higher the user bit rate, the shorter the range. The goal of third generation networks is to provide full coverage for low bit rates (<144 Kbps) and for higher bit rates a limited coverage may be acceptable. In Table 7.25 the maximum ranges for different bit rates are calculated assuming the same E_b/N_0 target of 4.0 dB for all services and the parameters in Table 7.23.

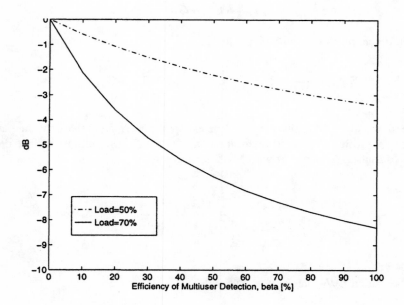

Figure 7.21 Savings in mobile transmission power with MUD.

Table 7.25
Maximum Range as a Function of User Bit Rate.

User bit rate	Relative maximum range (km)
8 Kbps	10.0
14.4 Kbps	8.5
64 Kbps	5.6
144 Kbps	4.5
384 Kbps	3.4
1 Mbps	2.6
2 Mbps	2.2

7.6.6 Summary

Coverage of a loaded DS-CDMA cellular network has been considered by analyzing the uplink range. Base station multiuser detection (MUD) has been proposed as an upgrade solution to provide good coverage even with high system load after the initial deployment. MUD has also been shown to decrease mobile station transmission power in a loaded network. Therefore, a network with base station MUD can be operated with a higher percentage of maximum load if the system capacity is limited by downlink.

Good coverage need not be sacrificed in order to use high system loads. For actual network planning we need to take into account the performance of MUD to be able to predict and plan the coverage. The impact of MUD into cell range depends also on the propagation environment (i.e., the percentage of the intracell interference from the total interference). The increase in the data rate will inevitably mean reduced range in the uplink as the transmission power is limited. Thus, in the cell design the coverage area for the low rate services is likely to be different than for high rate service. This effect has not been visible in the second generation networks since services are narrowband services with similar range performance.

REFERENCES

[1]   Proakis, J. G., *Digital Communications*, 3rd ed., New York: McGraw-Hill, 1995.

[2]   Xia, H. H., "Reference Models for Evaluation of Third Generation Radio Transmission Technologies," *Proceedings of ACTS Summit*, Aalborg, Denmark, October 1997, pp. 235–240.

[3]   Recommendation ITU-R M.1225, "Guidelines for Evaluation of Radio Transmission Technologies for IMT-2000," *Question ITU-R 39/8*, 1997.

[4]   Jakes, W. C., *Microwave Mobile Communications*, New York: Wiley, 1974.

[5]   Parsons, J., *The Mobile Radio Propagation Channel*, London: Pentech Press, 1992.

[6]   ETSI/SMG-5, Selection procedures for the choice of radio transmission technologies of the UMTS, UMTS 30.03, v. 3.0.0, May 1997.

[7]   Martin, G., and M. Faulkner, "1.9 GHz Measurement-based Analysis of Diversity Power versus the number of RAKE Receiver Tines at Various System Bandwidths," *Proceedings of PIMRC'97*, September 1997, Helsinki, Finland, pp. 1069–1073.

[8]   Otmani, M., and M. Lecours, "Indoor Radio Impulse Response Measurements with Polarization Diversity," *Proceedings of VTC'96*, Atlanta, Georgia, USA, May 1996, pp. 151–154.

[9]   CEC Deliv. No. R2020/TDE/PS/DS/P/040/b1, CODIT Final Propagation Model, June 1994.

[10] Rege, K., S. Nanda, C. F. Weaver, and W. C. Peng, "Analysis of Fade Margins for Soft and Hard Handoffs," *Proceedings of IEEE 6th Personal, Indoor and Mobile Radio Communications Conference (PIMRC'95)*, Toronto, Canada, 1995, pp. 829–834.

[11] Chopra, M., K. Rohani and J. D. Reed, "Analysis of CDMA Range Extension due to Soft Handoff," *Proceedings of VTC'95*, 1995, pp. 917–921.

[12] Andoh, H., M. Sawahashi, and F. Adachi, "Channel Estimation Using Time Multiplexed Pilot Symbols for Coherent Rake Combining for DD-CDMA Mobile Radio," *Proceedings of PIMRC'97*, Vol. 3, Helsinki, Finland, September 1997, pp. 954–958.

[13] Tiedemann, E. G., Jr., Y-C. Jou, and J. P. Odenwalder, "The Evolution of IS-95

to a Third Generation System and to the IMT-2000 Era," *Proceedings of ACTS Summit*, Aalborg, Denmark, October 1997, pp. 924–929.

[14]   Hämäläinen, S., P. Slanina, M. Hartman, A. Lappeteläinen and H. Holma, "A Novel Interface Between Link and System Level Simulations," *Proceedings of ACTS Mobile Communications Summit '97*, October 7–10, 1997, Aalborg, Denmark, pp. 599–604.

[15]   Hata, M., "Empirical Formula for Propagation Loss in Land Mobile Radio Services," *IEEE Transactions on Vehicular Technology*, August 1980, pp. 317–325.

[16]   Ariyavisitakul, S., "SIR Based Power Control in a CDMA System," GLOBECOM'92, Orlando, Florida, USA, December 1992.

[17]   Shapira J., and R. Padovani, "Spatial Topology and Dynamics in CDMA Cellular Radio," *Proceedings in VTC'92*, Denver, CO, May 1992, pp. 213–216.

[18]   Special Issue on Smart Antennas, *IEEE Personal Communications*, Vol. 5, No. 1, February 1998.

[19]   Ojanperä, T., K. Rikkinen, H. Hakkinen, K. Pehkonen, A. Hottinen and J. Lilleberg, "Design of a 3rd Generation Multirate CDMA System with Multiuser Detection, MUD-CDMA," *Proceedings of ISSSTA'96*, Vol. 1, Mainz, Germany, September 1996, pp. 334–338.

[20]   Nikula, E., A. Toskala, E. Dahlman, L. Girard and A. Klein, "FRAMES Multiple Access for UMTS and IMT-2000," *IEEE Personal Communications Magazine*, April 1998, pp. 16–24.

[21]   Jalali, A., and P. Mermelstein, "Effects of Diversity, Power Control, and Bandwidth on the Capacity of Micro Cellular CDMA System," *IEEE Journal on Selected Areas in Commun.*, Vol. 12, No. 5, June 1994, pp. 952–961.

[22]   Papoulis, A., *Probability, Random Variables, and Stochastic Processes*, 3rd ed., New York: McGraw-Hill, 1991.

[23]   Hämäläinen, S., H. Holma, and A. Toskala, "Capacity Evaluation of a Cellular CDMA Uplink with Multiuser Detection," *Proceedings of ISSSTA'96*, Mainz, Germany, September 1996, pp. 339–343.

[24]   Gass, J. H., D. L. Noneaker, and M. B. Pursley, "Spectral efficiency of a Power-Controlled CDMA Mobile Communication System," *IEEE Journal on Selected Areas in Commun.*, Vol. 14, No. 3, April 1996, pp. 559–569.

[25]   Padovani, R., "Reverse Link Performance of IS-95 Based Cellular System," *IEEE Personal Communications*, Vol. 1, No. 3, Third Quarter, 1994, pp.28–34.

[26]   Viterbi, A. M., and A. J. Viterbi, "Erlang Capacity of a Power Controlled CDMA System," *IEEE Journal on Selected Areas in Communications*, Vol. 11, No. 6, August 1993, pp. 892–899.

[27]   Ojanperä, T., P. Ranta, S. Hämäläinen, and A. Lappeteläinen "Analysis of CDMA and TDMA for 3rd Generation Mobile Radio Systems," *Proceedings VTC'97*, May 1997, Phoenix, Arizona, USA, pp. 840–844.

[28]   Holma, H., A. Toskala, and A. Hottinen, "Performance of CDMA Multiuser Detection with Antenna Diversity and Closed Loop Power Control," *Proceeding of VTC'96*, Atlanta, Georgia, USA, May 1996, pp. 362–366.

[29]   Parkvall, S., E. Ström, and B. Ottersen, "The impact of Timing Errors on the Performance of Linear DS-CDMA Receivers," *JSAC*, Vol. 14, No. 8, October

1996, pp. 1660–1668.

[30]   Buehrer, M., A. Kaul, S. Striglis, and B. D. Woerner, "Analysis of DS-CDMA Parallel Interference Cancellation with Phase and Timing Errors," *IEEE Journal on Selected Areas in Communications*, Vol. 14, No. 8, October 1996, pp. 1522–1535.

[31]   Simon, M., J. Omura, R. Scholtz, and B. Levitt, *Spread Spectrum Communications Handbook*, New York: McGraw-Hill, 1994.

[32]   Pizarroso, M., J. Jiménez (eds.), "Common Basis for Evaluation of ATDMA and CODIT System Concepts," *MPLA/TDE/SIG5/DS/P/001/b1*, MPLA SIG 5, September 1995.

[33]   Hämäläinen, S., H. Holma, A. Toskala, and M. Laukkanen, "Analysis of CDMA Downlink Capacity Enhancements," *Proceedings of PIMRC'97*, Helsinki, Finland, September 1997, pp. 241–245.

Chapter 8

HIERARCHICAL CELL STRUCTURES

8.1   INTRODUCTION

A third generation system must be able to support a wide range of services in different radio operating environments. Different types of cells are needed for different requirements: large cells guarantee continuous coverage, while small cells are necessary to achieve good spectrum efficiency and high capacity. Small range cells are used by low mobility and high capacity terminals, while high range cells serve high mobility and low capacity terminals. In addition, different cell types should be able to operate one upon another. Hierarchical cell structure (HCS) describes a system where at least

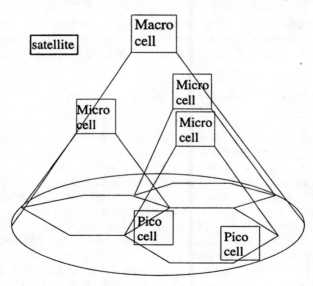

Figure 8.1   Hierarchical cell scenario.

249

two different cell types (e.g., macrocell and microcell) operate one upon another, as shown in Figure 8.1. Microcells are small cells covering areas with a radius of a few hundred meters. Low powered base stations are typically placed at lamp post level and they serve a street block. Macrocells, or umbrella cells, cover a radius of 1 km or more. Macrocells cover rural areas, provide continuous coverage of areas that are covered by microcells, and serve rapidly moving users. Picocells cover indoor areas with a cell radius of a few dozen meters. A low power base station usually covers an office, a floor of a high building, or a residence. Satellite cells introduce global continuous coverage by covering rural areas where terrestrial systems cannot be installed. Whenever possible, traffic should be directed to the smallest available cell, so that the system's spectral efficiency is improved. Particularly in the satellite segment, it is crucial to direct traffic to the macro- or microcells whenever possible [1].

In this chapter, two approaches for HCS design in CDMA are described, and advantages and disadvantages of the two methods are discussed. In the first approach, co-existing hierarchy layers operate in the same frequency band [2,3]. Users operating on different layers are separated by handovers and signal fading. In the second approach, different hierarchy levels are separated in the frequency domain. The aspects related to HCS are also relevant in a multioperator environment where an operator has to consider interference from the adjacent frequencies belonging to another operator. Network planning aspects such as guardbands related to multioperator environments are discussed in Chapter 11.

First, nonlinear power amplifiers are studied in Section 8.2 to establish an understanding of the adjacent channel interference mechanism in a system with different cell layers at different frequencies. A HCS system with micro- and macrocells in the same frequency, and at different frequencies, are studied in Sections 8.3 and 8.4, respectively. It should be noted that aspects related to interfrequency handover are not covered here but rather in Chapters 2 and 5. Section 8.4 evaluates link and system level performance of a cellular HCS network utilizing DS-CDMA and multiuser detection.

8.2   NONLINEAR POWER AMPLIFIERS

Even if hierarchy layers have different carrier frequencies, some adjacent channel interference (ACI) is generated between adjacent channel carriers. When bandlimited linear modulation methods are used, spectrum leakage between adjacent channel carriers depends on the linearity of the power amplifier (PA) [4]. Spectrum leakage to adjacent channels can be controlled by backing off the PA or by using some linearization method to equalize the nonlinearities of the power amplifier. Unfortunately, both methods decrease the achievable PA efficiency when compared to modulation schemes that have constant envelope and can be amplified with power efficient but nonlinear power amplifiers [4].

Real power amplifiers are always nonlinear. Linearity requirements of a CDMA transmitter are mainly determined by the spectrum ACI attenuation requirements, rather than from link level performance losses. If low ACI attenuations (i.e., larger spectrum spreading into adjacent carriers) can be tolerated, more efficient power amplifiers can be used in mobile terminals.

For link level simulations, a model of a nonlinear power amplifier is needed. In order to get real life amplifier model, a power amplifier IC was measured. The power amplifier IC is manufactured with the Gallium Arsenide heterojunction bipolar transistor (HBT) process, and the final stage of the IC is biased to class-AB.

8.2.1   Power Amplifier Characteristics

A typical power amplifier introduces both amplitude and phase distortion into the transmitted signal, resulting in AM-AM and AM-PM conversion effects. Figure 8.2 shows the AM-AM and AM-PM characteristics of the measured amplifier. The power amplifier amplifies input power linearly when input RMS power is −15 to 0 dBm. This is the linear area of the amplifier. When the input power is 0 to 5 dBm, the amplifier operates in its conversion or saturation area. The measured power amplifier has saturated output power at the level of 34 dBm. This value corresponds to 0 dB output backoff value [4]. Output backoff is defined as the difference between output power in the saturation point and output power in the operation point. Backing off the amplifier reduces output powers in the operation point and increases output backoff.

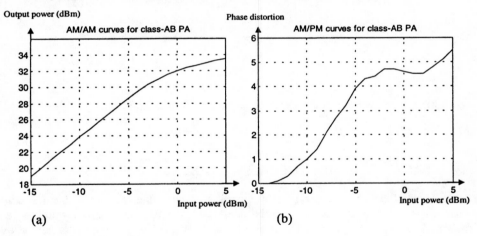

Figure 8.2   a) AM-AM curve and b) AM-PM curve for measured power amplifier IC.

8.2.2   Power Amplifier Efficiency

Efficiency of the power amplifier IC is measured using a constant envelope sine wave. Efficiency compared to output backoff is shown in Figure 8.3. At saturation power (0 dB output backoff), the power amplifier has slightly less than 50% efficiency. This means that if constant envelope modulation methods, such as Gaussian minimum shift keying (GMSK), were used, up to 50% efficiencies could be achieved. However,

252

because bandlimited linear modulation methods have nonconstant RF-envelope, output power needs to be lowered in order to avoid spectrum spreading [4]. The accurate efficiency of the power amplifier with bandlimited linear modulation methods (nonconstant RF-envelope) can be predicted by using the measured efficiency curve and the simulated pdfs of the used modulation method [4].

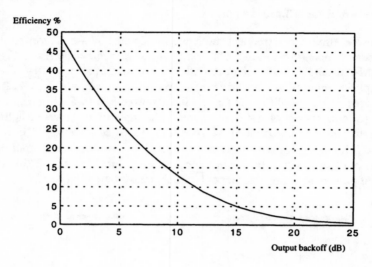

Figure 8.3   Power amplifier efficiency versus output backoff, class=AB PA.

8.3   MICRO- AND MACROCELLS AT THE SAME FREQUENCY

In the case of micro- and macrocells operating at the same frequency, the frequency reuse factor is set to one and spatial isolation is used for separating micro- and macrocell layers [2,3]. A reuse factor of one can be offered since processing gain allows users to experience interference originating from any cell layer. Intra-layer interference is controlled by power control and inter-layer interference by spatial isolation. In a generic flat-Earth model, the average transmission loss follows R^{-2} until a breakpoint that marks the separation between two segments. After the breakpoint, R^{-4} is followed. The location of the breakpoint depends on receiver and transmitter heights. As the base station is installed lower, the breakpoint occurs nearer the base station. Thus, the signal from a microcell base station attenuates faster than the signal from a macrocell base station. In Figure 8.4, spatial isolation is depicted for both hotspot case and for continuous coverage of microcells.

The problem with the presented method is that the mobile station should be instantly handed to the corresponding cell layer when it arrives to the intersection of micro- and macrocell attenuation curves. Soft handover eases the problem, but if the mobile station arrives in the microcell area, it is most often connected to both micro and macrocell base stations. This occurs since the handover region has to be rather large in order to avoid ping-ponging between cells (or cell layers), especially if the mobile stations are fast moving. Thus, the area where mobile stations are in a soft handover state is rather wide compared to the area of microcell.

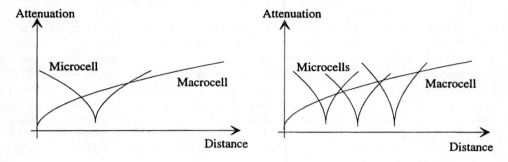

Figure 8.4   Spatial separation of umbrella cells and underlay microcells along [2]. The figure on the left depicts hotspot case covered by one microcell base stations, and the figure on the right depicts the full coverage of microcells.

Handover can never be very fast. If the number of candidate base stations is high, the pilot signal of a candidate base station can be measured only after some delay. In addition, this pilot has to be measured for a rather long period in order to filter fast fading. Handover can not be performed when the mobile station encounters a fading dip due to fast fading. After handover decision has been taken, signaling between the mobile station and the base station takes place. Before both the network side and the mobile side get all the handover related parameters, some delay has been experienced. The total delay due to pilot measurement, filtering, and handover signaling may be on the order of several seconds. If the mobile station runs with a 50 km/h speed, it moves almost 14m during 1 second. If the size of a microcell is 100m, only a 7 second delay on handover is needed to go to the center of the cell.

8.4   MICRO- AND MACROCELLS AT DIFFERENT FREQUENCIES

A system with different cell layers at different frequencies is easier to manage since cell layers do not interfere with each other as much as when they are at the same frequency. The main interference comes from the adjacent channel spill-over. A drawback of this approach is large spectrum requirements since each cell layer requires its own frequency. For example, as was discussed in Chapter 6 (Figure 6.2), the WCDMA scheme implements three layers within 15 MHz.

In this section, first ACI attenuation masks for the system level simulations are generated by using the characteristics of a measured power amplifier IC. The capacity for the uplink and downlink of a HCS network is then evaluated. Impact of power control, handover, and channel spacing on the system capacity are also discussed.

8.4.1 Adjacent Channel Interference and Link Level Performance

Adjacent channel interference and link level performance degradation determine the impact of nonlinear power amplifiers on overall spectrum efficiency. As shown below, the increased adjacent channel interference will dominate the degradation of spectrum efficiency due to nonlinear power amplifiers.

In order to simulate a HCS network with the system level simulator, adjacent channel interference has to be modeled. Here, adjacent channel interference is modeled so that the output power of an interfering user is reduced by *adjacent channel attenuation*. Adjacent channel attenuation describes how much the power of the interfered user is attenuated if it is received at the adjacent channel. Adjacent channel interference attenuations are simulated with OQPSK chip modulation, bandlimited with square-root raised cosine pulse shaping and using the described power amplifier model. Roll-off factor is selected to be 0.20 in the simulations. Because the system includes different chip rates for macro- and microcells (2.5 and 5.1 Mchip/s), two separate ACI masks need to be created for system level simulations. These masks are created by backing off the power amplifier by 3.5 dB and by measuring the spectrum spreading power with the receiver square-root raised cosine filter also having a roll-off factor of 0.2. By sliding the receiver filter in frequency domain, continuous ACI masks are created for system level simulations, as shown in Figure 8.5. The backoff used corresponds to about 31% power amplifier efficiency [4]. Because the measured amplifier effects are quite equal to OQPSK and QPSK modulation methods [4], the same ACI masks can be used for the downlink system level performance simulations.

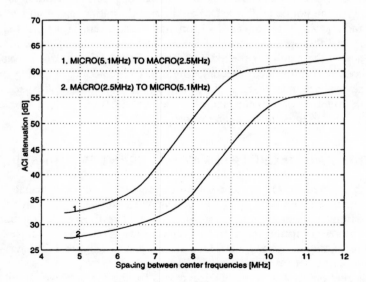

Figure 8.5 ACI masks for system level simulations.

Link level performance was also simulated with the same power amplifier model and with OQPSK and QPSK chip modulations, bandlimited with square-root raised cosine filter having a roll-off factor of 0.2. CDMA was found to be very robust against the nonlinearity of the power amplifier. Less than 0.2 dB link level performance losses occurred (both uplink and downlink) when the same output backoff (3.5 dB) as in the ACI mask definitions was used. This means that in a CDMA system, the spectrum mask requirements will limit the use of nonlinear amplifiers rather than link level performance losses. On the other hand, this means that if moderate ACI attenuation requirements can be tolerated in CDMA, the use of highly efficient power amplifiers becomes possible.

8.4.2   System Level Simulations of a HCS Network

In Figure 8.6, a HCS case with a hotspot of traffic in the middle of the area is shown. The high density of users located at the hotspot is served by microcells. The whole system area is covered by macrocells, which are used to provide continuous coverage.

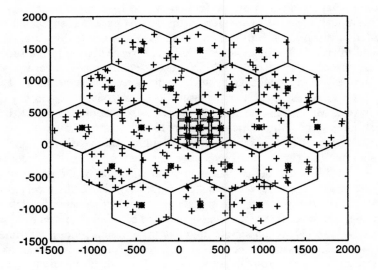

Figure 8.6   HCS environment where the macrocell base stations are on a hexagonal grid and the mobile stations (plus signs) are generated only to the streets. Eight microcell base stations are placed in the middle of the map (hotspot).

Figure 8.7 shows continuous coverage of microcell base stations. Here, both micro- and macrocell base stations provide continuous coverage. Fast moving mobile stations are handled by macrocell base stations, while high capacity terminals are served by microcell base stations.

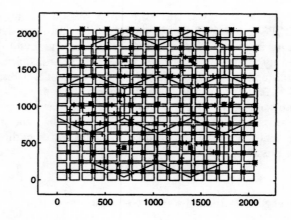

Figure 8.7   HCS environment with the full coverage of microcells.

System level simulations are performed at two HCS environments so that the hierarchy layers are separated in frequency. In the first case, the simulated system is covered by full coverage of macro- and microcell base stations. In this case, the number of macro- and microcell base stations are 7 and 128, respectively. In the second case, microcells are placed in a hotspot, while macrocells continuously cover the simulated area. The number of microcells is 8, and the number of macrocells is 19. In the simulation, macrocell base stations are placed in a hexagonal grid and the microcell base stations in a Manhattan-like grid. Microcell base stations are located in every second street intersection. All mobile stations are located in street canyons.

Instantaneous SIR is calculated by dividing the received signal by the interference and multiplying by the processing gain. In the uplink, it is assumed that a signal propagates from a mobile station to a base station via four equal strength Rayleigh faded paths. The observed base station adds together paths received from the observed user. It is supposed here that signals are combined coherently with maximal ratio combining. Maximal ratio combining is modeled by taking the sum of SNR values of each received path [5]. When simulating medium bit rate services, interference can be assumed to be Gaussian, since it is composed of transmissions to several independent users: signals from other users are just added together and seen as interference. If an interfering user was connected to the observed base station, the multiuser detection process is able to cancel the signal partially. The cancelled signal is multiplied by $(1-\beta)$, where β is *multiuser detection efficiency* and given as a parameter. SIR can be written as

$$SIR_{UL} = \frac{G_p \cdot S}{(1-\beta) \cdot I_{\text{intra}} + I_{\text{inter}} + N_0} \tag{8.1}$$

Since the system is interference limited, thermal noise N_0 is assumed to be small and neglected. External interference I_{inter} consists of intercell and inter-hierarchy level interference. Inter-hierarchy level interference is multiplied by *adjacent channel attenuation* δ. Thus, I_{inter} is

$$I_{inter} = I_{intercell} + \delta \cdot I_{interlayer} \tag{8.2}$$

In the downlink, signals from each base station to the mobile station propagate via four independent equal strength Rayleigh faded paths. Since macro diversity was considered at link level, the mobile station is able to receive all the paths from the base stations transmitting to it. Interference that propagates via the same paths than the desired signal is multiplied by an *orthogonality factor* α. This orthogonality factor states the orthogonality loss due to multipath propagation and energy losses due to limited receiver capacity. The orthogonality factor α is a simulation parameter since it varies according to radio environment, depending on the multipath delay spread (see Section 7.5.3). The transmission of the base stations that do not direct their transmission to the observed user contributes to the noise. SIR will then be

$$SIR_{DL} = \frac{G_p \cdot S}{(1 - \alpha) \cdot I_{intra} + I_{inter}} \tag{8.3}$$

In (8.3) I_{inter} consists of intercell and inter-hierarchy level interference, where interference from other hierarchy levels is multiplied by δ.

A performance measure for system simulations is outage percentage. A maximum of 5% of the obtained SIR values are allowed to be lower than the threshold SIR. SIR distributions are generated for both hierarchy layers and outage requirements that have to be fulfilled for both. Thus, the total outage requirement becomes even tighter than 5%. When the developed outage for one layer is 5%, the corresponding outage with the selected load will probably be less than 5% for the other layer. Simulation parameters are shown in Table 8.1.

8.4.3 Spectrum Efficiency Results

Spectrum efficiency results are shown in Table 8.2. Single layer micro- and macrocell capacities were simulated and used as references for HCS simulations. As seen previously, the capacity of the uplink with multiuser detection and antenna diversity is higher than the near orthogonal downlink capacity. In HCS simulations with full coverage of microcells, the microcell layer was loaded to 80% of a single layer microcell capacity. In case of a hot spot, macrocells were loaded to 80% of a single layer capacity. As seen from the results, co-existing layer capacity is almost the same as the single layer capacity, even when the interfering layer was loaded to 80% of its capacity. Even if the difference between pathloss of a microcell user and macrocell user may be several dozen decibels, power control with high dynamics makes the situation easy.

Capacities shown in Table 8.2 are obtained with high dynamic power control range. If the downlink power control range is set to 10 dB, a high difference between pathloss values cannot be compensated and outage becomes poor with the selected loading. Outage requirements can be fulfilled if the load is decreased or if channel spacing is increased. Very high power control dynamics may be difficult to implement in the downlink. Therefore, power control dynamics and channel spacing should be selected so that performance versus bandwidth usage is optimized and implementation requirements fulfilled. As an example, power control dynamics could be 10 dB and channel spacing 1.6 MHz higher than in the reference case. Then, on average, a 15 dB higher difference on pathloss values can be compensated. Another way to increase the ACI attenuation is to raise the output backoff of the power amplifier. Unfortunately, this decreases the achievable power amplifier efficiency quite fast, as shown in Figure 8.3. If the number of microcells is low, then the intercell interference ($I_{intercell}$) becomes low and the HCS microcell capacity can be even higher than the reference capacity. This is the case if microcells are in a hotspot. The same effect occurs for macrocell, as can be seen in the case where microcells cover a large area and the number of macrocell base stations is low (seven base stations).

Simulation results shown in this chapter are different from the results shown in Chapters 5 and 7, as the selected parameter set and channel models were different.

Table 8.1
System level simulation parameters.

Parameter	Uplink	Downlink	Unit
Chip rate: - macrocell	2.5	2.5	MHz
- microcell	5.1	5.1	MHz
Channel separation	4.6	4.6	MHz
Handover margin	3	3	dB
Active set size	3	3	-
MUD efficiency: - 144 Kbps, micro	60	-	%
- 12 Kbps, macro	40	-	%
Orthogonality, - macro	-	68	%
- micro	-	62	%
ACI attenuation (δ): - Macro-to-micro	28	28	dB
- Micro-to-macro	32	32	dB
Power control step size	1	1	dB
Power control dynamics	80	10 and 80	dB
E_b/N_o, - 144 Kbps, micro	0.4	6.4	dB
- 12 Kbps, macro	5.7	6.4	dB

Table 8.2
Simulation Results [Kbps/MHz/cell]

	Uplink		Downlink	
	Micro	Macro	Micro	Macro
Hot spot	1138	204	317	175
Full	634	271	216	276
Reference	792	236	274	218

In [6], the case where adjacent channels overlap (totally or partially) was neglected. From the simulations, it is expected that a CDMA-based network could operate so that both hierarchy levels use the same frequency. This, however, requires very fast power control and handover, and a very wide power control dynamic range for both transmission directions. Since the simulations were performed as snapshot simulations, the mobile station did not move during the simulation, and effects due to mobility were neglected. Mobility may destroy systems where the two hierarchy levels operate with the same frequency. Consider a case where a mobile station connected to the macrocell moves rapidly to the area of a microcell. The handover process is so slow that a mobile station runs deep inside microcell area before it is handed to the microcell base station. The only solution is very fast power control. If the mobile stations connected to the microcell base station are able to adjust their uplink powers fast enough, the system will not crash. The mobile stations connected to macrocells should have such a fast downlink power control that they can adjust downlink power in a way that the connection does not drop due to increased interference power from the microcell. In addition, the power control range for downlink should be so large that mobile stations can compensate for increased interference. The conclusion from system simulations is that hierarchical cell structure networks are feasible with CDMA, but some implementation problems may occur.

REFERENCES

[1] Andermo, P. G., "System Flexibility and its Requirements on Third Generation Mobile Systems," *Proceedings of PIMRC'92*, Boston, Massachusetts, USA, 1992, pp. 1–3.
[2] Shapira, J., and R. Padovani, "Spatial Topology and Dynamics in CDMA Cellular Networks," *Proceedings of VTC'92*, Denver, Colorado, USA, 1992, pp. 213–216.
[3] Shapira, J., "Microcell Engineering in CDMA Cellular Networks," *IEEE Transactions on Vehicular Technology*, Vol. 43, No. 4, November 1994, pp. 817–825.
[4] Lilja, H., "Characterizing the Effect of Nonlinear Amplifier and Pulse Shaping on the Adjacent Channel Interference with Different Data Modulations," *Licentiate Thesis*, Oulu University, Finland, 1996.
[5] Proakis, J. G., Digital Communications, New York: McGraw-Hill, 1995, pp. 778–785.
[6] Hämäläinen, S., H. Lilja and, J. Lokio, "Performance of a CDMA Based Hierarchical Cell Structure Network," Proceedings of PIMRC'97, Helsinki, Finland, September 1997, pp. 863–866.

Chapter 9

TIME DIVISION DUPLEX DS-CDMA

9.1 INTRODUCTION

In time division duplex (TDD), the uplink and downlink transmissions are time multiplexed into the same carrier, in contrast to frequency division duplex (FDD), where uplink and downlink transmissions occur in frequency bands separated by the duplex frequency. Figure 9.1 illustrates the principles of TDD and FDD.

Section 9.2 introduces second generation TDD systems: DECT, PHS, and CT2. Section 9.3 presents the motivation for using TDD for third generation systems.

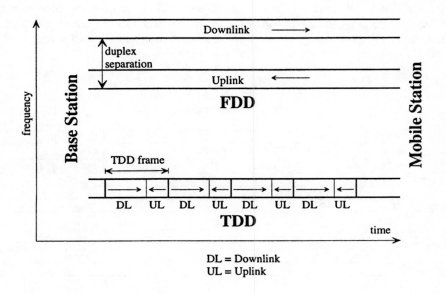

Figure 9.1 TDD and FDD principle.

261

In Section 9.4, interference aspects in TDD are considered. Intercell and intracell interference problems in asymmetric TDD-CDMA system are identified, as well as interference between operators. TDD-specific air interface design aspects, in particular the frame length and the receiver structure, are discussed in Section 9.5. A third generation TDD-CDMA system as an extension to an FDD-CDMA system is presented in Section 9.6. The performance of the system is analyzed for a WWW browsing traffic model. Section 9.7 summarizes other proposed wideband CDMA-TDD systems. TDMA-based third generation TDD proposals such as in [1] are described in Chapter 1.

9.2   SECOND GENERATION TDD SYSTEMS

Examples of second generation TDD systems are Digital European Cordless Telephone (DECT), Personal Handy Phone System (PHS), and CT2. The main parameters of these systems are listed in Table 9.1. These systems are intended for a low tier radio environment, mainly for indoor operation. A common feature of second generation TDD systems is no or very little channel coding, which restricts the performance to low mobility radio environments. TDD systems have not gained as much market support as the second generation FDD technologies (GSM, IS-95, PDC, and US-TDMA). The main reason for this seems to be the limited mobility and coverage provided by the TDD systems.

The DECT standard was adopted by ETSI in 1992, and during 1993 the first generation of DECT products were commercially launched. In addition to speech services, DECT provides data services with bit rates up to 512 Kbps half duplex and 256 Kbps full duplex. DECT also provides interworking with GSM.

PHS was developed in Japan in the early 1990s. Commercial operation was launched in 1995. An interesting feature specific to PHS is that direct mobile-to-mobile calls are allowed using the lower end of the frequency band (1895 to 1898 MHz). In the United States, PHS was merged with the WACS system in 1994. This formed Personal Access Communications System (PACS), which has also an FDD mode.

Table 9.1
TDD System Parameters

	DECT	PHS	CT2
Multiple access	FDMA/TDMA	FDMA/TDMA	FDMA
Frequency band	1880 – 1900 MHz	1895 – 1918 MHz	864 – 868 MHz
Carrier bandwidth	1.728 MHz	300 kHz	100 kHz
Carrier bit rate	1.152 Mbps	384 Kbps	72 Kbps
Frame length	10 ms	5 ms	2 ms
Timeslots/frame	24	8	2
Modulation	GMSK	DQPSK	GFSK
Speech coding	ADPCM 32 Kbps	ADPCM 32 Kbps	ADPCM 32 Kbps
Max. transmitter power	240 mW	10 mW	10 mW
Power control	No	Yes	No

9.3   REASONS TO USE TDD

This section discusses the reasons for selecting TDD as a duplex scheme for a third generation CDMA system.

9.3.1   Spectrum Allocation

Providing high bit rate services, even with high spectral efficiency, requires a large bandwidth. Additionally, the services required beyond the year 2000 are not yet clearly defined. Efficient and flexible utilization of all the available bandwidth, including the TDD band, is essential for viable third generation mobile radio systems. Figure 9.2 and Table 9.2 show spectrum for IMT-2000 according to WARC'92. Since the IMT-2000 spectrum allocation is asymmetric, it supports both FDD- and TDD- based systems. In Europe, ERC has designated these frequencies for UMTS [2]. Even though not specifically decided, frequency bands 1900 – 1920 and 2010 – 2025 MHz are unpaired bands and could be used for TDD applications. Japan has adopted a similar frequency plan for IMT-2000.

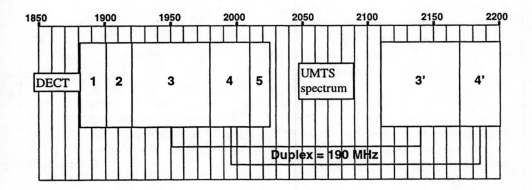

Figure 9.2   Proposed spectrum allocation for UMTS. Frequency band 1 is currently used by DECT.

Table 9.2
Proposed Spectrum Allocation for UMTS

Band	Frequency (MHz)	Bandwidth (MHz)	Allocation
2	1900 – 1920	20	UMTS terrestrial applications (TDD)
3	1920 – 1980	60	UMTS terrestrial applications (FDD)
4	1980 – 2010	30	UMTS satellite component (FDD)
5	2010 – 2025	15	UMTS terrestrial applications (TDD)
3'	2110 – 2170	60	UMTS terrestrial applications (FDD)
4'	2170 – 2200	30	UMTS satellite component (FDD)

9.3.2 Asymmetric Services

Services such as the Internet, multimedia applications, and file transfers often set different capacity requirements for the uplink and downlink. The utilization of a TDD frequency band is not fixed between the uplink and downlink (unlike with FDD) and this flexibility in resource allocation can be used if the air interface design is flexible enough. This is one motivation for considering the TDD extension for FDD-CDMA–based full coverage third generation mobile radio systems.

9.3.3 Reciprocal Channel in Uplink and Downlink

In FDD operation, the uplink and downlink transmissions are separated by a duplex separation. Since fast fading due to multipath propagation depends on the frequency, it is uncorrelated between the uplink and downlink. The FDD transmitter cannot predict the fast fading that will affect its transmission.

In TDD operation, the same frequency is used for both the uplink and the downlink. Based on the received signal, the TDD transmitter is able to know the fast fading of the multipath channel. This assumes that the TDD frame length is shorter than the coherence time of the channel. This assumption holds if TDD mobiles are slowly moving terminals. The reciprocal channel can then be utilized for:

- Open loop power control;
- Spatio-temporal transmission diversity (adaptive antennas for transmission [3], pre-RAKE [4–6]).

With open loop power control, the need for power control signaling is reduced compared to closed loop power control. Closed loop power control signaling also introduces some delay and is subject to errors, which is not the case with open loop power control. In order to have fast enough open loop power control, the TDD frame must be short enough. According to [7], Doppler frequencies up to 80 Hz (43 km/h at 2-GHz carrier frequency) can be supported with a very small degradation if the uplink part of the TDD frame length is 1.5 ms. If the TDD system is intended only for slowly moving terminals, then longer TDD frames could also be used. With open loop power control, the interference situation at the receiver is not known by the transmitter, only the signal level is known. Changes in the interference level must be signaled.

Transmission diversity can be utilized with diversity antennas (space domain diversity) or with pre-RAKE (time domain diversity). In selection diversity combining, the receiver measures the received signal from diversity antennas and selects the best antenna for reception. Antenna diversity techniques are easily applied at the base stations but those receiver techniques are not suited for small handheld terminals. In order to achieve antenna diversity in the downlink, transmission diversity is utilized at the base station. Based on uplink reception, the best antenna can be selected for downlink transmission in TDD.

In FDD-CDMA transmission, a RAKE receiver is used to collect the multipath components and to obtain multipath diversity. The optimal RAKE receiver is a matched filter to the multipath channel. If the transmitter knew the multipath channel, it could apply RAKE in the transmitter (pre-RAKE). Transmission would be such that the multipath channel would act as a matched filter to the transmitted signal. In the receiver no RAKE would be needed (i.e., multipath diversity could be obtained with a one-finger receiver). However, it should be noted that only a little multipath diversity may be available in indoor propagation environments, as shown in Chapter 7 with ITU channel models. Indoor and microcell environments are the most probable application areas for TDD communication. Therefore, antenna diversity transmission will be a more attractive diversity technique for TDD operation than the multipath diversity technique.

If such a TDD proposal, which has different solutions for the uplink and downlink (where uplink uses a single wideband carrier and downlink a multicarrier approach) is applied, channel reciprocity cannot be utilized. The effect of such a structure depends on the environment, but, as calculated in Chapter 7, the coherence bandwidth cannot be guaranteed to be so high that similar fading characteristics could be obtained for both uplink and downlink.

9.4 PROBLEMS WITH TDD-CDMA

This section introduces the problems encountered by TDD-CDMA systems and presents the disadvantages of using such a duplex method.

9.4.1 Interference From TDD Power Pulsing

If fast power control frequency with open loop is desired to support higher mobile speeds, then short TDD frames must be used. The short transmission time in each direction results in the problems listed below:

- Audible interference from pulsed transmission both internally in the terminal and to the other equipment. Generated pulsing frequency in the middle of voice band will cause problems to small size speech terminal design where audio and transmission circuits are relatively close to each other and achieving the needed isolation is costly and requires design considerations. At high power levels this may not be achievable at all.
- Base station synchronization requirements are tight and more overhead must be allocated for guard times and also for power ramps as EMC requirements limit the ramping speed.
- Fast ramping times set tighter requirements to the components (e.g., to the power amplifier).

Lower pulsing frequency, say, 100 Hz (i.e., a TDD frame of 10 ms), results in less audible pulsing but limits the maximum tolerable mobile speeds. In the TDD-CDMA in [8,9], the uplink slot and the downlink slot are both 0.625 ms, resulting in an audible interference at 800 Hz.

9.4.2   Intracell and Intercell Interference Between Uplink and Downlink

In CDMA systems, the SIR may be quite low (e.g., below −15 dB) at carrier bandwidth. After despreading, the SIR is improved by the processing gain. In TDD systems, a transmitter located close to a receiver may block the front end of the receiver, since no RF filter can be used to separate uplink and downlink transmission as in FDD operation. This blocking may happen even if the transmitter and receiver are not operating in the same frequency channel but if they are operating in the same TDD band. In that case, the processing gain at baseband does not help since the signal is already blocked before baseband processing. These interference problems with TDD operation are considered in this section.

Within one TDD-CDMA cell, all users must be synchronized and have the same time division between uplink and downlink in order to avoid interference between uplink and downlink. This time division is based on the average uplink and downlink capacity need in that particular cell. Each user then applies multirate techniques to adapt its uplink and downlink capacity needs to the average need in that cell. The same time division must be applied to all carriers within one base station. If the base station transmits and receives at the same time as adjacent carriers, it would block its own reception.

Asymmetric usage of TDD slots will impact the radio resource in neighboring cells. This scenario is depicted in Figure 9.3 and the resulting signal to adjacent channel interference ratio is calculated in Table 9.3. Intercell interference problems occur in asymmetric TDD-CDMA if the asymmetry is different in adjacent cells even if the base stations are synchronized. MS2 is transmitting at full power at the cell border. Since MS1 has different asymmetric slot allocation than MS2, its downlink slots received at the sensitivity limit are interfered with by MS1, causing blocking. On the other hand, since BS1 can have much higher effective isotropically radiated power (EIRP) than MS2, it will interfere with BS2 receiving MS2. It is difficult to adjust the asymmetry of an individual cell in a network due to interference between adjacent cells. If TDD-CDMA cells are located adjacent to each other, offering a continuous coverage, then synchronization and asymmetry coordination between these cells is required. This ensures that the near-far problems of interference between mobiles in adjacent cells can be controlled. Another scenario, where the previously described blocking effect clearly exists, is if TDD operation were also allowed in the FDD band.

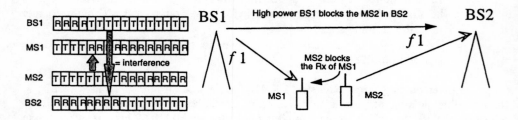

Figure 9.3   TDD interference scenario.

Table 9.3

Adjacent Channel Interference Calculation

BTS transmission power for MS2 in downlink 1W	30 dBm
Received power for MS1	−100 dBm
Adjacent channel attenuation due to irreducible noise floor	50 to 70 dB
Signal to adjacent channel interference ratio	−60 to −80 dB

9.4.3 Inter-Operator Interference with Continuous Coverage

If there are several operators offering the service in the same geographical area in the TDD band, base station synchronization for the different operators is required and asymmetry flexibility between uplink and downlink becomes considerably more difficult. Asymmetric TDD-CDMA systems are therefore not well suited if several operators share the spectrum and the same area. In those cases, TDD-TDMA systems are better off since the signal is concentrated in the time domain. In TDD-TDMA, interference can be averaged with time hopping or avoided with dynamic channel allocation techniques not applicable in CDMA systems. A difficult problem for operation in an unlicensed band is the case where different operators use different multiple access techniques in the same TDD band.

Adjacent channel interference between operators may also cause problems. As the terminals have limited dynamic range and neighboring channel filtering capability, adjacent channel interference in this kind of uncoordinated operation may prove to be very severe. The power differences between the transmission of the desired base station and the interfering mobile transmitting at the same time as adjacent carriers can block the receiver terminal's A/D converters and can also cause problems in the RF components.

Assuming the 15-MHz spectrum for uncoordinated TDD, a chip rate of 4.096 Mcps can then accommodate only three carriers. If symmetric 2.0 Mbps should be supported with a TDD network, a double chip rate of 8.192 Mcps would be required.

9.4.4 Synchronization of Base Stations

Intercell and inter-operator interference problems are present in a TDD-CDMA system if the base stations are not synchronized. Synchronization is therefore desirable for TDD-CDMA. Synchronization accuracy must be at the symbol level but not at the chip level. It can be achieved with, for example, GPS receivers at the base station or by distributing a common clock with extra cabling. These methods increase the cost of the infrastructure.

9.5 TDD-CDMA AIR INTERFACE DESIGN

The specific aspects of a TDD-CDMA system air interface design are presented in this section, with particular attention to frame design.

9.5.1 TDD Frame Length

Since within coherence time the channel for uplink and downlink is the same, there is no need for closed loop power control. However, the frame structure limits the maximum command rate for the power control. The mobile station measures the received power during the downlink transmission and determines · the uplink transmission power, which is applied during the uplink transmission. Thus, if the frame length is 10 ms, the command rate is 100 Hz. For power control, no feedback can be provided during the uplink part: thus, it is desirable that the channel is constant during the uplink part. If symmetric division between uplink and downlink is assumed, the uplink part is 5 ms in this example. Coherence time depends on the Doppler frequency and is shown in Table 9.4 for 2-GHz carrier frequency. Coherence time should be clearly longer than the uplink part of the TDD frame for power control to perform adequately. With a 10 ms TDD frame (i.e., 5 ms uplink part in symmetric allocation), open loop power control is effective for mobile speeds of up to about 10 to 20 km/h.

Table 9.4
Coherence Times for Different Mobile Speed

Mobile station speed (km/h)	Doppler frequency at 2 GHz (Hz)	Coherence time (ms)
5	9	108
10	18	54
20	37	27
30	56	18
40	74	14
50	93	11
80	148	6.8
100	185	5.4

9.5.2 Hardware Requirements

Since the channel is reciprocal, it is possible to simplify the receiver while still maintaining effective diversity techniques against fading. In TDD-CDMA, pre-RAKE could be applied either to downlink or to uplink. If pre-RAKE is utilized in uplink, simple TDD base stations could be built. This option is attractive if all the terminals support both FDD and TDD operation, and they must therefore have a RAKE receiver for FDD operation. In that case, TDD base stations could be made very simple when no RAKE is needed. Also, multiuser detection algorithms at the base station would be simpler if there is only one multipath component per user to be tracked. If TDD-only terminals are used, then pre-RAKE could be applied to the base station to reduce the complexity of TDD terminals. Pre-RAKE cannot be applied to both transmission directions at the same time.

Downlink performance could be improved with transmission antenna diversity at the base station. The downlink transmission antenna is determined based on the uplink reception. Also, downlink beamforming with adaptive antennas is easier to utilize in TDD than in FDD, but in typical TDD environments (like indoor), users may be difficult to separate by the direction of arrival.

In a dual-mode FDD/TDD terminal, additional complexity is concentrated in the RF filtering, compared to a FDD only terminal. Modifications at the baseband parts are minor compared to changes in the RF parts.

9.6 TDD-CDMA EXTENSION TO FDD-CDMA SYSTEM

A TDD-CDMA extension to a FDD-CDMA based third generation cellular system, as presented in [10], is considered in this section. A stand-alone TDD-CDMA system providing full coverage is not considered because of possible interference problems in TDD operation. A system scenario with both TDD and FDD is presented, where frequency bands allocated to TDD operation can be effectively utilized in complementing the services and increasing the capacity provided by a wide area coverage FDD system.

It is assumed that the FDD-CDMA system is used to provide both wide area speech and data services. The FDD-CDMA operator offers public access to the network for all IMT-2000 users. The TDD-CDMA with limited coverage is used as an extension to the FDD-CDMA. The TDD system typically covers hotspots with high capacity requirements, such as office buildings, airports, and hotels. These TDD systems are operated either by the same public operator as the FDD system or by private operators such as companies in their own premises or service providers providing the access in office buildings shared by several companies. The TDD system is not intended to offer continuous coverage, and systems in different locations can operate independently of each other. The system scenario is shown in Figure 9.4.

9.6.1 TDD-CDMA System Description

The TDD-CDMA extension is based on the FDD-CDMA system. When the parameters of the TDD extension (bandwidth, frame length, chip rate) are aligned with the FDD, the implementation of FDD-TDD terminal is easier.

It is assumed here that CDMA terminals support both FDD and TDD operation. By employing pre-RAKE in mobile stations, the complexity requirements of TDD-CDMA base stations are reduced (note that mobile stations already have RAKE receiver due to the FDD operation). The simplified architecture for the TDD-CDMA mobile terminal is shown in Figure 9.5. During the downlink (three switches shown in Figure 9.5 are up), the received signal is despread and this correlation process is followed by a RAKE filter. Pilot symbols within the TDD frame are used to form a channel estimate. During the uplink (switches down), the channel estimate obtained from the downlink is used in the pre-RAKE filter, and the channel effectively forms a matched filter to the transmitted signal.

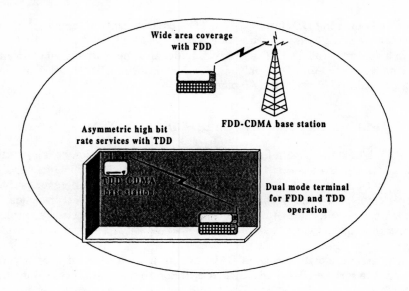

Figure 9.4   TDD-CDMA extension to full coverage FDD-CDMA system.

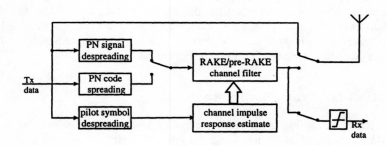

Figure 9.5   Simplified block diagram of TDD-CDMA mobile terminal architecture.

The capacity of the TDD system is divided asymmetrically between the uplink and downlink to support multimedia services. This asymmetry can be adjusted to match the uplink and downlink requirements for the TDD-CDMA cell. All downlink channels will be synchronized to obtain good orthogonality between users. The TDD-CDMA asymmetric frame structure is shown in Figure 9.6. The simulation results in this chapter are obtained with a frame length of 20.5 ms, and it comprises an uplink burst, two guard time bits, and a downlink burst. The guard times in a TDD system must accommodate the maximum round trip delay time in the cell as well as the hardware delays.

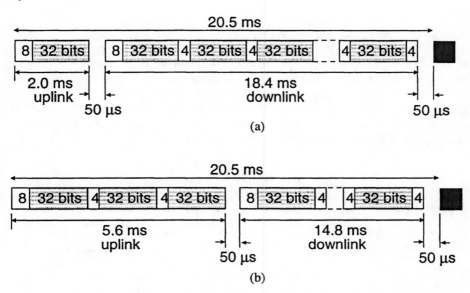

Figure 9.6   TDD-CDMA asymmetric frame in pre-RAKE simulation: (a) uplink/downlink load = 1/10 and (b) uplink/downlink load = 3/8.

A frame which has a downlink capacity of 10 times that of the uplink is shown in Figure 9.6(a) and corresponds to the WWW browsing model described in Table 9.5. The TDD-CDMA air interface is flexible and will allow extra timeslots to be added to the uplink at the expense of downlink timeslots, as shown by Figure 9.6(b).

Different TDD asymmetries between uplink and downlink in near or adjacent cells can lead to high intercell interference levels. To combat this, such cells must either have the same asymmetry and be synchronized, or be synchronized with certain timeslots being replaced by guard times (which is effectively a loss of capacity). This is based on the assumption that the cell reuse efficiency would be 100%.

The beginning of both the uplink and downlink for each frame contains an 8-bit header. Four of these bits are pilot bits, and four are for control. Higher level control can be provided by arranging the frames into superframes. The frame is divided into eleven 32 bit timeslots, each being preceded by four pilot bits. On the synchronous downlink, the pilot bits are transmitted at 100% of the transmitter power and provide training for RAKE (and pre-RAKE); they also aid the synchronization process. Pilot overhead is equivalent to 12.5% of total capacity. One of the downlink control bits can be used for closed loop power control. Only a single bit per frame is necessary to provide fine adjustments. The fast open loop power control is achieved due to the correlated nature of the uplink and downlink impulse responses, but the maximum power control frequency is limited to 50 Hz by the TDD frame rate. Thus, fast moving terminals are not supported by the TDD-CDMA system since they will increase the near-far effect.

9.6.2   WWW Browsing Session Performance with TDD Extension

The simulation parameters for the TDD-CDMA extension are given in Table 9.5. Simulations have assumed perfect chip synchronization. The frame structure shown in Figure 9.6(a). is used in the simulations. The WWW model gives rise to a nonuniform document inter-arrival rate, and so, transmission buffers must be used. We assume that the document arrival at the transmitter takes zero time, but that its transmission is limited by the data rate provided by the communication link.

Table 9.5
TDD-CDMA System Simulation Parameters

Max. symbol rate	20 Kbps
Nominal uplink bit rate	1.56 Kbps
Downlink bit rate/channel	15.61 Kbps
Chip rate	5.12 Mcps
Code length	256
Modulation	BPSK
Channel coding	None
Doppler rate	10 Hz classic
Multipaths	4 path (equal powers)
Antenna configuration	Single antenna
RAKE/pre-RAKE	Implemented without pre-RAKE prediction
Multiuser detection	Not implemented
Traffic parameters	
Mean downlink WWW document arrival rate	60 seconds (Poisson distribution)
Mean document size	30 Kbyte (Rayleigh distribution)

Figure 9.7 shows theoretical results based on the downlink for an uncoded system with a perfect channel estimate applied to a four-tap RAKE filter. The number of users for this system is not the same as the number of channels utilized. The average number of channels required per user for the WWW model is about 1/4 so there is only one channel required for about every four users. The "user SNR" refers to the mean received SNR (pre-correlation) seen by each user, excluding any multiuser interference. The processing gain of code length of 256 is equal to 24.1 dB.

Figure 9.8 shows the simulated error rates for a different number of users. Here the RAKE filter uses the channel impulse response estimate obtained by averaging the four pilot bits provided for each of the downlink timeslots. No other filtering or signal processing was applied to this signal and the performance could be further improved. The last channel estimate provided by the downlink is used for the pre-RAKE filter since this will be highly correlated with the uplink impulse response at low Doppler spreads. No RAKE filter is implemented in the base station, and only a single antenna is considered in these simulations.

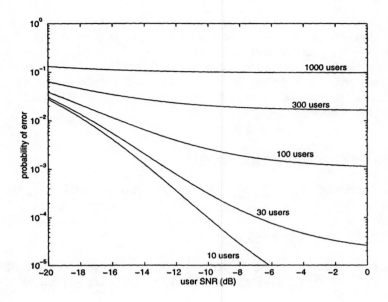

Figure 9.7   Error probability for downlink with perfect channel estimate and uniform channel loading. Processing gain is 24.1 dB.

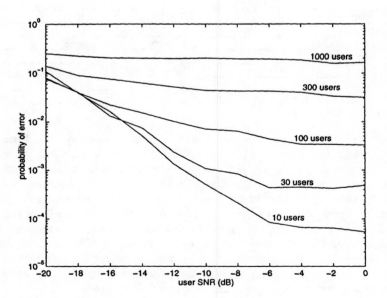

Figure 9.8   Simulated downlink error probability for the WWW service model using channel estimates derived from the pilot bits. Processing gain is 24.1 dB.

Clearly, the simulated results are poorer than the theoretical ones, which represent a lower bound. There are two reasons for degradation. First, theoretical results assume a constant load for each user, whereas the load in the simulated results is governed by the Poisson document inter-arrival process. This will cause some peaks in the load and corresponding peaks in error rate. Secondly, theoretical results assumed perfect channel estimation, whereas the simulated results must derive the channel estimate from the pilot symbols.

9.7 WIDEBAND TDD-CDMA SYSTEMS

In this section Cylink TDD-CDMA and third generation TDD-CDMA proposals are presented.

9.7.1 Cylink S-CDMA TDD for PCS

One of the first concepts to use TDD operation with CDMA is Cylink's S-CDMA [11,12] intended for the US ISM band at 902-928 MHz. The concept has the parameters listed in Table 9.6.

Table 9.6
Cylink S-CDMA TDD for PCS system concept parameters

Modulation	DS/QPSK
Voice coding	32 Kbps ADPCM
Data transmission	4.8 Kbps signaling channel
Processing gain	32
Synchronization	Network, both downlink and uplink
Users per carrier	Max 16
QPSK Symbol burst rate	1.536 Msymbol/s

Source: [13].

This proposal contains the basic features of a cellular system. Base stations transmit broadcast information as part of their TDD burst and thus allow the mobiles to acquire frequency and time synchronization and receive general base station information. The concept uses sequences optimized in terms of autocorrelation, and all users use the same sequence within the cell. For this kind of operation, in addition to power control, timing control by the base station towards the mobiles is also required to maintain the achievable (near) orthogonality between users. For both network and user, synchronization accuracy of the order of one chip interval (651 ns) is required. The operation environment is considered to be indoor and indoor-to-outdoor with velocity below vehicle speeds (pedestrian environment), where delay profiles remain relatively small and do not change very rapidly due to low mobile speeds.

Base stations use preamble (pilot) data in the beginning of the common broadcast part to allow synchronization and acquisition of the signal. The use of base station transmission diversity is also included.

9.7.2 Third Generation Wideband CDMA TDD Schemes

A 5-MHz TDD-CDMA cellular system with synchronized base stations has been proposed in [8,9]. This system has also been adopted in ARIB as part of the wideband CDMA proposal and has also been the basis for the WCDMA TDD mode in ETSI and cdma2000 TDD mode in TR45.5 in the United States. The ETSI TDD scheme has also been influenced by the TD-CDMA scheme described in Chapter 1. We expect further harmonization between the ARIB and ETSI wideband CDMA TDD modes in the future.

The system parameters are described in Table 9.7. Time division between uplink and downlink is proposed to be fixed (i.e., asymmetric capacity allocation is not supported). Diversity scheme in base stations is used both in uplink and downlink by utilizing the reciprocal channel. Because TDD uses only one radio frequency band and transmission diversity at the base station can be employed, space diversity is not that important at the mobile station and thus the mobile station can be made smaller. TDD-CDMA systems can also perform effective transmission power control with only the use of open loop control. A single antenna and no duplexer are the merits of TDD-CDMA for terminal size and battery life. This system is especially effective in an environment such as microcellular systems or indoor office systems, where the spreading bandwidth is narrower than the coherent bandwidth, and path diversity gain by itself is small.

Table 9.7
System Parameters of Wideband TDD-CDMA

Access method	DS-CDMA/TDD
Minimum frequency band	5 (10/ 20) MHz
Chip rate	4.096 (8.192/16.384) Mcps
Frame length	10 ms = 0.625 ms × 8 slots × 2 (TDD)
Slot length	0.625 ms
Transmission power control	Open loop control + slow compensation control by control channel
Diversity	Mobile station: RAKE Base station: transmission/reception space diversity + RAKE
Modulation and Demodulation	Data: QPSK/pilot symbol coherent detection Spreading: BPSK
Error correction	Inner: K = 7 R = 1/2 convolutional code + Viterbi soft decoding Outer: R = 4/5 Reed-Solomon code
Inter-BS synchronization	Synchronization
Maximum user information transmission rate	144 Kbps (5 MHz bandwidth, vehicular) 2 Mbps (20 MHz bandwidth, indoor office)

For wideband TDD-CDMA, experimental equipment has been developed in order to evaluate such fundamental characteristics as reception diversity at base station, transmission diversity at base station, and open loop power control. The specifications of the TDD-CDMA testbed are shown in Table 9.8. Spreading bandwidth is 5 MHz for chip rate of 4.096 Mcps. The maximum number of space diversity branches is four at base station, and the maximum number of RAKE fingers is six. This testbed consists of one base station and two mobile stations (8-Kbps transmission user and 144-Kbps transmission user). Simultaneous communication of 8 Kbps transmission user and 144 Kbps transmission user is possible.

Table 9.8
Main Testbed Specifications

Chip rate	4.096 Mcps (5-MHz bandwidth)
Power control	Open loop control + slow compensation control by control channel Closed loop control
Diversity	Mobile station: 1 branch antenna and 6 fingers RAKE combining Base station: 1 to 4 branch antenna diversity and 6-finger RAKE combining
FEC	Inner: K = 7, R = 1/2 convolutional code Viterbi soft decoding 10/40/80/160/320 ms interleaving Outer: R = 9/10 Reed-Solomon code 40/80/160/320 ms interleaving

9.7.2.1   IMT-2000 TDD Proposals

Four wideband CDMA TDD schemes have been submitted to ITU RTT evaluation: ARIB WCDMA TDD, ETSI WCDMA TDD, cdma2000 TDD, and a time division synchronous code division multiple access (TD-SCDMA) proposal from China. The ARIB WCDMA TDD scheme is very similar to the above described wideband CDMA TDD scheme.

ETSI WCDMA TDD: In the ETSI wideband CDMA TDD mode, a wideband CDMA carrier with chip rate of 4.096 Mcps is divided between uplink and downlink in time. The frame length is as in the FDD mode, currently 10 ms, and the number of time slots per frame is 16. Figure 9.9 shows an example TDD frame structure with one switching point for uplink / downlink separation within a frame. Another option under study is to use multiple switching points. As shown in Figure 9.9, a burst consists of three parts (data block - midamble - data block). The TDD mode uses QPSK data modulation and currently employs fixed spreading factor. In addition, variable spreading factor is under study.

Cdma2000 TDD: The frame length of the cdma2000 TDD mode is 20 ms and it is divided into sixteen 1.25-ms slots. Every other slot is for TX and every other for RX. Similar to the FDD mode, the TDD mode has multicarrier and direct spread option.

TD-SCDMA: The TD-SCDMA concept has a chip rate of 1.1136 Mcps and a carrier spacing of 1.2 MHz. The frame length is 5 ms and each frame is divided in 8 time slots (4 for the uplink and 4 for the downlink). One time slot can accommodate 16 CDMA codes.

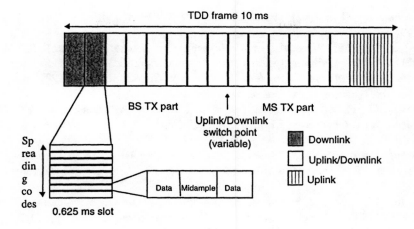

Figure 9.9   An example TDD frame structure.

REFERENCES

[1] Klein, A., R. Pirhonen, J. Sköld, and R. Suoranta, "FRAMES Multiple Access Mode 1 – Wideband TDMA with and without Spreading," *Proceedings of Personal, Indoor and Mobile Radio Communications (PIMRC'97)*, Vol. 1, September 1997, Helsinki, Finland, pp. 37–41.

[2] European Radiocommunications Committee ERC Decision of 30 June 1997 on the frequency bands for the introduction of the Universal Mobile Telecommunications System (UMTS), ERC/DEC/(97)07, can be found from ERO website http://www.ero.dk/.

[3] Litva, J., T. Lo, *Digital Beamforming in Wireless Communications*, Boston: Artech House, 1996.

[4] Povey, G. , "Frequency and Time Division Duplex Techniques for CDMA Cellular Radio," *Proceedings of 3rd International Symposium on Spread Spectrum Techniques & Applications (ISSSTA'94)*, Oulu, Finland, 1994, pp. 309–313.

[5] Povey, G., "Capacity of a Cellular Time Division Duplex CDMA System," *IEE Proc.-Commun.*, Vol. 141, No. 5, October 1994, pp. 351–356.

[6] Esmailzadeh, R., and M. Nakagawa, "Pre-RAKE Diversity Combination for Direct Sequence Spread Spectrum Mobile Communications Systems," *IEICE Transactions on Communications*, Vol. E76-B, No. 8, August 1993, pp. 1008–1015.

[7] Esmailzadeh, R., M. Nakagawa, and E. Sourour, "Time-Division Duplex CDMA Communications," *IEEE Personal Communications*, April 1997, pp. 51–56.

[8] Hayashi, M., K. Miya, O. Kato, and K. Homma, "CDMA/TDD Cellular Systems utilizing a Base-Station Based Diversity Scheme," *Proceedings of 45th Vehicular*

Technology Conference (VTC'95), July 1995, Chicago, Illinois, USA, Vol.2, pp. 799–803.

[9] Miya, K., M. Watanabe, M. Hayashi, T. Kitade, O. Kato, K. Homma, "CDMA/TDD Cellular Systems for the 3rd Generation Mobile Communication," *Proceedings of 47th Vehicular Technology Conference (VTC'97)*, May 1997, Phoenix, Arizona, USA, Vol. 2, pp. 820–824.

[10] Povey, G., H. Holma, and A. Toskala, "TDD-CDMA Extension to FDD-CDMA Based Third Generation Cellular System," *Proceedings of ICUPC'97*, San Diego, California, USA, October 1997, pp. 813–817.

[11] Omura, J., and P. Yang, "Spread Spectrum S-CDMA for Personal Communication Services," *Proceedings of Milcom'92*, October 1992, San Diego, California, USA, pp. 269–273.

[12] Omura, J., "Spread Spectrum Radios For Personal Communication Services," IEEE ISSSTA'92, Yokohama, Japan, November/December 1992. pp. 207–211.

[13] Simon, M., J. Omura, R. Scholtz, and B. Levitt, *Spread Spectrum Communications Handbook*, New York: McGraw-Hill, 1994.

Chapter 10

IMPLEMENTATION ASPECTS

10.1 INTRODUCTION

This chapter concentrates on the implementation aspects of a wideband CDMA transceiver. The goal is to help the reader understand the basic analog and digital structure of a wideband CDMA terminal and get an idea of the problems encountered in real implementation.

Section 10.2 highlights different optimization criteria (power consumption, cost, and size) that should be taken into account while designing a CDMA transceiver. The task is certainly not easy because the number of degrees of freedom is large. The main concern is on the mobile terminal side, but in the future power consumption, cost, and size will become important also for base stations.

In Section 10.3, the transceiver modularity concept is discussed in conjunction with wideband CDMA systems. A wide set of defined third generation system services calls for terminals capable of adapting to different data rates and quality of services. In practice this makes the physical terminal design much more complicated because many functions have multiple sets of operating parameters.

Baseband signal processing is investigated in Section 10.4. The emphasis is on the basic building blocks, but attention is also paid to more advanced techniques like interference cancellation. First, the different processing units are considered from the functional architecture point of view. This is followed by a discussion of how to make the mapping from the functional description to a physical implementation. The problem is not, however, unambiguous, making an optimal solution difficult to find. In all cases, digital application specific integrated circuit (ASIC) or digital signal processing (DSP) - based implementation is assumed although some baseband processing like matched filter could be performed in an analog domain. At the end of the section, different control loops for AGC, AFC, and power control are highlighted.

Section 10.5 presents well-known RF architectures suitable for a wideband CDMA transceiver. The RF section is considered to incorporate everything from radio

279

frequency signal to baseband, independent of analog or digital implementation. Complexity of the front-end is increased compared to second generation systems mainly because of the need for higher linearity, better selectivity, and larger dynamic range requirements. Most radio frequency processing is still done in the analog domain, while in the future the digital border is moving towards the antenna.

Section 10.6 describes the concept of software configurable radio and section 10.7 illustrates an example configuration of a wideband CDMA mobile terminal architecture, integrating the different parts.

Section 10.8 discusses implementation of multimode terminals. The different implementation requirements between spread and nonspread, between slotted and continuous, between circuit and packet switched, and finally between FDD- and TDD-based systems are discussed.

10.2   OPTIMIZATION CRITERIA

In the design of a wideband CDMA terminal and base station, critical factors to be optimized are power consumption, cost, and size. Each of these can be evaluated by qualitative criteria such as input power requirements, number of components, number of instructions, or ASIC gate primitives. Figure 10.1 illustrates a generic wideband CDMA transceiver partitioned into RF, IF, and baseband sections. Traditionally, RF and IF sections have been realized by analog technology, and baseband by digital technology. However, the interface between analog and digital sections depends on the selected receiver and transmitter architectures, as discussed in Section 10.5.

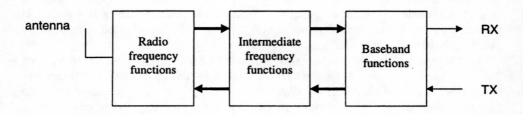

Figure 10.1   Block diagram of a CDMA transceiver.

10.2.1 Power Consumption, Cost, and Size

Due to the radical evolution in the digital signal processing field in terms of DSP processors and new ASIC technologies (see Table 10.1), the analog RF section is becoming the major power consumer in normal speech terminals. On the other hand, more complex baseband functions are being required (e.g., multiuser detector), and if the digital-analog border shifts towards antenna (see Section 10.5.2), the baseband may still form a significant part of the power consumption.

Table 10.1
Development of DSP and ASIC Technologies

	1992	1994	1996	1998	2000
Gate density (gates/mm^2)	2000	5000	15,000	30,000	50,000
Processing power (MIPS)	20	50	80	120	200
Relative power consumption (1/MHz)	100	30	10	3	1
Relative power consumption (max MIPS)	100	75	40	18	10

Source: [1].

On the RF side, the dominating current consumers are the power amplifier and frequency synthesizers. The former is seen as a critical component in the third generation systems due to high linearity requirements likely causing a loss of efficiency. It should, however, be noted that due to smaller cell sizes (pico/microcells), output power is reduced, and hence, the overall power consumption can be kept acceptable. The power consumption of a synthesizer is directly related to the activity factor and is thus larger for systems such as CDMA, which transmit continuously. In addition, two synthesizers are needed in CDMA terminals since transmission and reception take place simultaneously in contrast to TDMA terminals, which need only one synthesizer.

The transceiver cost factor depends quite evenly on every required component. So far, the division between analog RF and digital baseband sections has been approximately fifty-fifty. However, the trend leans toward the analog side because the integrated digital processing side is shrinking into a single chip.

The actual analog and digital data processing functions (encoding/decoding, filtering, mixing, equalization) are not very critical in terms of the total terminal size. This may have been the case in the past, but today the dominant components are the battery and the user/external interfaces. On the data path side, some RF section components such as the duplex filter and possibly the antenna may, however, occupy a large amount of space.

The development of the first third generation mobile terminals is a difficult task. They should have all the features of the second generation systems and include new features related to higher bit rate services. The fact that consumers also expect to be able to connect to the existing systems calls for a dual/multimode engine.

10.2.2 Evaluation Methods

There are a few quantitative parameters that can be used to evaluate the transceiver power consumption, size, and cost. This section explains what should be taken into account when assessing analog and digital section complexities. Radio frequency parts are seen as analog, and baseband functions as digital. IF processing lies somewhere in between.

On the analog RF side, power consumption can be assessed by estimating the average input power requirements in both talk and idle modes. The power consumption is a function of user data rate, activity factor, required effective radiated power, and transmission spectral mask (linearity requirements). Emphasis is on deriving the power amplifier efficiency in different situations. The activity factor of a synthesizer has a large impact on the power consumption. Therefore, effective power-down/up procedures should be used to turn off the synthesizer whenever possible.

Cost and size of the RF section depends strongly on the number of individual components. The major problems relate to isolation requirements and integration of different semiconductor processes, which prevent a single chip RF front-end implementation. Increased integration, especially on the intermediate frequency side, has already been achieved (possibly joined with some baseband blocks), but still there are a few discrete components such as channel selection and duplex filters that must be implemented externally.

Evaluation on the digital processing side is a much more straightforward task because an analytical approach can be followed. Each function can be divided into instruction and/or gate primitive operations, which can further be mapped to real DSP processors and ASIC processes to give direct estimates for power consumption, cost, and size. The most problematic task here is to compare DSP- and ASIC-based implementations. A good method for doing this is to take the word length requirements into account when counting the MIPS requirements. This leads to a concept called normalized MIPS (nMIPS), which states the number of 1-bit operations needed to execute within a second. The example below compares the complexities of a RAKE receiver correlator and complex phase rotation primitives. The former is a clear ASIC-oriented function, while the latter is more ideal for a DSP processor.

Example. The despreader is for a QPSK chip modulated signal with an input sample rate of 4.096 Msps and input word length of 4 bits. The spreading ratio is 64. The same spreading code is used in both I and Q branches, so only two correlations are needed. The phase rotator takes a complex 10-bit input vector and multiplies that with another 10-bit vector. The operation rate is 64 Ksps. In this case, multiplication is assumed to take (Wlin · WLin) and addition (Wlin + WLin) primitive operations. Thus,

$$\text{nMIPS(despreader)} = 2 \cdot F_s \cdot [WLin \cdot WLc + ((WLin \cdot WLc) + WLin2)] = 148 \text{ nMIPS}$$

where F_s = sampling frequency = 4.096 Msps, WLin = input sample word length = 4, WLc = reference code word length = 1, and WLin2 = integrator second input word length = 10.

$$nMIPS(rotator) = Fs \cdot [4 \cdot WLin \cdot WLin + 2 \cdot (WLin + WLin)] = 28 \text{ nMIPS}$$

where Fs = sampling frequency = 64 Ksps and WLin = input sample word length = 10.

10.3 MODULARITY CONSIDERATIONS

Third generation wideband CDMA terminals will provide a set of different services varying from a low rate speech service to a high rate, high quality service for data. When designing RF and baseband architectures this should be taken into account. The more adaptable the architecture, the better the available processing resources can be utilized to provide this variety of operation modes.

In wideband CDMA systems, the higher data rates can be obtained by decreasing the spreading factor or by introducing more parallel code channels (see Section 5.8). The latter option, especially in the case of only a few parallel channels, results in a high peak-to-average power requirement and thus to high backoff. Therefore, it is not very good from the transmitter power amplifier point of view. It may be best to use parallel channels in the downlink to obtain good modularity and variable spreading factor in the uplink to avoid parallel transmission in the mobile station.

The baseband section chip rate elements have a fairly fixed complexity as a function of data rate. This is because the data rate is mainly defined by the correlator dumping period, not the chip/sample rate. Multiple code channels, however, create extra hardware requirements, but still many blocks (like receiver front-end, multipath delay estimation, and RAKE finger allocation, as well as complex channel estimation units) may not need to be duplicated for different codes. The critical parts from the complexity point of view are the transmitter pulse shaping filter, additional RAKE receiver fingers, and code generators. The complexity of a pulse shaping filter depends heavily on the input word length, and in the case of several parallel code channels plus high adjacent channel attenuation requirements (long filter), implementation becomes difficult. Each parallel code channel used requires code generator(s) for spreading and despreading. With short M-sequences (simple shift registers), the complexity increase can be tolerated, while with the long Gold and Kasami codes, which require several shift registers, the silicon area consumption will grow significantly. Although a RAKE finger itself is a reasonably simple device, the total receiver complexity increases because each multipath component requires as many despreaders as there are parallel code channels.

The main effect of increasing user bit rate is, however, seen on the narrowband side because processing requirements at the symbol rate are closely tied to the user data rate. This mainly affects the channel encoding, decoding, and interleaving functions. The key aspect here is how to make the architecture as scalable as on the wideband side.

The required scalability can be achieved in two ways. Extra processing power is achieved by installing several processing units operating in parallel; or existing unit throughputs are increased by using a higher clock rate. The former seems justified from the ASIC technology evolution point of view because the available silicon area is constantly expanding. Clock rate increase can be used to save silicon but on the other hand, may cause hotspots. However, it is not desirable to have high rate system clocks at the mobile station.

Another thing to consider is the specified set of different coding schemes for several bearer services. It is obvious that data services require better BER than speech services. This must be taken into account while designing the encoder/decoder architecture in order to provide maximum resource utilization. The more common lower level processing primitives that can be found, the higher the resource sharing that can be provided. This is easier with encoders, while decoders tend to have more individual constructions. A good example is to use the same physical shift register for convolutional and Reed-Solomon encoders. Software oriented implementation would be a good solution here, but for high data rates the processing power requirements may be too high and too expensive.

10.4   BASEBAND SECTION

Baseband processing implements all the data path functions operating on a signal whose spectrum center is located at zero frequency. In wideband CDMA, the baseband section can be divided into bit/symbol and chip/sample domains, each having a different operation rate. This section describes the basic baseband functions (data path/control) required for a W-CDMA terminal implementation. In all cases, the physical implementation is assumed to take place in the digital domain (not counting the converters).

10.4.1 Baseband Receiver

Figure 10.2 illustrates the generic block of a direct sequence CDMA receiver. After analog-to-digital (A/D) conversion and filtering, the wideband I/Q sample stream is fed into a correlator bank that performs signal transform (despreading) into the narrowband domain where the multipath diversity combiner collects the channel energy from different RAKE fingers. Channel delay profile estimation is required to resolve the multipaths and to set the RAKE fingers to track a specific code phase. After combining, the bits are deinterleaved and decoded to produce transmitted information bit stream. Multiuser detection (MUD) after the despreading and before the multipath combining can be used to increase the system performance. MUD is primarily base station-oriented function and can be implemented in the wideband domain as well.

10.4.1.1   Analog-Digital Conversion

In a wideband CDMA system, because of the wideband signal, a high sampling rate A/D converter is required. Power consumption and cost are not, however, radically increased because the processing gain obtained by spreading allows usage of lower dynamic range converters than in narrowband systems. The direct sequence signal as such is robust against ADC nonlinearities. For a brief discussion of the aberrations, refer to [2].

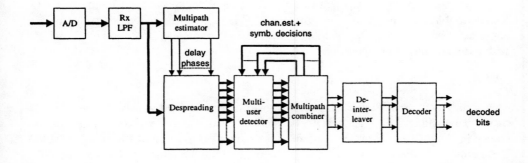

Figure 10.2 Generic receiver block diagram with optional interference cancellation stage.

As stated by Nyquist, the sampling rate must be at least twice the input signal bandwidth. Due to practical reasons the sampling rate should, however, be a multiple of the chip rate (if, for example, 4.096 Mchip/s takes 5-MHz band, it is better to use a rate of 4 × 4.096 Msamples/s than 2 × 5 Msamples/s). Furthermore, a higher oversampling rate means that a better multipath resolution can be obtained without utilizing interpolation techniques, which may be quite complex from the RAKE receiver point of view.

The required dynamic range of an A/D converter should be selected in such a way that the converter does not saturate with proper AGC setting and still quantization noise does not contribute to the overall link performance. The CDMA signal differs in this respect from nonspread systems because the actual information bearing signal is buried in noise. For example, with a spreading ratio of 128 and symbol-energy-to-noise density ratio (E_s/N_0) of 8 dB, the carrier-to-interference ratio (CIR) is only −13 dB. In practice, 4-to 6-bit converters have been noted to provide adequate performance [3].

10.4.1.2   Baseband Signal Filtering

An optimal receiver filter is matched to the transmitter pulse shaping filter to provide maximum SNR. A purely digital implementation is too complex; a better choice is to find a compromise between analog IF/BB and digital filtering. In addition, the latter should be designed in such a way that it is able to compensate for the possible signal distortion caused by the transmitter/receiver analog filters and other data path non-linearities.

The main problem with the receiver filter compared to the transmission filter is the higher input dynamic range (4 to 6 bits) and sampling rate requirement. This means that long filters cannot be provided practically, which is the case especially with the mobile terminal. A shorter nonoptimal filter can, however, be used because it deteriorates the signal quality only slightly. For filter implementation examples in the IS-95 system, refer to [4].

It should still be noted that filtering must be performed to both I and Q signals, which doubles the processing requirements. One option for reducing the hardware complexity is to use only high throughput filters with two delay lines, and clocking this with twice the sampling rate.

10.4.1.3   Signal Despreading

The purpose of signal despreading is to transform the received wideband signal into a narrowband signal. This can easily be implemented using the despreading circuits shown in Figure 10.3. (See Section 5.6.2 for discussion on different spreading circuits.) The number of required correlators depends on the modulation method, possible multi-code reception, and the number of multipath taps to be despread per code channel. The dumping rate of the correlators is set according to the spreading ratio. The wideband clock may be set equal to the chip clock given that the delay phase is correct. This can be achieved by utilizing an interpolator before each RAKE finger, or a higher accuracy clock on the multipath estimator side. In the latter case, the despreader wideband clock is divided from the higher rate clock, and the phase of the chip rate clock must be adjustable.

It should be noted that by using BPSK spreading, the number of correlators can be halved. The case is the same with QPSK spreading if the cross-correlations between the I and Q branch codes are zero. Still another way of reducing the number of correlators is to use the same spreading code for both I and Q signals. This approach is, however, vulnerable to phase errors and requires more accurate channel estimation.

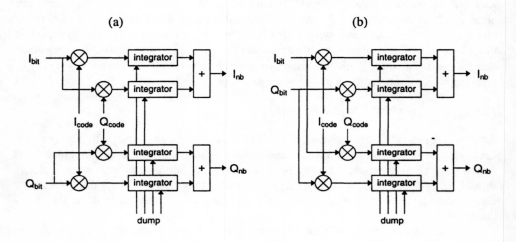

Figure 10.3   Despreader architectures for QPSK chip modulated signal: (a) dual-channel despreading and (b) complex despreading.

Especially in the downlink, flexible data rates can be implemented by using multiple parallel code channels. Extra correlators are needed for each new code, and if the number of codes increases to several tens, mobile terminal implementation becomes cumbersome. Phase and delay estimation can, however, be shared since each code channel passes through the same multipath, and hence, some alleviation can be obtained in terms of implementation.

Higher data rates implemented with the variable spreading factor (VSF) scheme result in a receiver implementation whose hardware complexity is fairly constant as a function of the data rate. Only despreaders for the multipath taps of a single code channel are needed, and the different data rates are implemented by changing the correlators' dumping periods. The integration unit must be designed according to the largest spreading ratio, which defines the worst case register word length.

An important aspect is the number of bits required for the wideband signal presentation. The implementation complexity and cost of the receiver depend heavily on the required wideband signal dynamic range. The input signal consists of additive white Gaussian noise mainly due to the receiver front-end and different user signals propagated through multipath channels. Receiver gain control is used to adjust the input signal level suitable for the A/D-converter dynamic range. By noting that quantization noise depends on the desired input signal level with respect to the converter input range, an estimate for the needed word length can be obtained. Figure 10.4 shows signal-to-noise degradation with respect to the effective quantization word length.

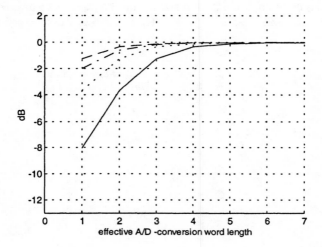

Figure 10.4   SNR degradation as a function of quantization word length. The four curves are with different Gaussian noise variances δ_n into ± 1 A/D input (dashed: $\delta_n = 0.5$, dashdot: $\delta_n = 0.375$, dotted: $\delta_n = 0.25$, solid: $\delta_n = 0.125$).

288

10.4.1.4   Channel Estimation

The channel estimation block estimates multipath delays and complex tap coefficients (phase and amplitude). There are numerous different algorithms and approaches to solve the estimation problem. Some of the more frequently used ones are discussed below, emphasizing the actual implementation.

The main target of the multipath delay estimation unit is to provide accurate enough delay estimates at a fast enough rate. Traditionally, the implementation has utilized a sliding correlator that calculates one delay tap power per dumping period (see, for example, [5-6]). The more parallel correlators available, the faster the multipath window can be scanned. A good criteria for finding the "real" taps and rejecting the false ones is to employ so-called two-dwell approach [7]. The procedure has two parts: first, the candidates are found, then a verification is performed. A more exotic approach is to use the subspace concept in which the received signal is projected against the desired user code (vector spanning a subspace) in different delay phases [8]. Should there exist a multipath tap, the projection value will exceed a threshold.

The accuracy of the delay estimate is generally defined by the sampling rate. If, however, interpolation techniques are used, better accuracy can be obtained at the expense of increased complexity. Figure 10.5 shows an example of SNR degradation due to an erroneous delay estimate. It is interesting to note that the filtered input signal is more robust against delay errors that a nonfiltered square wave. The solid line presents correlation with raised root cosine (RRC) filtered reference sequence (input is also RRC filtered); the dotted line is with rectangular input; and the dashed line is with both rectangular input and reference signals. There are four samples per chip in each case.

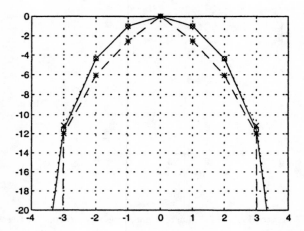

Figure 10.5   SNR after despreading as a function of timing error.

A commonly adopted approach, shown in Figure 10.6, is to use a coarse delay estimation unit, which triggers delay locked loops (DLL) connected to the despreader fingers. The DLL forms a signal proportional to the code phase error, which is used to adjust the code generator(s). The time constant is made long to provide a good enough estimate without excess phase jitter. Implementation of the sliding corrector is simple, but extra complexity is introduced because of decentralized control. In real conditions, multipath taps tend to merge, locking two physical fingers to the same tap. This calls for an indication signal from each RAKE finger to the multipath estimator control, carrying the corresponding finger code phase value.

The system in Figure 10.6 operates as follows. First, the coarse multipath estimator scans the defined channel impulse response window defined by the maximum delay spread. By averaging, the decision logic selects the highest peaks and allocates available RAKE fingers to corresponding code phases. Each RAKE finger's DLL locks to the phase and starts following the drifting multipath tap. From time to time, in addition to a "synch.lost"-indication, RAKE fingers send code phase information to the allocation control, which checks that no peaks are tracked by two or more units. In addition, the multipath control must continue updating the channel impulse profile to detect possible new peaks.

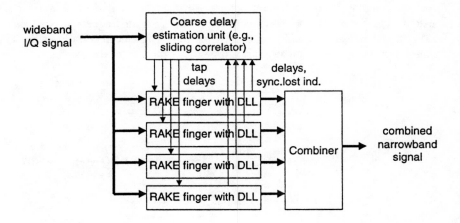

Figure 10.6   Traditional DLL-based receiver.

Today's ASIC technologies are at the point that a full digital code matched filter can be implemented without great effort. In practice this means that the delay estimation unit can be made fast and accurate, suggesting that the DLLs are no longer needed. Impulse response averaging must naturally be applied in order to prevent RAKE allocation due to a noise event. The advantage is a centralized control, which determines both RAKE finger allocation and the releasing strategy that can be optimized according to changing channel conditions and available hardware (number of fingers).

Complex amplitude estimation (i.e., phase and amplitude estimation) performed for individual channel taps is required for coherent detection. Complex channel estimation is performed with the help of a pilot channel consisting of known transmitted symbols (see Section 5.2.2). The accuracy of the estimate is essential from the link performance point of view, and it depends on the pilot channel energy, the algorithm used, and environment conditions. Variable mobile speed grades require a lot from the algorithm and in the extreme case, call for an adaptive solution.

The pilot channel can be provided in two basic ways. In the continuous pilot case, there is one physical code channel dedicated fully to constant pilot symbol transmission (code multiplexed pilot), while another option is to insert pilot symbols into the data stream (time multiplexed pilot). The former is widely used in the downlink transmission, while the latter approach is needed to implement a coherent uplink. The former argument is not, however, true if adaptive downlink beamforming is applied. Here each MS faces a different channel, which is also the case with uplink transmission.

Figure 10.7 shows one possible phase estimation architecture based on a continuous pilot channel. Each RAKE finger comprises extra correlators for despreading the pilot channel in the same code phase as the data channel tracked by the finger. The obtained instantaneous phase estimate is filtered with a suitable time constant, which can also be made adaptive to the Doppler frequency. Note that before passing the signal to the combiner, a maximal-ratio phase correction is made. The pilot channel correlator can utilize any spreading ratio regardless of the data dumping period.

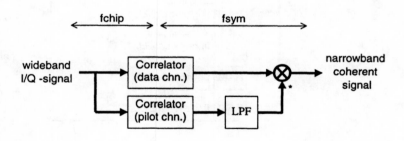

Figure 10.7   Continuous pilot signal-based phase estimation.

In case the pilot signal is transmitted as symbols in conjunction with the data channel, there is no more need for extra pilot correlators and code generators. Two well-known methods of performing the estimation for these are based on either interpolation or decision feedback techniques.

The decision feedback approach shown in Figure 10.8 performs hard decisions for the nonpilot symbols and removes the data modulation. Hence, a continuous pilot can be constructed and later filtered with a low-pass filter. Pilot symbols within the data are mainly used for taking care of phase ambiguity testing.

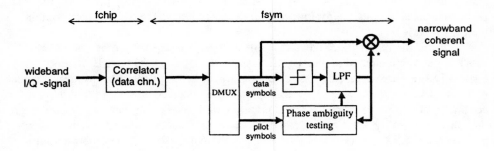

Figure 10.8   Complex phase estimation using decision feedback techniques.

If low rate coding, resulting in very small symbol energy, is used, the decision feedback performance starts to degrade because of increased symbol errors. In this case the interpolation technique may suit better because it uses only the pilot symbol energy for the estimation. The pilot energy can be increased by applying more symbols or by using a higher transmission power. The interpolating system faces difficulties with high mobile speeds because in order to get a good enough estimate a long time constant must be used and thus, the system cannot track the channel. Another problem here is the introduced data path delay due to noncausal estimation. Figure 10.9 shows an example architecture for a RAKE finger with an interpolating estimator.

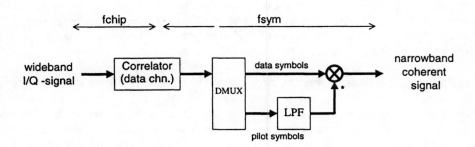

Figure 10.9   RAKE finger with interpolating phase estimator.

10.4.1.5   Multipath Combiner

A multipath combiner takes the narrowband outputs from the different RAKE fingers and combines them with each other. Maximal ratio combining produces the best performance. It weights the multipath symbols according to the SNR before combining them.

From the receiver architecture point of view, the tap weighting (maximal ratio case) can be implemented in a straightforward manner inside the RAKE fingers. This is especially the case with coherent detection where each finger incorporates a signal rotation. The rotation is done by multiplying the received signal with the complex conjugate of the phase error vector having length 1. Weighting can easily be merged into this process by setting the length of the vector according to the corresponding multipath tap SNR. In practice this means that the only task for the combiner unit is to add the coherently detected and weighted signals together.

Each RAKE finger does not always process a strong multipath component (e.g., due to fading). If the SNR of a tap is low enough, it is better to leave it out of the combining process. This is because it might have a destructive effect on the total signal quality. Some SNR level value (possibly adaptive) can be used to decide which taps are taken into the combining process.

10.4.1.6 Multiuser Detection/Interference Cancellation

The principles of multiuser detector/interference cancellation algorithms were presented in Chapter 5. A large number of different suboptimum multiuser detection approaches exists. This section concentrates on describing the best known suboptimal approaches, namely, decorrelating, MMSE, parallel interference cancellation, serial (successive) interference cancellation, decision feedback detectors, and neural network-based detectors from the implementation point of view.

Interference cancellation is a baseband processing function, which is one of the main advantages over the other capacity enhancement approaches such as adaptive antennas. State-of-the-art digital DSP and ASIC technology already provide an opportunity of building some kind of interference cancellation unit, although optimal implementation (MLSE) is still far too complex.

Decorrelating Detector. A decorrelating detector is depicted in Figure 10.10. A matrix inversion is used for the cross-correlation matrix, and the K signals from the traditional RAKE receiver branches are filtered with this R^{-1} operator. From the implementation point of view, the system becomes complex because the matrix inversion must be calculated every time the users' mutual time delays change. The Cholesky decomposition [9] can be used to perform the matrix inversion faster and hence ease the processing. The matrix is not always invertible (i.e., the matrix is singular) although according to [10] this should not happen in most cases. From the computation point of view, the precision required to perform the inversion is also critical.

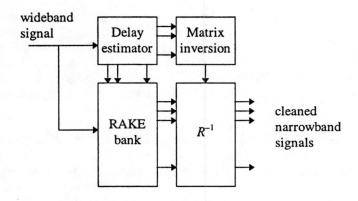

Figure 10.10   Decorrelating detector.

A decorrelating detector can also be implemented in the wideband domain. In this case, the traditional type correlator receiver is used while replacing the reference chip sequence with a signal orthogonal to all the other received users [11]. From the complexity point of view, the difference is the multilevel nature of the despreading signal and the processing element to calculate this sequence. The latter causes a radical complexity increase if operated dynamically, and hence, a suitable application would be a single user detector applied to a synchronous system.

In practice, the implementation complexity of the direct matrix inversion might be too high, especially for a large number of users and multipath components. Therefore, as was discussed in Chapter 5, iterative algorithms such as the conjugate gradient (CG) method have been proposed for implementation of the decorrelator.

Minimum Mean Square Estimate (MMSE) Based IC. From the performance point of view, a decorrelating detector has the disadvantage of noise power enhancement in the matrix inversion process. An MMSE-criteria-based interference cancellation receiver overcomes this by taking the noise power into account. The matched filter output is fed through a linear transform derived according to MMSE criteria [12]. The linear operator is obtained by taking the inversion of the cross-correlation matrix that is conditioned with received user signal powers and estimated noise power on the diagonal. Performance improvement comes, however, with extra detector complexity. This is due to the need to estimate individual user signals and noise power levels. The MMSE-based receiver performance is similar to that of the conventional matched filter with low SNRs, while with high SNRs, the performance approaches the decorrelating detector.

The MMSE receivers have attracted interest because of their applicability to decentralized adaptive implementation. Adaptive implementation does not require information of the interfering spreading codes, and thus, the adaptive MMSE receivers could be used in the mobile station as well. However, they require the signal to be cyclostationary (i.e., periodic). Therefore they can be used only with short spreading codes.

Matrix inversion turns out to be the main source of implementation complexity. Because the matrix is symmetric and still positive definite, Cholesky's method can be used to ease the calculation. It should still be noted that as in the case of the decorrelating detector, the inversion must be done every time the mutual user multipath delays change. In addition, the update must be performed every time the user or noise signal powers vary. Hence, similar to the decorrelator, iterative algorithms should be used.

Parallel Interference Cancellation. Parallel interference cancellation (PIC) cancels the interference of all users simultaneously. The multistage PIC shown in Figure 10.11 suppresses the multiple access interference in multiple consecutive steps. Stage n cancels the interference by utilizing the hard symbol decision from stage n - 1. The cancellation process can be performed either in narrowband (NB) or wideband (WB) domains. For the corresponding detectors we use notations NB-PIC and WB-PIC. The cancellation operation itself requires knowledge of the user codes, the relative code phases, complex channel estimates for the multipath taps, and finally hard symbol decisions for each user.

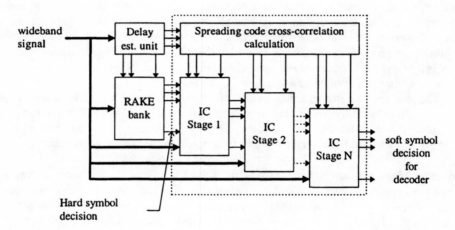

Figure 10.11   Multistage PIC detector with two cancellation stages.

Multistage PIC [13] operating in the narrowband domain performs computation with a symbol rate. Hard symbol decisions from the preceding stage are used in conjunction with complex channel estimates and mutual normalized cross-correlations between the spreading codes to generate aggregate interference estimates for each despread multipath component of interest. Multiple access interference (MAI) cleaned multipath signals are combined for each user, and more accurate hard symbol estimates are made from these. Most of the gain can already be obtained with two cancellation stages [13].

From an architecture point of view, implementation of the NB-PIC detector is a straightforward procedure because it can be fitted fairly directly into a traditional CDMA receiver between the correlators and the deinterleaver/decoder. Multipath combining and hard symbol decisions are inherently included in the interference cancellation stages. The actual cancellation operation is simple because it takes place at the symbol rate. The total processing requirements become high, however, for two reasons. First, the spreading code cross-correlations must be updated every time the mutual tap delays change. In the extreme case, if long spreading codes are used, the whole matrix must be recomputed for every symbol. In addition, the aggregate interference estimates depend on the cross-correlations and differ for each multipath. Hence, we need to compute at symbol rate $K{\times}L$ interference estimates each consisting of $(K{\times}L - 1)$ additions and multiplications (K = no. of users, L = no. of multipath taps per user).

Physical implementation of the NB-PIC detector depends on the symbol rate, spreading factor, type of spreading, number of stages, and the total number of multipaths involved in the IC process. The interference estimation is likely to need an ASIC-based solution, while the actual cancellation could be done with DSPs. Multiple stages can be implemented using either serial computation with one physical instance or if the processing rate is too high, with parallel instances in pipeline mode.

Should long spreading codes be used, the receiver complexity due to the constant need for cross-correlation updating is increased vastly. A wideband version of PIC, also called regenerative PIC, would be a better choice in this case. Finally, it should still be perceived that the narrowband multistage detector architecture relies strongly on unified code channel symbol rate, making it more suitable for systems utilizing parallel code channels instead of variable spreading to provide variable bearer services.

A major problem with the NB-PIC detector is that each multipath component, due to different mutual cross-correlations, sees a different interfering signal on the narrowband side. This can be prevented by conducting the cancellation in the wideband domain. WB-PIC, shown in Figure 10.12, is based on the regeneration principle, which is used to calculate the so-called residual signals. Regeneration, which requires knowledge of complex channels, hard symbols decisions, tap delay estimates, and the spreading code, is done for every multipath component taken into the cancellation process. The combined regenerated signal estimate is subtracted from the received wideband signal to get the residual signal. This is added to the regenerated individual multipath components in order to get a cleaned wideband signal for the despreaders.

WB-PIC has the advantage that it can be utilized with both short and long code based systems. In addition, the residual signal provides an interference resistant input for multipath profile estimator. Although the actual interference cancellation operation requires a higher operation rate than in the NB-PIC case, the lack of cross-correlation calculation makes the total processing/resource requirements smaller. The complexity of the regeneration part depends heavily on the pulse shaping filtering scheme used and hence on the sampling frequency in conjunction with the word length. The type and length of the spreading codes also have a noticeable effect on the total system complexity.

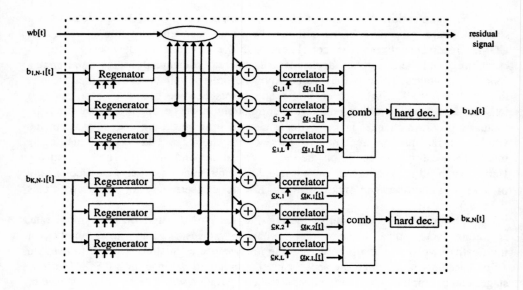

Figure 10.12 Wideband domain parallel interference cancellation (WB-PIC) stage.

Multiple simple parallel operations in WB-PIC functional architecture suggest clearly an ASIC implementation. A fully parallel architecture with multiple physical stage instances provides the lowest latency, while one WB-PIC operated sequentially would require less gate area. It would also be easy to perform serial userwise, or even pathwise, processing to further minimize the area. WB-PIC indeed suits many different processing schemes, which makes it a potential candidate for a real application.

In the previous text concerning multistage detectors it has been assumed that the interference cancellation comprised all the users with their multipath components. In practice the receiver could be made adaptive in such a way that only the strongest users/multipaths are canceled. This would require extra system/control resources for decision making but otherwise would reduce the actual PIC complexity. The system performance may, however, degrade slightly due to the selective cancellation, but in some cases it may even increase. The latter is the case when weak users are left out from the IC process. The above idea seems like a good one, but we must remember that in DS-CDMA the power control tries to keep the receiver signal levels alike. On the other hand, this is not actually the case when the VSF scheme is used for higher bit rates or when signal power control is used to adjust the quality of service.

Successive Interference Cancellation [14]. Successive interference cancellation (SIC), shown in Figure 10.13, ranks the users according to their power levels. First, the user with the strongest signal is detected, and the MAI estimate for the following users is updated with this knowledge. Next, the second strongest user is processed continuing to the weakest one. From the implementation, and also from the performance point of view, the process can be stopped after cancellation of a subset of users (the most interfering ones). The cancellation can be done on the wideband or narrowband side. The former uses the canceled user symbol hard decision, signature waveform, and amplitude and phase estimates to regenerate the wideband signal. This is then subtracted from the received wideband input stream, which is later despread by the rest of the users. Another method is to use cross-correlations between the spreading codes of the strong user and of the "cleaned" users, and do the cancellation after the despreading operation. It should be noted that the cleaned wideband canceled signal can be used in the multipath channel estimator, thus making it MAI resistant.

(a)

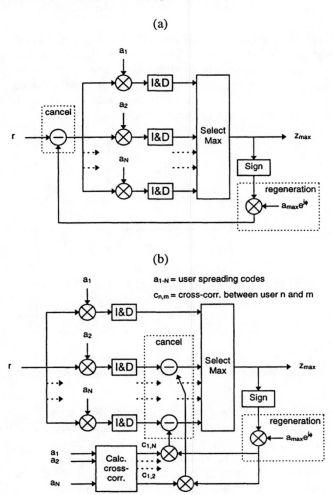

(b)

Figure 10.13   Wideband- and narrowband- based successive cancellation receivers (*From*: [15]).

Decision Feedback (DF) Detectors. DF detectors combine linear and successive interference cancellation. They take hard symbol decisions for all, or a subset of active users, and feed this value back to estimate the MAI. The receiver generally consists of a feed-forward cancellation part relying on the previous stage symbol decisions, and a feedback part utilizing the following stage "cleaned" hard symbols in estimating the MAI [12]. The times of signal arrivals play an important role here by deriving the order of cancellation. A conventional detector is used as the first stage. The performance can, however, be improved by introducing an extra feedback part already to the first stage [16]. In case of severe near-far reception, a good approach is to rank the users according to the power levels and perform the intra-stage cancellation starting from the strongest user [17].

The decision feedback detectors are very similar to the already-discussed multi-stage ones. With respect to the former, the main difference is that within a single stage, the users are detected sequentially in time (according to power levels, or time of arrival), and hence the hard decisions based on readily "cleaned" user symbols can be utilized with subsequent detection of the same stage users (see Figure 10.14). This could also be called a serial-parallel interference cancellation.

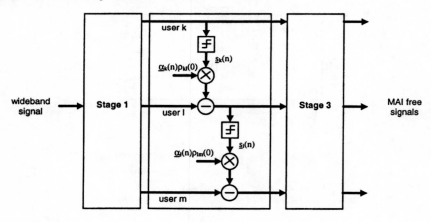

Figure 10.14   Decision feedback detector with three synchronous users.

Neural Network-Based Detectors. Neural networks (NN) provide a means of solving optimization problems with computational elements called nodes that are connected to each other with weighted (possibly adapted) interconnections. The effectiveness of neural networks is due to parallel computation taking place in numerous simple processing elements. In multiuser detection, neural nets can be utilized to find the most likely transmitted bit sequence that produced the matched filter output. There are many different neural network architectures [18], while the most commonly used for MUD realization are the normal feed-forward multilayer perceptron [19] and a simpler Hopfield net shown in Figure 10.15 [20].

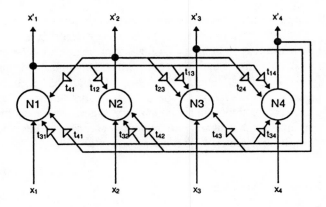

Figure 10.15   Hopfield neural net with four nodes.

A major disadvantage of the feed-forward structure is the prerequisite for training. In addition, the number of nodes increases exponentially with respect to users to be detected. Because of these drawbacks of the feed-forward structure, the Hopfield network is seen as a better neural network choice for multiuser detection problems. Its complexity expands with the square of the number of users, which will also lead to a large system in case of many active users. Training is not needed, while the processing burden comes from the iteration rounds (depends on the number of users). Therefore, neural networks cannot be used to process the whole set of received users, and the number of inputs should be reduced. Figure 10.16 shows an approach utilizing a reduced detector (RD) in conjunction with Hopfield net. The reduced detector implemented an iterative algorithm focusing only on users with nonconfident parameters (reduced search space) [21]. If the iteration turned out to be irreducible, a Hopfield net was used to perform a more extensive search.

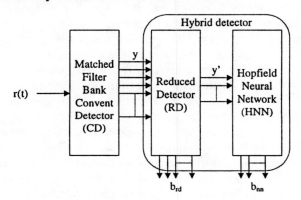

Figure 10.16   Hybrid matched filter, reduced receiver, Hopfield neural net detector (*From*: [21]).

From the implementation point of view, the neural networks are naturally ASIC oriented (many parallel elements). In the past, analog techniques were applied to provide topologies with up to 50 neurons. Digital approaches tend to be problematic for larger nets with nonbinary inputs and interconnections, although this is about to change due to even higher integration levels. Multiplication, addition, and some nonlinear functions are the primitive operations to be provided. The number of multiplications depends on the interconnections, and when binary inputs are used, these are fairly simple. Additions, on the other hand, cause major problem here because the required adder trees call for primitive adders with nonbinary inputs.

10.4.1.7   Decoding Architectures

Forward error correction (FEC) and detection coding are used to protect the transmitted information bits and provide the specified quality of service. The most frequently used FEC schemes are based on convolutional and block coding. The Viterbi algorithm is used to perform convolutional decoding, while simple shift registers can be applied for block codes. Third generation cellular systems create many new challenges for the terminal decoders. These are due to the increased number of different bearer services with different data rates, BERs, and delay requirements.

The different operation modes and coding schemes call for modular and flexible decoder architectures. For pure speech terminals the physical implementation can rely on DSPs, while the higher data rates certainly need an ASIC accelerator. The key goal here is to realize which primitive operations are common to the different coding schemes and how these can be effectively utilized to provide variable data rates.

Block coding schemes in conjunction with error detection coding (CRC) can easily make use of the same physical processing resources. This is because the decoding can be implemented using shift register chains, and the different schemes can be supported with variable polynomial lengths and changeable polynomial connections. Constant processing clock rate with alternating activity periods provide a good means for adapting the decoder to different data rates. The modest complexity of the shift register depends naturally on the polynomial degree.

The Viterbi decoder is built on top of a trellis tree consisting of stages and transitions. The basic operation consists of partial metric calculation, trellis updating based on path selection, and back-tracing. The partial metric processing involves calculation of 2^K values (K = constraint length) for each received bit. The total number of the stages is $2^{(K-1) \times L}$ where L is the so-called truncation length. After L bits have been updated to the tree, back-tracing is performed providing the decoded bits. From the complexity point of view, K determines the required processing rate, and, incorporated with L, the total decoder memory requirements can be derived. The truncation length also defines the decoding delay. Based on simulation results, L equal to $5 \times K$ is an adequate selection [22]. So far, constraint lengths less than or equal to 7 have been widely used. It is likely, however, that the third generation cellular standards include specification for larger constraint lengths (e.g., nine). This, in conjunction with high bit rates, means that decoder processing requirements increase remarkably compared to second generation systems.

From the physical architecture point of view, a key primitive with the Viterbi decoder is the so-called add-compare-select (ACS) primitive that is used to select the partial metrics that have lower values. For fast decoding requirements an ASIC implementation is a necessity, while low rate systems may incorporate a software solution (assuming that the selected DSP can provide an ACS command). The best Viterbi decoders at the moment are capable of providing throughputs of 1 Gbps [23]. Sub-optimal Viterbi decoders are mainly based on reducing the number of states in the tree. M- and t -algorithms are examples of these [24]. The latter could also be classified as an adaptive decoder because the number of states is variable. For a comprehensive description of the Viterbi algorithm see, for example [25,26].

Finally, it should be noted that new decoding approaches based on iterative algorithms have emerged. These are referred to as Turbo decoding, and they have claimed to possess better performance compared to convolutional schemes. On the other hand, their main disadvantage is an increased decoder complexity.

10.4.2 Baseband Transmitter

The transmitter baseband section is mainly responsible for channel coding, including both error detection and protection functions. In the case of third generation systems, several different schemes are required to fulfill many service qualities and user data rates. Encoding functions are considered as bit level procedures.

Signal spreading specific to CDMA systems takes place on the baseband side. This operation causes the input bandwidth to widen according to a defined spreading factor or processing gain. Signal spreading is purely chip-level processing.

Digital filtering is commonly used to shape the transmitted signal spectrum. The filtering must be performed for a wideband signal and becomes complex due to higher clock rates. There exist, however, good architectures providing implementation-efficient solutions.

10.4.2.1   Channel Encoding

Error correction algorithms can be divided into block- and tree-based schemes, convolutional coding being the best known from the latter category [27]. Recursive convolutional codes, also called Turbo codes, have been under intense research recently.

Traditionally, the implementation of channel encoders has concentrated on finding good solutions for some constant set of parameters like input data rate or coding parameters. This is not the case anymore due to diverse requirements and the new emphasis on flexibility. The encoders for the third generation systems must be capable of operating with different data rates and coding parameters. In the far extreme, the same physical encoder hardware should even support several coding algorithms. Additional problems still arise since the coding scheme might have to be switched on the fly, the requirement of which is due to packet access and link adaptation.

In the second generation systems, to attain flexibility channel encoding has been normally implemented in software mainly because the data rates have been reasonable low. Third generation systems will place much tougher requirements on the encoders due to radically increased user data rates up to 2 Mbps. Fortunately, the well-known encoding algorithms can be efficiently implemented in ASICs. However, ASIC implementation has the problem of not providing high flexibility, but we should remember that most of the encoding algorithms can be implemented using simple shift register chains. This is also the case with Turbo encoders [5].

Hardware encoder throughput can be adjusted by three different approaches (see Figure 10.17). The first option is to provide parallel hardware, and activate necessary resources when needed. This suits well systems with data rate expansion using parallel "low-rate" code channels. In the other options the same physical encoder core is used, but the operating cycle, or the clock rate, is made variable. The latter may be problematic from a mobile terminal point of view.

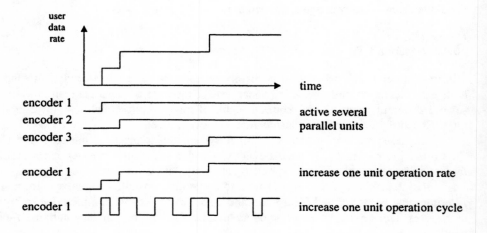

Figure 10.17   Different options to increase the encoder throughput.

10.4.2.2   Data Modulation and Spreading

After channel coding, data modulation is performed. Usually coherent phase shift keying (BPSK or QPSK) data modulation, which provides good spectral efficiency, is applied. The physical implementation of the bits-to-symbol mapping is a very simple operation.

A signal with a wideband spectrum is obtained by multiplying the transmitted symbols by a faster rate chip stream. The operation is highly hardware oriented due to its simplicity (binary multiplication) and its high frequency. Depending on the modulation method, the spreader may include only in-phase or both quadrature branches. In addition, the spreading method defines how the chip sequence(s) is multiplied with the symbol stream. As shown in Figure 10.18, for an I/Q-signal this can be done individually for both branches (dual-channel spreading) or by using a complex multiplication. The latter option means an increase in processing requirements due to extra multipliers and adders.

(a)   (b)

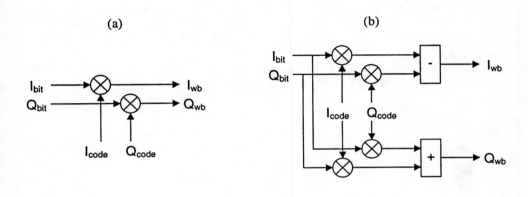

Figure 10.18   (a) Dual-channel and (b) complex spreader architectures.

From the overall transceiver point of view, the spreader procedure has only a minor effect on the total complexity. Some increase, however, may be seen in the baseband power consumption and silicon area, mainly due to fairly long code generators.

10.4.2.3   Baseband Pulse Shaping Filtering

Due to bandwidth-limited output signals, the output chip stream from the spreading modulator is filtered using either a digital or analog filter (pulse shaping filter). A digital filter with linear phase can be designed more accurately but will require a larger dynamic range from the DACs following it. Still, the implementation of long filters (close to 100 taps) may turn out to be too complex. Square RRC-type impulse response is generally used due to the Nyquist no ISI sampling criterion. The roll-off factor defines the sharpness of the spectrum.

For digital implementation, the filter length may have to be truncated in order to minimize complexity. Radical truncation effects on the spectral shape can be alleviated to a certain degree by using a windowing function [28]. Hamming, Hanning, Bartlett, and Kaiser functions are examples of the best known. The filtered output signal must have a high enough word length to keep the quantization noise due to D/A conversion below spectral mask requirements. As a rule of thumb, one extra bit increases the signal-to-quantization noise ratio by 6 dB [28].

The oversampling factor, which is also an important design parameter, defines the sample-and-hold (S/H) boxcar "filter" and, more importantly, the reconstruction filter requirements. The higher the oversampling factor is, the further apart the copied spectral components appear, which relaxes the post-DAC filter specification. Analog filters, like Bessel or Chebyshev, with two to five taps would be suitable in many cases. Care must be taken, however, to prevent the filter phase response from distorting the signal. Mixed-mode simulations must be performed if tight transmitter specifications are to be fulfilled.

The two plots in Figure 10.19 show the effect of D/A conversion following pulse shaping. The RRC filter with roll-off 0.35 was employed. The signal spectrum is copied into frequencies, which are multiples of the sampling rate.

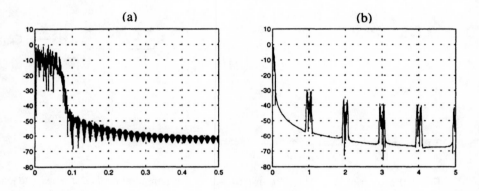

Figure 10.19 Effect of D/A conversion box-car filter on the signal spectrum: (a) signal spectrum after RRC filter and (b) signal spectrum after D/A-conversion.

10.4.2.4 AFC, AGC, and Power Control

In order to make the transceiver work, there are three important functions still to be performed. Automatic frequency control (AFC) is needed in a mobile terminal to synchronize its local oscillator to the base station oscillator. Automatic gain control (AGC) on the receiver side is responsible for preventing the ADC from saturating, while at the same time trying to provide a maximum available dynamic range. The transmitter side still incorporates a power control (PC) adjustment, which is particularly important in CDMA systems, as discussed in Sections 5.11.6 and 7.3. Figure 10.20 depicts the control loops. The dotted lines represent optional connections. For example, in the case of AFC, the carrier frequency can be changed by adjusting the reference oscillator or possibly the digital synthesizers on both the TX and RX sides.

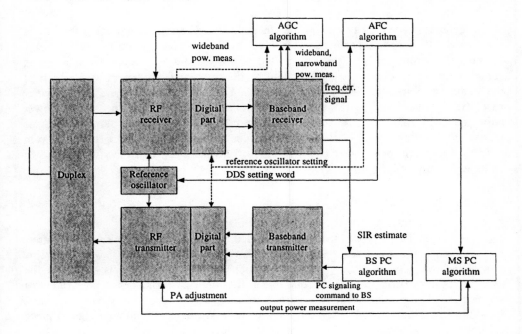

Figure 10.20 Mobile station internal control loops (optional ones with dotted lines).

Automatic frequency control aims at bringing the mobile up/down-conversion frequencies as close as possible to the base station ones. This can be done either by adjusting the MS reference oscillator or, if utilized, the digital frequency down-conversion. Generally, the system digital clocks are also tied to the reference clock, and by performing AFC with the reference, the side effect also provides more synchronized sample/chip/bit clocks. The AFC loop can be made very slow because the oscillator drifting in time is also a slow time varying process. Frequency control could also be performed to compensate for the Doppler effect. This is, however, a slightly different problem compared to slow AFC because on the receiver side the compensation must be done in the opposite direction, which requires a fast absolute frequency error estimate.

The received signal envelope has a large dynamic range. Gain control is required to adjust the amplitude suitable for the A/D converter input range, which tries to provide the highest dynamic range. Finding the optimal operation point is not an easy task because it depends on many aspects like modulation, fading, multipath channel, and, in DS-CDMA systems, on other user signals. AGC loop adjustment is based on the received signal power (RSSI), which can be measured from both wideband and despread narrowband signals. The wideband power can be measured either directly from the RF unit or after the receiver A/D converter.

CDMA systems are vulnerable to erroneous transmission power control because any error will introduce extra cell interference and hence reduce the capacity. The power control loop must be fast enough to react to abrupt signal changes and accurate enough not to generate undesired power on the channel. The criterion for adjusting the power level should be based on needed service quality; SIR is a good option for this purpose. In the open loop approach, the transmitted power is adjusted according to received signal; while in the closed loop power control, the commands are transmitted as part of the signaling information (e.g., every 5 ms with 1 dB up/down command). In the latter case, the time between the RX power measurement (by BS) and the actual TX power adjustment (by MS) must be kept as brief as possible. In a real implementation the power amplifier's varying characteristics in time should be taken into account. From time to time some power measurement value from the PA output should be used to calibrate the transmission.

10.5   RF SECTION

This section reviews a few of the best known RF transmitter and receiver architectures. In a wideband CDMA transceiver the RF must be designed to suit a wideband low power spectral density signal. Furthermore, transmission is continuous unlike in the case of time division systems. Requirements for high dynamic range, accurate fast power control loop cause most of the problems compared to second generation systems. In addition, the linear modulation and multicode transmission (multilevel signal) place challenges on the power amplifier linearity and efficiency design.

10.5.1 Linearity and Power Consumption Considerations

Third generation systems place much more difficult linearity and efficiency requirements for the RF front-end. The linearity constraint is due to tighter output spectral mask specification, higher signal envelope variations (linear modulation), and, in the case of the PA, the need to keep the operation level near the compression point in order to achieve a high enough efficiency. In addition, when multicode transmission is applied, more backoff is needed causing a loss of efficiency. Adjacent channel interference (ACI) levels of −30 dB are likely to be specified, while with large umbrella cells and flexible channel allocation the requirements may increase to −60 dB.

The transmitted signal spectrum at the antenna defines how much energy is emitted to the adjacent and alternate channels (by alternate channels we mean those after the adjacent ones). Interference is increased due to the spill-over of the signal into adjacent channels, and the network capacity decreases accordingly. As was discussed in Chapter 8, systems incorporating hierarchical cell layers become very difficult because without highly orthogonal spectra, one umbrella-cell user may entirely block microcell traffic due to its radically higher transmission power. Output transmission spectrum depends mainly on the modulation method, multirate solution (multicode vs. variable spreading), transmission filtering, and RF nonlinearities. In the last case, the power amplifier has the largest contribution to the signal distortion.

It should be noted that the overall transmitter linearity is an optimization problem between PA gain characteristics and transmission filtering. The former makes the spectrum narrower but at the same time increases the signal crest factor (= peak-to-mean signal ratio). The higher the crest factor, the more the signal is distorted by the power amplifier. However, it should be noted that examining only the crest factor may be misleading. From the generated adjacent channel interference point of view, a high crest factor is tolerated if the occurrence of peak amplitudes is low. Hence, instead of crest factor, the power amplifier input signal power histogram should be investigated, as seen in Figure 10.21.

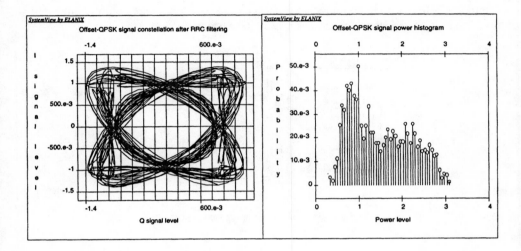

Figure 10.21   Constellation and histogram plots for RRC-filtered O-QPSK signal (roll-off = 0.35).

Power amplifier linearity depends on the operating class. The class can be selected by using different biasing levels. A-class amplifiers are linear at the expense of low power efficiency because they consume power regardless of the input signal level. C-class amplifiers, on the other hand, are power efficient with the introduction of higher nonlinear characteristics. PA linearity can be defined by input amplitude versus output amplitude (AM-AM), and input amplitude versus output phase (AM-PM) relationships. The most nonlinear operation takes place within low/high input signal regions. These characteristics still vary within time due to aging and supply voltage, output load, and temperature change.

The PA can be made linear by applying higher biasing (to A/AB-class), by increasing input signal backoff, or by using linearization techniques. The two former solutions are not very suitable for handheld terminals because the efficiency is decreased accordingly. Linearization techniques are thus seen as the key solution to overcome the tightened spectral mask requirements in conjunction with acceptable amplifier efficiency. It should be noted here that the linearization techniques do not ease the multicode transmission, which will in any case call for extra backoff.

Linearization techniques can be divided into four main categories: (1) feed-forward, (2) feedback, (3) envelope elimination and restoration, and (4) predistortion

[29,30]. Each of these have a set of variants providing different implementation complexity, ACI improvements, and bandwidth/convergence rates. Figure 10.22 shows an example of PA architecture linearized with the complex gain predistortion technique. This approach provides a fairly simple implementation with high bandwidth/convergence rate [29,31]. The input signal power is used to address the predistortion vector stored in RAM. The predistortion vector is updated according to changing environment parameters. Because the number of entries is fairly small, the adaptation is fast. It is likely, however, that the high dynamic range complex multiplier turns out to be the most critical component in the system.

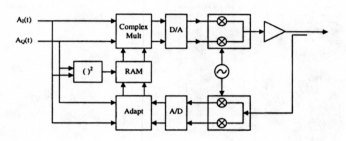

Figure 10.22   Power amplifier linearization with complex gain predistortion.

10.5.2 Receiver Architectures

While the transmitter is facing optimization between linearity and efficiency, the receiver is concerned with providing high selectivity with low noise figure (NF). The non-linearities must be handled in order not to cause excessive signal distortion. Noise figure tells how much the receiver deteriorates the incoming SNR. While more optimal technologies and components have emerged, lower noise figure RX chains can be implemented. Good receivers provide NFs as low as 5 to 7 dB. High selectivity is desired in order to prevent adjacent/alternate channel energy from getting to the input of the ADCs, and decreasing the dynamic range requirements. The former is critical especially when operating in a hierarchical cell system where the nearby channel power may be several tens of decibels higher. A low sensitivity receiver would cause signal saturation in the A/D input.

Linearity of the signal path on the receiver side becomes more important in the case of linear, multilevel modulation or multicode transmission schemes. Still, if the system is operating close to narrowband systems with high power spectral densities, the intermodulation products may distort or even block the whole reception.

The desired channel signal strength may vary for several reasons. The receiver must ensure that if the signal goes up rapidly the ADC is not saturated, or if there is fast power level degradation the signal quality does not pass below an acceptable level. Fast and high dynamic range (up to 80 dB) AGC is responsible for adjusting the receiver variable gain amplifiers and attenuators in order to feed the ADC with as optimal a signal level as possible. In CDMA systems the adjustment can be done every symbol time without regard to slot boundaries. This is because the spread bits inherently include a training sequence, and channel estimation/equalization takes place on the symbol level.

10.5.2.1   *Super-Heterodyne Receiver*

The super-heterodyne architecture shown in Figure 10.23 is the most well-known receiver to implement good selectivity. The RF signal is filtered, amplified, and converted to an intermediate frequency in which the channel filtering is performed. The IF frequency is kept constant, while channel selection is done by changing the RF mixer frequency.

Several stages with different IF frequencies are needed to get the desired performance at the expense of higher receiver cost. A single conversion receiver is the simplest option and is commonly utilized.

From the total receiver noise figure point of view, the low noise amplifier (LNA) preceding the RF mixer is the key component. A typical LNA has a noise figure below 4 dB, while the best offer an NF around 1 dB with power gains up to 30 dB.

Good channel selectivity with feasible linearity can be achieved by using surface acoustic wave (SAW) filters at the IF frequency. The size of the filter depends on the channel bandwidth and IF frequency and hence can be implemented smaller for wideband CDMA systems. The problem with the SAW filter is its high insertion loss and relative cost. As a result, solutions to get rid of it have been searched for (direct conversion and digital receivers being such examples).

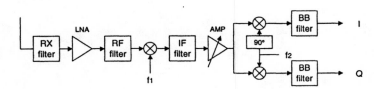

Figure 10.23   1-stage super-heterodyne receiver architecture.

As was the case with the IF-based transmitter, digital direct synthesis (DDS) can be utilized in super-heterodyne receivers to perform I/Q separation (see Figure 10.24). By adjusting the digital down-conversion frequency, DDS can also be applied in the AFC procedure. It should be remembered, however, that the analog channel selection filter defines the adjustment margin. Thus, it might be useful to define its bandwidth wider than the final desired channel BW and perform the final selection digitally. If, on the other hand, a wide IF filter with high dynamic range ADC is used, the DDS can make the actual frequency channel selection. This may not be as useful with wideband systems as with narrowband ones and would clearly be a base station oriented solution.

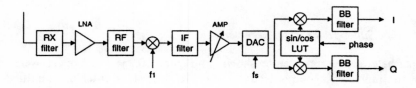

Figure 10.24   1-stage super-heterodyne receiver with digital I/Q splitting.

10.5.2.2   *Direct Down-Conversion*

The direct down-conversion receiver shown in Figure 10.25 mixes the receiver RF signal straight into I/Q baseband components for which the channel filtering is performed using simple, cheap low-pass filters. The benefit of this architecture is the reduced complexity and performance, especially in terms of the SAW filter (high insertion loss), and may in the end lead to a single chip transceiver including also the transmitter section.

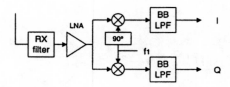

Figure 10.25   Direct-conversion receiver with baseband channel filtering.

There are, however, a few problems related to the direct conversion receiver technique. One aspect is the DC component, which may appear because of self-mixing the local frequency and/or nearby strong users. A CDMA signal inherently is resistant against narrow band interferers, but high level DC offset could saturate the input ADCs and destroy the reception.

The key component here is the quadrature local oscillator, which should provide good phase and gain matching in order to provide pure I/Q-branch sine waves with equal amplitude and 90-degree phase shift. Still, the isolation from the mixer input must be high enough to avoid the above-mentioned self-mixing. The mixer itself should possess good linearity to suppress intermodulation products caused by adjacent/alternate channel transmission.

Finally, it should be noted that the direct conversion receiver can be fit into a single chip by using, for example, the silicon bipolar process (in the future, a CMOS process could also be applied). The low pass filters are excluded from this and are put into the baseband side.

10.5.2.3   *Harmonic IF Sampling*

So far it has been assumed that the receiver I/Q baseband signal is converted into the digital domain using a sampling frequency twice as high as the highest spectral component in the input. The Nyquist criteria states, however, that it is enough to use a sampling frequency twice the desired signal bandwidth, which suggests that it is fully possible to utilize the actual signal alias term. In the frequency domain, the desired band is translated to another phase according to the selected sampling rate. Hence, the sub-sampling can also be considered as a down-conversion process (see Figure 10.26). Before feeding the signal to an A/D converter, bandpass channel selection filtering must be done for the desired signal to prevent other channels to cause their aliases within the desired frequency band.

The ADC can be used as such if the sampling rate suits the signal central frequency. If not, a separate track/hold function with a low pass filter should be applied to relax the ADC requirements. Maximum input signal frequency to the sampling device depends on the track and hold device bandwidth. In practice, several hundreds of megahertz can already be handled.

From the CDMA point of view, sub-sampling receiver does not give any benefits over traditional narrowband systems. From a implementation point of view, however, only one ADC with fast T/H instead of two ADCs is required. Complexity on the digital side is increased due to I/Q separation that must be done at a high sampling rate.

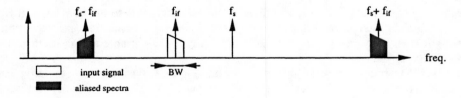

Figure 10.26   Signal spectrum after performing subsampling.

10.5.3 Transmitter Architectures

This section presents the basic RF transmitter architectures suitable also for wideband CDMA transceiver implementations. Critical aspects are highlighted and, in conjunction with each approach, advantages and disadvantages are revised.

The baseband signal to be transmitted is converted to the final carrier either directly or through some intermediate frequency. The latter approach has been more common in the past, while present RF technology is capable of following the former approach.

In direct sequence spreading systems, phase and/or amplitude shift keying chip-level modulation is generally utilized. A single sideband signal with suppressed carrier is generated using an I/Q-mixer with 90-degree phase difference in the frequency between in-phase and quadrature local inputs.

10.5.3.1   Traditional IF-Based Up-Conversion

Transmitted signal conversion to intermediate frequency before the final carrier provides a few advantages. First, different stages are easier from a component matching point of view. An I/Q modulator, for example, operating with low frequency can offer better phase and amplitude balance. One important aspect is that channel filtering can also be done using IF passband filters and consequently relax the processing power requirements of the digital parts.

The main sources of noise come from the nonideal local frequencies. This may be problematic if there are several IF stages because each local oscillator (LO) adds to the total transmitter noise contribution.

CDMA systems demand tough requirements for the dynamic range, accuracy and speed of the transmission power control. In practice, the amplification/attenuation is done in many phases to fulfill these requirements and an IF-based transmitter has an intrinsic feature to support gain control in several stages.

A functional block diagram of a 1-stage IF transmitter is shown in Figure 10.27. The I/Q -signal is up-converted to a predefined first IF frequency that is likely to be constant. This is followed by passband filtering and an IF amplifier. Finally, the signal is translated to carrier frequency, amplified by the PA, and input to the transmission RF filter. A duplex filter is required in case of simultaneous transmission and reception. The highest up-conversion frequency is traditionally made variable and is used in the transmission channel selection.

DDS techniques can be utilized with an IF-based transmitter. In the simplest form, DDS replaces the I/Q mixer or even the analog RF synthesizer. The disadvantage of this approach is that the generated frequencies are not as clean as the analog counterparts. The advantage is the flexibility in terms of frequency and settling speed.

By moving the digital-analog border closer to the antenna, the arrangement shown in Figure 10.28 is achieved. Here the I/Q -modulation is done digitally before converting to analog format. Good amplitude and phase balance can be obtained, but the disadvantages noted before still hold. Digital-to-analog converter (DAC) also causes distortion due to the S/H-function, which must be taken into account when designing the reconstruction filter.

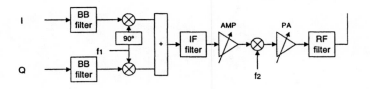

Figure 10.27 Typical IF transmitter architecture.

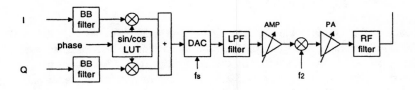

Figure 10.28 IF based transmitter with digital I/Q -modulation.

10.5.3.2 Direct Up-Conversion

The main disadvantage of a several-stage transmitter is the number of different components required, which increases the cost of a portable terminal. In a direct conversion-type transmitter (Figure 10.30), the signal is transferred directly from baseband to RF band. There is a need only for one synthesizer whose output is taken to the I/Q -modulator. Care must be taken to provide good enough 90-degree phase offset and gain match between the quadrature branches in order not to create too low an unwanted sideband suppression.

Channel filtering in the case of a direct up-conversion transmitter must be performed on the baseband before I/Q mixing. Digital filtering with linear phase, and hence constant group delay, is a good option for the purpose. The analog low pass filters following the DAC could also be used in the process.

Power control requirements are slightly more difficult to fulfill in case of direct up-conversion. Some alleviation can, however, be provided by implementing higher

314

dynamic range digital parts and especially the D/A-converter. From the implementation point of view, higher dynamic range digital parts, and especially the D/A-converter, are needed.

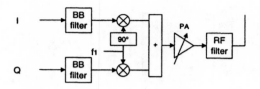

Figure 10.29   Direct conversion transmitter.

10.6   SOFTWARE CONFIGURABLE RADIO

The most flexible implementation for a communication terminal would incorporate a software configuration. Such an approach could provide an access to several different wireless systems with a single hardware unit simply by downloading a new configuration internally or from a network. The idea of a software radio has originally been connected to military applications. Such a device, however, would provide flexible roaming between today's multitude of commercial cellular systems. Good articles about software and multimode radios can be found in [32–34].

Figure 10.30 illustrates the concept of the software radio. There are a few basic features that relate to the software radio. The RF section must support transmission and reception of differently modulated signals within several frequency bands. This calls for a linear power amplifier operating over a wide frequency range with acceptable efficiency. Similarly, antennas must provide low loss and uniform gain across the same range. After down-conversion to IF or baseband initial channel, selection prior to the A/D conversion must be performed. This is necessary in order to restrict the signal dynamic range to within acceptable limits. If a large dynamic range A/D converter and a wide enough filter are used, the actual channel filtering can be made digitally. This may not, however, be enough if the nearby channel powers are radically larger than the desired channel. One solution here is to introduce a filter bank and select one bandwidth according to the operations mode.

The baseband section of a software radio consists of programmable hardware and a powerful DSP processor(s). In the former case, the configuration can be done by employing field programmable gate array (FPGA) technology and well defined hardware description languages like VHDL (Very High Speed Integrated Circuit Hardware Description Language) or VERILOG. Depending on the access method, different baseband functions need to be supported. The main concern here is in the channel equalization and forward error-correction coding/decoding. The baseband architecture must be designed in such a way that both burst- (TDMA) and symbol-type (DS-CDMA) equalization approaches can be supported in a straightforward manner. When it comes to channel encoding/decoding there are numerous methods available, each providing certain coding rates and service qualities. Software implementation is fine with low effective bit rate systems, but high rate systems call for hardware accelerators.

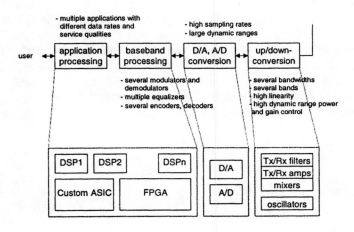

Figure 10.30   Mobile station software radio top-level architecture.

The application side of a mobile terminal has so far been mostly speech oriented and several complex source codecs have been implemented into the handsets. New applications related mainly to video coding and other data processing are emerging, and hence, the requirements for the software radio application support become greater.

An important issue with the software radio is related to flexible usage of the available processing resources. The system clock rate cannot be increased easily in a mobile terminal, and hence, one of the key solutions is to utilize configurable silicon. Higher processing capabilities are obtained by creating and activating extra processing units on a need basis. This could be interpreted as a "breathing" parallel processor that expands as needed. Power consumption is occasionally increased and must be taken into account while designing the device power management.

10.7   TYPICAL WIDEBAND CDMA MOBILE TERMINAL ARCHITECTURE

This section presents an architecture for a wideband CDMA mobile terminal integrating the different functions presented earlier. For the RF section, a traditional down-conversion receiver architecture with two IFs was selected, and on the transmitter side, direct up-conversion will be employed. The baseband side consists of all the bit- and chip-level processing primitives needed to provide all bearer service quality classes, band spreading/despreading, coherent multipath diversity detection, and some extra functions for link maintenance.

Figure 10.31 shows the mobile transmitter section. On the transmitter side, the information-bearing user data is obtained from outside the radio modem part. The stream is coded and interleaved according to selected QoS. Two basic encoding schemes, namely convolutional and Reed-Solomon, are available. The rate of the user data is matched to the air interface by choosing some of the available spreading ratios and unequal symbol repetition. Frame control header information about the frame data rate and coding scheme is encoded and interleaved to the quadrature branch. In addition, the power control and the required amount of reference symbols are placed into the transmitted stream. Before D/A conversion, the physical in-phase and quadrature branches are spread with two PN codes and passed through a pulse shaping filter to obtain the specified transmission spectrum.

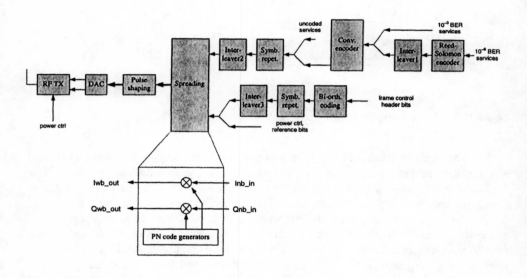

Figure 10.31   Mobile transmitter section.

The receiver side down-converts the incoming signal to the baseband before A/D conversion. Figure 10.32 depicts the mobile terminal receiver section. Channel filtering and AGC have been implemented on the analog RF side. In addition, a loop back from the baseband channel estimator provides an error signal to adjust the reference oscillator for clock synchronization purposes. The I/Q signal is converted to the digital domain using a 6-bit, four times oversampling ADC. Digital filtering following the ADC is matched to the transmitter filter, thus improving the SNR. The digital filter can also be used to compensate for nonlinearities caused by the analog side filters. Wideband power measurement is performed for the post-ADC signal and is used in the receiver AGC loop. The multipath channel estimator utilizes a full matched filter to find multipath components quickly and to perform accurate delay estimation without any DLL. RAKE fingers consisting of only two physical correlators perform the signal despreading for each multipath component and parallel 20 ksymbol/s channel. A maximal ratio combiner estimates complex channel taps from the despread pilot channel, rotates/weights the RAKE output signals, and sums them together. In addition, the combiner outputs signals for frequency error, narrowband signal power, and SIR. These are used in AGC, AFC, and PC-loop, as described earlier. The selector picks up the code channels (I/Q branches) owned by the specific user and routes to de-interleavers/channel decoders.

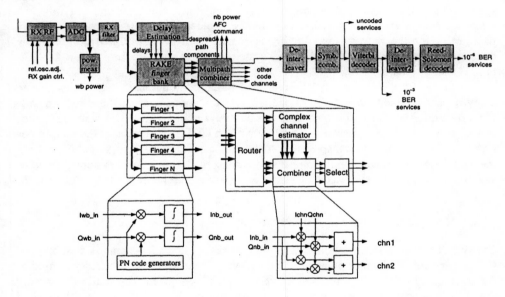

Figure 10.32   Mobile terminal receiver baseband section.

10.8   MULTIMODE TERMINALS

Since at the beginning of wideband CDMA system deployment there will still be several second generation systems in operation, dual-mode or multimode terminals can be used to provide seamless service for the users. However, from the transceiver point of view, this may be cumbersome because the same physical terminal must provide technical solutions to support potentially dissimilar radio technologies.

The main second generation cellular systems are GSM, IS-136, IS-95, and PDC. Furthermore, PHS and DECT are second generation low-tier wireless systems. Third generation wideband CDMA systems will be implemented together with some of these second generation systems. The most likely combinations are GSM/WCDMA, PDC/WCDMA, and IS-95/cdma2000.

Different systems place different requirements on different parts of the transceiver. On the other hand, the right selection of transceiver architecture can ease the implementation of dual mode terminals. In the RF section, the duplexer is impacted by the fact that second and third generation systems have different frequency bands. RF and IF filter bandwidths are different due to different bandwidths of the systems. Intermodulation and phase noise requirements are also most likely different. Synthesizer settling times vary, as well as A/D and D/A converter requirements. Since different systems have different symbol rates and other clocking requirements, possibly two reference oscillators are required. This could be avoided by proper selection of system parameters for the new system. In the baseband, receiver algorithms are somewhat different. For example, a TDMA-based system requires an equalizer and a CDMA-based RAKE receiver.

Since there are so many different dual-mode combinations, it is impossible to analyze each of them in detail. Instead, several commonalties can be observed, and implementation of dual-mode terminals can be analyzed in general. The main difference of the above-listed systems is the spreading used in CDMA systems and the continuous transmission compared to nonspread, slotted signal of TDMA-based systems. Therefore, the text below assumes separation between nonspread and spread systems, between slotted and continuous systems, and according to a duplex scheme.

10.8.1 Nonspread Versus Spread Systems

The difference in the implementation of nonspread and spread systems is certainly not straightforward. The major parameter is the channel bandwidth, and if these are similar for both the systems, the actual transceivers seem fairly unified. The major differences are on the baseband transmitter and receiver front-ends. In the former case, the spread system faces a multilevel signal due to parallel code channels, which increaseases transmission filter and D/A converter requirements. On the receiver side, the nonspread equalizer (DFE or MLSE) is also replaced with a RAKE receiver, or even a multiuser detection unit.

If the bandwidths are dissimilar, like with wideband CDMA and GSM, the situation is different. First, the IF/RF section will require new filters on both TX/RX sides. For a dual-mode terminal, there should be several filters from which the correct one is selected. Another choice is to implement an adjustable filter, which has the disadvantage that it cannot always be adjusted the most optimal way.

Channel raster is directly related to the bandwidth used. If the specified rasters are not factors of each other and analog synthesizers are used, there is a problem with the reference frequency. In order to provide simple and fast PLLs, the reference frequency should equal the channel raster. For a multimode terminal this would mean a compromise between complexity (possible several references) and speed (lowest common factor frequency).

On the baseband side the processing requirements between spread and non-spread systems can be considered fairly similar. A wideband signal calls for a higher sampling rate, while at the same time it can be presented with less bits/sample. In favor of nonspread systems, signal spreading/despreading adds some extra complexity to the terminal. For long channel delay spreads, a RAKE receiver tends to be simpler than an optimal MLSE equalizer.

10.8.2 Slotted Versus Continuous Systems

Continuous systems better suit ASIC implementation because most of the functions operate constantly. Selecting a software approach for a continuous system would cause high control overhead per received symbol because each one would be processed independently by generating an interrupt for a DSP. This could be overcome by buffering some number of symbols prior to transmission and reception. The problem here, however, is the increased data path latency.

Slotted systems on the other hand suit software processing fairly well. This is because symbol detection may start only when the whole/partial burst has been received. This is especially the case when the channel estimation is performed from the burst midamble sequence. Overhead due to control is minimized, and the number of instructions per processed symbol becomes feasible.

From a slotted systems' RF point of view, the power amplifier sees discontinuous transmission with high peak power. With continuous systems, the PA is active constantly, and, if the average transmission power is assumed the same, the peak power is reduced respectively. Continuous systems may, however, act as discontinuous systems if DTX mode is applied. This is common especially with speech services.

10.8.3 FDD Versus TDD Systems

Continuous systems generally use frequency duplex division for providing up and downlink channels. From an implementation point of view, this requires synthesizers on both the transmitter and receiver side due to simultaneous operation. In addition, a duplex filter must be applied to prevent TX signal leaking to RX side.

In slotted systems, another possible method is to utilize time division duplex separation. The idea is to transmit up and downlink channels with the same frequency but at different times. The main benefit here is that the terminal is not transmitting/receiving simultaneously, and hence only one synthesizer and no expensive duplex filter is needed.

On the baseband processing side there are few physical resources that could be shared between the transmitter and receiver. DS-CDMA code generators are such an example (polynomials, and states reloaded upon TX/RX switching). In addition, the multipath channel estimation could be made easier because the channel is reciprocal (i.e., the channel estimate obtained for one direction is valid within coherence time for the other direction).

REFERENCES

[1] Neuvo, Y., "Wireless Will Catch Up Wireline and More", *Keynote speech in PIMRC'97*, Helsinki, Finland, September 1997.

[2] Bernel, D. A., and T. C. Hofner, "Digital Comms. Require Dynamic Performance," *Electronic Engineering*, September 1997, pp 67–71.

[3] Gaudenzi, R. D., "The Influence of Signal Quantization on the Performance of Digital Receivers for CDMA Radio Networks," *ETT*, Vol. 8, No. 1, January - February 1997, pp. 89–97.

[4] Do, G. L., "Efficient Filter Design for IS-95 CDMA Systems," *IEEE Trans. on Consumer Electronics*, Vol. 42, No. 4, 1996, pp.1011–1019.

[5] Dixon, R. C., *Spread Spectrum Systems with Commercial Applications*, New York: John Wiley & Sons Inc., 1994.

[6] Pickholz, R. L., D. L. Schilling, and L. B. Milstein, "Theory of Spread-Spectrum Communications – A Tutorial," *IEEE Trans. on Comm.*, Vol. 30, No. 5, May 1982, pp. 855–884.

[7] Polydoros, A., and C. L. Weber, "A Unified Approach to Serial Search Spread-Spectrum Code Acquisition – Part 2: General Theory," *IEEE Trans. on Comm.*, Vol. 32, No. 5, May 1984, pp. 550–560.

[8] Bensley, S. E., and B. Aazhang, "Subspace-Based Channel Estimation for Code Division Multiple Access Communication Systems," *IEEE Trans. on Comm.*, Dec. 1994, pp. 1009–1020.

[9] Kreyszig, E., *Advanced Engineering Mathematics*, 7th ed., New York: John Wiley & Sons Inc., 1993.

[10] Lupas, R., and S. Verdú, "Linear Multiuser Detectors for Synchronous Code-Division Multiple Access Channels," *IEEE Trans. on Information Theory*, Vol. 35, No. 1, January 1989, pp.123–136.

[11 Paris,] B. P., "Finite Precision Decorrelating Receivers for Multiuser CDMA Communication Systems," *IEEE Trans. on Comm.*, Vol. 44, No. 4, April 1996, pp. 496–506.

[12] Xie, Z., R. T. Short, and C. K. Rushforth, "A Family of Suboptimum Detectors for Coherent Multiuser Communications," *IEEE Journal on Selected Areas in Communications*, Vol. 8, No. 4, May 1990, pp. 683–690.

[13] Varanasi, M. K., and B. Aazhang, "Multistage Detection in Asynchronous Code Division Multiple-Access Communications," *IEEE Trans. on Comm.*, Vol. 38, No. 4, April 1990, pp. 509–519.

[14] Holzman, J. M., "DS/CDMA Successive Interference Cancellation," *Proceedings of ISSSTA'94*, Oulu, Finland, July 1994, pp. 69–78.

[15] Patel, P., and J. Holtzman, "Analysis of a Simple Succesive Interference Cancellation Scheme in a DS/CDMA System," *IEEE Journal on Selected Areas in Communications*, Vol. 12, No. 5, June 1994, pp. 796–807.

[16] Giallorenzi, T. R., and S. G. Wilson, "Decision Feedback Multiuser Receivers for Asynchronous CDMA Systems," *Proceedings of Globecom'93*, Houston, Texas, USA, November 1993, pp. 1677–1682.

[17] Duel-Hallen, A., "A Family of Multiuser Decision-Feedback Detectors for Asynchronous Code-Division Multiple Access Channels, " *IEEE Comm*, Vol. 43, No. 2/3/4, February/March/April 1995, pp. 421–434.

[18] Lippmann, R. P., "An Introduction to Computing with Neural Nets," *IEEE ASSP Magazine*, April 1987, pp. 4–19.

[19] Paris, B. P., B. Aazhang, and G. Orsak, "Neural Networks for Multi-User Detection in CDMA Communications," *IEEE Trans. Comm*, Vol. 40, July 1992, pp. 1212–1222.

[20] Kechiotis, G., and E. S. Manolakis, "Hopfield Neural Network Implementation of the Optical CDMA Multiuser Detector," *IEEE Trans. Neural Networks*, January 1996, pp. 131–141.

[21] Kechiotis, G., and E. S. Manolakis, "A Hybrid Digital Signal Processing-Neural Network CDMA Multiuser Detection Scheme," *IEEE Trans. on Circuits and Systems-II*, Vol. 43, No. 2, February 1996, pp. 96–104.

[22] Proakis, J. G., *Digital Communications*, 3rd ed., New York: McGraw-Hill, 1995.

[23] Dawid, H., G. Fettweis, and H. Meyr, "A CMOS IC for Gb/s Viterbi Decoding: System Design and VLSI Implementation," *IEEE Trans. on VLSI Systems*, Vol. 4, No. 1, March 1996, pp. 17–31.

[24] Anderson, J. B., and E. Offer, "Reduced-State Sequence Detection with Convolutional Codes," *IEEE Trans. on Information Theory*, Vol. 40, May 1994, pp. 944–955.

[25] Forney, G. D., "The Viterbi Algorithm," *Proc. of IEEE*, Vol. 61, March 1973, pp. 268–278.

[26] Viterbi, A. J., and J. K. Omura, *Principle of Digital Communication and Coding*, New York: McGraw-Hill, 1979.

[27] Blahut, R. E., *Theory and Practice of Error Control Codes*, Reading, MA: Addison-Wesley Publishing Company, 1993.

[28] Oppenheim, A. V., and R. W. Schafer, *Discrete-Time Signal Processing*, Prentice Hall International Inc., 1989.

[29] Sundström, L., "RF Amplifier Linearisation Unsing Digital Predistortion," *Thesis for the degree of Teknisk Licentiat*, Lund University, Sweden, June 1993, pp. 8-17.

[30] Johansson, M., and M. Faulkner, "Linarization of wideband RF power amplifiers," *Proceedings of Nordic Radio Symbosium*, April 24–27, 1995.

[31] Cavers, J. K., "A Linearising Predistorter with Fast Adaptation," *Proceedings of 40th IEEE Vehicular Technology Conference*, May 1990, pp.41–47.

[32] Special issue on software radios, *IEEE Communications Magazine*, Vol. 33, No. 5, May 1995.

[33] Kuisma, E., "Technology Options for Multi-Mode Terminals," *Mobile Communications International*, February 1997, pp. 71–74.

[34] Edmond, P., "Towards a multi-mode mobile telephone design," *Electronics Engineering*, October 1996, pp. 67–70.

Chapter 11

NETWORK PLANNING

11.1 INTRODUCTION

When using a cellular phone, you have most likely experienced several undesired effects such as dropped calls when you move from one location to another, busy network when attempting to make a call, or just poor speech quality. The grade of service wireless subscribers experience is dependent on the quality of the radio network planning. Thus, network planning is one of the key competence areas for any operator in order to satisfy customers and thus make a profit.

Network planning covers two major areas: radio network planning and network dimensioning. Radio network planning includes the calculation of link budget, capacities, and thus the required number of cell sites. Furthermore, radio network planning includes detailed coverage and parameter planning for individual sites. In network dimensioning, the required number of channel elements in the base station, the capacities of transmission lines, the number of base station controllers, switches, and other network elements are calculated. This chapter presents the main concepts of third generation network planning; for details of IS-95 network planning refer to [1].

A large number of different bit rates and the diversification of services complicate the network planning process for third generation systems, as compared to the second generation systems. It is more difficult to predict traffic and usage patterns of different services. Which unit to use to measure traffic density is our first topic in this chapter. Next, we introduce performance measures, such as outage and blocking, which are used to measure the network quality. High bit rates services can be offered either with uniform coverage over a cell or with smaller data rates at the cell edge. Furthermore, either continuous coverage over a wide area or only hotspot coverage could be provided. A careful assessment is required to understand what is the impact of different deployment strategies on the demand of services.

The radio network planning process can be divided in three phases: preparation, estimation of cell count, and network optimization. Each of these is discussed in depth.

In the network optimization phase, the following aspects are discussed: detailed characterization of the radio environment for individual cell planning purposes, CDMA control channel power planning, pilot pollution, planning of soft handover parameters, interfrequency handover, iterative network coverage analysis, and radio network testing. The impact on the network planning of microcell and indoor planning are discussed, as are sectorization and beamforming.

The main emphasis in the section on network dimensioning is on the calculation of the number of channel elements in the base station.

The co-existence of wideband CDMA air interface with already deployed air interfaces will be an important factor in the beginning of wideband CDMA system deployments. The last sections discuss intermodulation interference, guardband and zones, and transition considerations.

11.2   TRAFFIC INTENSITY

The starting point for network planning is the assessment of individual user traffic and traffic intensity (offered traffic) in a given area. To quantify the traffic intensity, different units can be used. Most often, traffic intensity is measured in *Erlang*, where 1 Erlang is equivalent to one circuit in use for 1 hour (3600 seconds).

$$\text{Traffic intensity in Erlang} = \frac{\text{Number of calls per hour} \times \text{average call holding time (seconds)}}{3600} \quad (11.1)$$

For example, if a user makes one call per hour and the average call duration of calls is 120 seconds, it produces 120/3600 = 33 mErlang of traffic. Given the user density in a geographical area, the traffic density can be calculated and is expressed in Erlang per square-kilometer (Erl/km^2).

Since third generation systems will have a very large variety of services, a single traffic measure might not be suitable to all cases. The *equivalent telephony Erlang* (ETE) defines traffic with relation to basic telephone calls [2]. However, this definition is dependent on the transmission rate of the basic telephone service. For data services, traffic measured in Mbps/km^2 will better characterize the traffic density.

11.3   PERFORMANCE MEASURES

The performance measures for the grade of service are *area coverage probability* and *blocking*. The area coverage probability is related to the quality of the radio planning and the radio network capacity. Blocking is related to the available hardware, for example, to the base station channel cards. The term *soft blocking* is used when a user is blocked because the cell runs out of capacity (i.e., a user is denied access to the network in order to ensure the quality of already admitted users). First, the network is designed for the desired area coverage probability. Within this area coverage probability, then, certain blocking requirements, usually 2%, need to be fulfilled.

Area coverage probability can also be defined by *outage*. Outage is the probability that the radio network cannot fulfill the specified quality of service target. Several definitions exist for the measure of outage. In general, it is defined as a drop in the required quality below a prespecified target value. However, it needs to be defined for how long the quality must be below the target value to create an outage. Furthermore, quality criteria for outage must be defined. We can separate between a *signal outage*, when the signal-to-noise ratio (SNR) drops below the target value, and *interference outage,* when the signal-to-interference ratio (SIR) drops below the target value. The outage is impacted by several factors, such as shadowing, pathloss model, outside cell interference, handover type, and power control dynamics, and is usually evaluated by system level simulations.

If the system is *coverage limited*, outage can then be defined as the probability that pathloss and shadowing exceed the difference between the maximum transmitted power and the required received signal level. Since shadowing is log-normally defined, very low outage probability leads to very high power margin and thus small cells and an expensive network. This is illustrated in Figure 11.1, which shows a plot of a log-normal distribution with standard deviation of 8 dB. The area of the tail of the distribution gives the outage probability. In order to achieve 10% outage probability (i.e., 90% *availability* or coverage probability), a 10.3-dB shadowing margin is required. For 5% outage target, a 13.2-dB margin is required, and for 1% outage target, an 18.6-dB margin. Thus, we can see that increasing the outage target leads to very high margins and thus reduced range. We also plot shadowing with the standard deviation of 10 dB. The increased shadowing requires higher margins since it results in higher probability of large shadowing values.

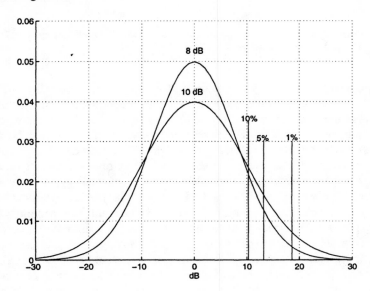

Figure 11.1 Log-normal distribution and shadowing margin.

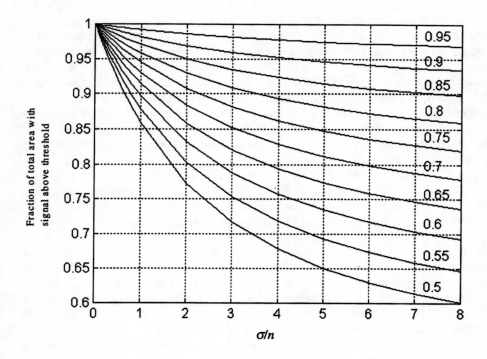

Figure 11.2 Curves relating fraction of total area with signal above threshold as a fuction of probability of signal above threshold on the cell boundary.

The coverage probability can be defined as a cell edge probability or a cell area probability. The cell edge probability refers to a probability that the mobile would receive a signal above a threshold at the cell edge, for example, RF signal strength. The threshold value chosen need not be the receiver noise threshold but may be any value that provides an acceptable signal under Rayleigh fading conditions [3]. The cell area probability means the percentage of locations within a circle of radius R in which the received signal from a radiating base station antenna exceeds a particular threshold value. The relation between area and cell edge probability has been derived in [3] (pp. 126-127) and is illustrated in Figure 11.2, where n is the pathloss exponent and σ is the standard deviation of the shadowing given in decibels. In a macrocell environment, a typical value for the pathloss exponent is 3.6 and for standard deviation is 8 dB. Thus, σ/n is 2.22, and a boundary coverage probability of 85% would give 95% area coverage.

Blocking is defined as the probability of a blocked call on a first attempt. A blocked call occurs due to a lack of system resources (i.e., if all network resources are busy, a user will be denied access to the system). The reasons for blocking can be shortage of channel elements in the base station or shortage of fixed network resources such as transmission lines, BSC, or switching capacity. Thus, each network element needs to be dimensioned appropriately for the traffic demands. Usually, a blocking probability of 2% is used when dimensioning cellular systems.

Let N be the total number of channels and T be the offered traffic in Erlang. Then, the probability of all channels being busy is obtained from a Poisson distribution:

$$P(N;T) = \frac{T^N e^{-T}}{N!} \tag{11.2}$$

where $P(N;T)$ is the blocking rate. Thus, if the traffic density (offered traffic) and blocking probability are known, the required number of network resources (e.g., channel elements in base station or capacity of the transmission lines) can be calculated using the Erlang tables. The Poisson distribution has the effect of achieving better trunking efficiency as the number of channels is increased. Trunking efficiency, also called channel utilization efficiency, is defined as

$$\text{Efficiency}(\%) = \frac{\text{Traffic in Erlang}}{\text{Number of channels}} \times 100\,\% \tag{11.3}$$

Given the blocking probability, (11.2) can be used to calculate the traffic in Erlang for different numbers of channels. Thereafter, (11.3) gives the trunking efficiency. In Figure 11.3, we have plotted the trunking efficiency for a blocking probability of 2%. We can clearly see that the efficiency increases as the number of channels increases.

11.4 RADIO NETWORK PLANNING PROCEDURE

The prerequisite for good radio network planning is the know-how of the radio environment. Rough network planning and deployment plans can be made based on general radio channel models. However, since the radio environment is highly variable, even within the area of one cell, detailed measurements and optimization need to be performed for each individual cell.

The radio network planning process can be divided into three phases:

- Preparation;
- Estimation of cell count;
- Detailed network planning.

11.4.1 Preparation Phase

In the preparation phase, coverage and capacity objectives are established, network planning strategy defined, and initial design and operating parameters determined.

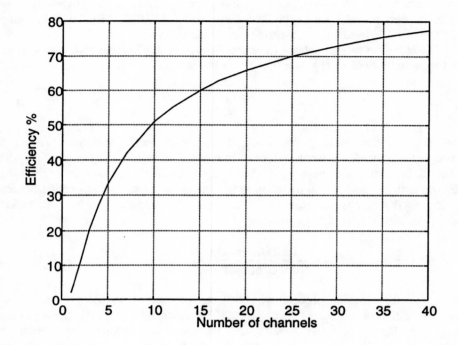

Figure 11.3 Trunking efficiency.

11.4.1.1   Coverage and Capacity Objectives

Coverage and capacity objectives are a trade-off between desired quality and overall network cost. A smaller signal outage probability means smaller cells and thus higher overall network costs; and smaller interference outage probability means smaller capacity and thus also higher cost. A typical outage probability target is 5% to 10%, corresponding to 90% to 95% availability/coverage probability. The coverage probability could be different for different services as discussed below.

11.4.1.2   Network Planning Strategy

The network planning strategy includes issues like microcell deployment, provision for indoor and high bit rate coverage, and migration from second generation systems. Several factors need to be considered for the most feasible network planning approach. These include cost of fixed line transmission, how easily cell sites can be acquired, and at what cost cell sites can be acquired. Furthermore, environmental issues such as cell tower appearance will impact where a base station can be deployed. Traffic distribution will of course impact the deployment strategy.

One deployment strategy could be to use macrocells for outdoor coverage and picocells for indoor coverage in office buildings. In addition, macrocells would be used to fill the gaps in the indoor coverage. This is because extensive indoor coverage is most likely required in any case. Therefore, it might be wiser to build additional capacity by increasing the number of indoor cells rather than trying to provide indoor coverage from outdoor cells and, thereby, be forced to introduce microcells earlier due to capacity restrictions. If coverage in indoors is provided by outdoor base stations, the building penetration margin, typically 10 to 20 dB, needs to be taken into account in the link budget calculations.

Another approach is to use microcells extensively from the beginning and to provide indoor coverage from them. This might be feasible in dense urban areas.

High bit rates can be provided either uniformly over the cell area or the data rate at the cell border could be smaller than when close to the base station to allow a larger cell range. This depends on the nature of the high bit rate services. For services which utilize available bit rates (i.e., services that do not require any quality of service guarantees), nonuniform coverage might be acceptable, but for applications that require maximum bit rate such as video transmission, uniform coverage is required.

If an operator has a deployed second generation network, migration aspects need to be considered in the network planning strategy. These include reuse of existing cell sites, handovers between the new and old systems, and co-existence requirements (see Section 11.9).

11.4.2   Estimation of Cell Count

The number of wireless users in a given area is obtained by multiplying the population by the penetration. The number of users and offered traffic per user determine the overall offered traffic. When cell capacity and range are known, a rough number of cells can be determined. Figure 11.4 illustrates the procedure for the estimation of cell count.

11.4.2.1   Offered Traffic

In the calculation of the number of users that need to be served, at least the following factors should be accounted for:

- Population living in given area;
- Population working in given area;
- Vehicle traffic;
- Special events, and use of recreational areas.

All the above aspects contribute their share to the total number of users that might use the cellular network. When the total number of users is known, we need to estimate what percentage of the total number of users has a wireless terminal (i.e., what is the penetration?).

The offered traffic per user depends on what services are used and how often they are used. The prediction of service usage is especially difficult for third generation for which many new services will be introduced. Different services produce different bit rates. For data services, we need to estimate the average requirement of megabit per second per user. The average calling time for circuit switched users depends on the service (e.g., speech, video), customer types (business, residential), and tariffing. Typical values for the average calling time for speech service range from 120 to 180 seconds. For other services, the average calling time is more difficult to define since there is less experience. Since the average calling time has a major impact on the generated network traffic, it is very important to be able to predict it with reasonable accuracy.

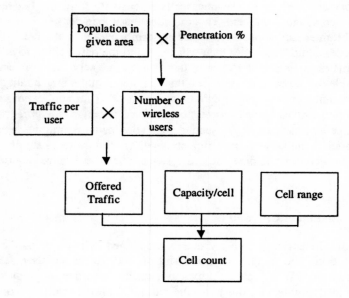

Figure 11.4 Estimation of cell count.

Since the demand for services also varies according to the time of the day, the network capacity must be dimensioned according to busy hour traffic (i.e., when the traffic demand is highest). Typically, busy hour is during business hours. However, new user types (e.g., young people) and a more widespread use of entertainment services through wireless connections might change this.

Example: The total user density is 10,000 user/km^2, 50% of users are using speech and 10% data. Thus, the number of users requiring speech services is 5000 user/km^2 and data services 1000 user/km^2. A speech user generates 30 mErlang traffic and a data user 100 mErlang. Thus, the total offered traffic for speech is 150 Erlang and for data 100 Erlang. If a data call requires on average 100 Kbps, the total offered data traffic is 10 Mbps.

11.4.2.2   Estimate Cell Capacity

As described in Chapter 7, cell capacity can be estimated based on simulations or analytical formulas. User data rate, traffic characteristics (variability, burstiness), quality of service requirements (delay, BER/FER), and outage probability are the main factors that determine the supported capacity. The higher the bit rate, the less users can be supported. The smaller the interference outage probability (i.e., the better the network quality), the smaller the provided capacity. It can be easily understood that provision for better quality radio connections requires more radio resources, and thus fewer users can be supported within a fixed amount of spectrum.

A possible impact of sectorization or adaptive beamforming, also called spatial division multiple access (SDMA), on spectrum efficiency needs to be estimated. It depends on specific radio environment and antenna equipment. For example, in microcells the signals tend to propagate along the street corridors no matter to what direction they were originally transmitted. Thus, the maximum sectorization in the downlink gain is less than in a more open space environment.

A CDMA radio network cannot be operated at a so-called *pole capacity*. Pole capacity can be defined as the theoretical maximum capacity. The *load factor* defines how close to the maximum capacity the network can operate. When determining a suitable load factor, a number of factors need to be taken into account. These include traffic characteristics, radio resource management algorithms, and even the planning capabilities of the network operator. Loads over 75% have been found to cause instability in the system [4].

Example 1: Assume 64-Kbps data service. Based on the network simulations performed in Chapter 7, we conclude that with full load the spectrum efficiency is 100 Kbps/MHz/cell (i.e., 500 Kbps/cell given 5-MHz bandwidth). Assuming 50% load, the actual spectrum efficiency is 250 Kbps/cell (i.e., slightly under 4 Erlang).

Example 2: Assume 8-Kbps speech users. Based on the network simulations performed in Chapter 7, we get spectrum efficiency of 108 kbps/MHz/cell (i.e., 33.75 Erlang with 5-MHz bandwidth and 75% load). Assuming each user generates 30 mErlang traffic, one cell can support 1125 users.

11.4.2.3   Estimate Maximum Cell Coverage

The link budget is calculated according to principles presented in Chapter 7. In addition to the basic assumptions such as data rate and E_b/N_0 performance, the equipment-specific factors such as cable losses, antenna gain, and receiver noise figure need be taken into account.

Soft handover gain has a large impact on link budget. The soft handover gain is the gain brought by handoff at the boundary between two or more cells, where there is equal average loss to each of the cells. The soft handover gain can be obtained by first calculating the log-normal fade margin (i.e., the margin required to provide the specified coverage probability at the border of a single isolated cell, and then the corresponding margin required at the boundary between two or more cells). The soft handover gain is given by the difference (in decibels) between the two different margins. The soft handover gain depends on the shadowing correlation and coverage probability. The larger the coverage probability requirement (i.e., smaller outage probability), the larger the required margin. In addition to the gain brought by cell selection, soft handover brings the macro diversity gain through increased diversity (see Section 7.2.5). The actual gain depends on the radio environment, and number of RAKE fingers.

Since each radio environment has its own characteristics, for more detailed coverage prediction some correction factors for the pathloss models presented in Chapter 4 are required. Field measurements can also be used.

For the uplink, the impact of load factor η in the link budget, the interference margin I_m (in decibels), can be determined from

$$I_m = 10\log\left(\frac{1}{1-\eta}\right) \tag{11.4}$$

and is illustrated in Figure 11.5. As can be seen, the interference margin increases, and thus, the range would decrease with increasing load factor. As was discussed in Chapter 7, multiuser detection can reduce the impact of load factor on the cell range.

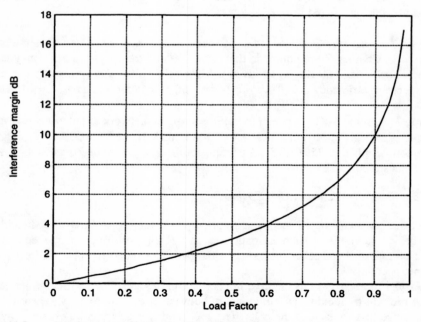

Figure 11.5   Interference margin as a function of load factor.

Asymmetric traffic has to be considered in the link budget calculations. CDMA can trade the uplink system load for coverage. This is useful since usually the mobile station transmission power limits the maximum cell range. If there is three times more downlink than uplink traffic, the uplink may only be loaded to 25%, when the downlink is loaded to 75%. Consequently, the uplink coverage will improve by 4.8 dB compared to the downlink. However, the downlink coverage can be compensated for by increasing the transmission power of the base station.

In Table 11.1, an example of link budget calculation is given. See Chapter 7 for a detailed explanation how to calculate the link budget. Note that in this calculation the carrier frequency and antenna height have been taken into account. We see that the uplink and downlink have different link budgets and thus different ranges. In the last column, we have different bit rates (144 and 28.8 Kbps) and unequal loading in the uplink and downlink. In practice, the downlink link budget is matched to the uplink by adjusting the base station transmission power. This has been done in the last case by increasing the downlink transmission power.

11.4.2.4 Estimate Cell Count

The number of cells required to cover a given area can be calculated based on the capacity and link budget. A network can either be *coverage* or *capacity limited*. Capacity limited means that the maximum cell radius cannot support the total offered traffic. In this case, the cell count is obtained as follows. By dividing the number of users, one cell can support by the number of user/km^2, we obtain the cell area in square kilometer. The total number of cells is obtained by dividing the total area by the cell area. Coverage limited means that there is enough capacity in a cell to support all traffic (i.e., the cell size could be larger from a capacity point of view, but the maximum cell range limits it). In this case, the maximum cell area is used to find out the required number of base stations.

Example: The area to be covered is 100 km^2 and the number of speech users is 5000/km^2. The cell area 23.32 km^2 is obtained from Table 11.1. One cell can support 1041 users with a 50% load, corresponding to 3-dB interference margin. Thus, the cell area is 1041/5000 = 0.2082 km^2 and altogether 100/0.2082 ≈ 481 cells are required to cover the area. If the user density was 30 user/km^2, then the network would be coverage limited since one cell could cover 37.4 km^2, from the capacity point of view. However, this exceeds the maximum cell area of 23.32 km^2. The required number of cell sites would in this case be 100/23.32 ≈ 5.

11.4.3 Detailed Network Planning

After the cell count has been obtained, detailed radio network planning, taking in account the exact radio environment where each cell is deployed, can start. Due to cost reasons, zoning laws, building restrictions, or some other impediment, it is not possible to obtain optimum cell sites in a real network. This will impact the initial coverage plan. For the detailed network planning a network planning tool is used. The previously described process of obtaining rough cell count could also be included in a network planning tool. A network planning tool has a digital map of the area to be planned. Building heights and antenna patterns are also modeled. Since each vendor has its own tools with slightly different capabilities, we try to reflect here the general principles and not the detailed procedures used in real assessment. The optimization process of the radio network coverage includes:

- Detailed characterization of the radio environment;
- Control channel power planning;
- Soft handover parameter planning;
- Interfrequency handover planning;
- Iterative network coverage analysis;
- Radio network testing.

Table 11.1
Link Budget Calculation

Service	Speech		Data		Data		
	Downlink	Uplink	Downlink	Uplink	Downlink	Uplink	Unit
Average TX power/TCH	30	24	30	24	30	24	dBm
Number of code channels	1	1	1	1	1	1	
TX power/code channel	30.00	24.00	30.00	24.00	37.14	24.00	dBm
Max TX power/TCH	30.00	24.00	30.00	24.00	37.14	24.00	dBm
Max total TX power	30.00	24.00	30.00	24.00	37.14	24.00	dBm
Cable etc losses	2	0	2	0	2	0	dB
Antenna gain	13	0	13	0	13	0	dBi
TX EIRP/TCH	41.00	24.00	41.00	24.00	48.14	24.00	dBm
Total TX EIRP	41.00	24.00	41.00	24.00	48.14	24.00	dBm
RX antenna gain	0	13	0	13	0	13	dB
Cable and connector losses	0	2	0	2	0	2	dB
Receiver noise figure	5	5	5	5	5	5	dB
Thermal noise density	-174	-174	-174	-174	-174	-174	dBm/Hz
Interference margin	3	3	3	3	6	1.25	dB
Tot. noise + interf. density, No+Io	-166	-166	-166	-166	-163	-167.75	dBm/Hz
Information rate	8	8	144	144	144	28.8	kHz
10log(Information rate Rb)	39.03	39.03	51.58	51.58	51.58	44.59	dBHz
Eb/No (incl. macro div. gain)	8	6.6	8	6.6	8	6.6	dB
Receiver sensitivity	-119.0	-120.4	-106.4	-107.8	-103.4	-116.6	dBm
Handoff gain (selection gain)	5	5	5	5	5	5	dB
Expl. diversity gain	0	0	0	0	0	0	dB
Other gains / losses	0	0	0	0	0	0	dB
Body loss	3	3	3	3	3	3	dB
Log normal fade margin	11.3	11.3	11.3	11.3	11.3	11.3	dB
Maximum path loss	150.67	146.07	138.12	133.52	142.26	142.26	dB
Carrier frequency f	2000.00	2000.00	2000.00	2000.00	2000.00	2000.00	MHz
Base station antenna height, hB	15.0	15.0	15.0	15.0	15.0	15.0	m
Range	3.97	3.00	1.84	1.39	2.37	2.37	km
Cell area	40.96	23.32	8.80	5.01	14.62	14.62	km^2

11.4.3.1   Detailed Characterization of the Radio Environment

For optimization purposes, the radio environment has to be characterized in detail by actual field measurements or other techniques such as ray tracing [2]. Characteristics such as pathloss, shadowing, and delay spread are studied. Furthermore, a RF survey can be performed using spectrum analysis. RF measurements reveal competing cellular operators and background interference and possible intermodulation products.

11.4.3.2   Control Channel Power Planning

The transmission of control channels reduces the overall capacity of the network. The total transmission power depends on how the power is divided between pilot, synchronization, paging, broadcast, and traffic channels. Power allocation principles are related to coverage.

In FMA2 and wideband IS-95, a common pilot channel is used for handover measurements and coherent detection (see Chapter 6). In Core-A, a BCCH channel is used for handover measurements only (see Chapter 6). The coverage of these channels must be larger, when compared to traffic channels, in order for the mobile station to be able to decode other base stations before entering the soft/softer handover zone. Roughly 5% of total base station power will be allocated to the synchronization channel. Since the broadcast channel including the cell information has to be decoded before the mobile station enters the coverage area of a cell, for example 3 dB more power than for a 8-Kbps speech traffic channel should be allocated. The paging channel(s) can manage with about the same power as a 8-Kbps speech traffic channel.

11.4.3.3   Pilot Pollution

By pilot pollution it is meant that there are a number of pilot signals but none of them are dominant enough to allow the mobile station to start a call. The symptoms of such a situation are typically good mobile received power, poor E_c/I_0, and poor forward BER. To prevent pilot pollution, network planning needs to create a cell plan where a dominant pilot exists. This can be done by scaling pilot powers, down-tilting antennas, or increasing coverage of certain sectors or cells.

11.4.3.4   Soft Handover Parameter Planning

Soft handover performance impacts the required fade margin against shadowing and the number of users in soft handover. As was shown earlier, the fade margin is one factor in the link budget and thus will influence the number of base stations that have to be deployed. Users in soft handover require additional channel elements and backhaul connections (see Section 11.8). The network operator's goal is to minimize the fade margin and number of users in soft handover, while maintaining a satisfactory quality of service.

The number of users depends on handover thresholds and service type. It is possible that packet services do not use soft handover at all. Also, antenna tilt and orientation impact the percentage of cell coverage in soft handoff and should be considered in the handover parameter and region planning. The optimum active set size depends on the handover thresholds, number of available RAKE fingers, and radio environment. Typically, more than three base stations in the active set do not give significant soft handover gain increase.

In the following we review the soft handover planning based on the IS-95 soft handover algorithm. The main parameters for soft handover are handover thresholds, handover timers, and pilot search window at the mobile station. The parameters used in the active set updating are the following:

- T_{ADD}: this threshold indicates the point when a pilot should be added to the candidate set. The mobile station is measuring the pilot E_c/I_0.
- T_{DROP}: this threshold indicates when a pilot should be dropped from the active set if pilot strength has dropped below T_{DROP} for T_{TDROP} seconds.
- T_{COMP}: the mobile station reports that a candidate set pilot is stronger than an active set pilot only if the difference between their respective strengths is at least $T_{COMP} \times 0.5$ dB.

A typical value of T_{ADD} is between −12 and −16 dB [5]. Increasing the T_{ADD} threshold reduces the number of users in the soft handover and thus reduces the required equipment overhead. However, reducing it too much increases interference and thus reduces performance [5]. Since the time required to detect a new pilot should be minimized, filtering in the searcher for pilot in neighboring sets should be minimized. The T_{ADD} threshold should be high enough to prevent false alarms due to large noise in pilot measurements.

T_{DROP} should be low enough to prevent the loss of a good pilot. Typical values for T_{DROP} are between −17 and −20 dB. If the T_{TDROP} timer is too short, unnecessary handovers might occur, so T_{TDROP} should be in the order of the time required to establish handover. If T_{ADD} and T_{DROP} are too close to each other and the T_{TDROP} timer is very short, a handover ping-pong effect (i.e., adding and dropping the same pilot consecutively) might occur.

T_{COMP} should be set to a value that would prevent the mobile station from continuously sending active set update messages as a consequence of small changes in the strengths of pilots in the active set and the candidate set. Too large values, however, would introduce too long a delay before a pilot strength measurement message is issued, delaying the handover procedure [6].

The impact of T_{COMP}, pathloss slope and standard deviation of shadowing (log-normal fading), and different radio environments have been studied in [7]. The following results were reported: the higher the log-normal standard deviation, the higher the shadow margin required to combat fading. Higher pathloss slope causes better isolation between adjacent cells and hence a lower margin is required. T_{COMP} did not impact performance very much in [7]. The required fade margin increases with delay in establishing handover. A typical handover delay in today's systems is 2 to 5 seconds [8].

In the future, other criteria such as uplink interference might be required in the handover process. Furthermore, different services with different quality of service requirements might require different criteria for handover and even different active set sizes. This increases the network planning complexity.

Since soft handover gain depends on the correlation between the signal from different cell sites and on the correlation length of the shadow fading from the same cell site, field measurements are required to quantify the exact gain and thus the network capacity. Lower correlation between sites leads to higher call reliability. Furthermore, a large correlation length of the log-normal fading increases the fading margin [8]. The impact of correlation length depends on the mobile speed: the faster the mobile moves, the smaller the impact, since the correlation coefficient for the log-normal fading is smaller.

The searcher window determines the number of chip delays for which a mobile is instructed to look for pilot. A large window helps to collect all the multipath energy but leads to a slow measurement process, which might be detrimental for high-speed mobiles. A too narrow window leaves some of the significant multipaths unnoticed and thus degrades performance. Obviously, the window size is a function of delay spread and should be optimized for each cell, taking the expected mobile speed distribution into account.

11.4.3.5   Interfrequency Handover

The compressed mode is one alternative for seamless interfrequency handover (see Chapter 5). The main parameter of interfrequency handover is the frequency at which the base station commands the mobile station to measure neighboring carriers. The stronger the serving cell, the more seldom the mobile would check the neighboring channels. At cell edge, the mobile should measure pilot channels more often in order to be prepared for interfrequency handover.

The use of slotted frames leads to performance degradation. This degradation depends on the number of slotted frames. Assuming that the performance of a slotted frame is 2 dB worse compared to continuous frame, the overall degradation depends on the number of slotted frames. For example, if every 10th frame is slotted, the degradation would be 0.2 dB.

In case there are both micro- and macrocell coverage, then the criteria for which each layer is selected has to be decided. In general, if a fast moving mobile is detected in microcells it should be handed over to macrocells. Therefore, the speed of mobile stations needs to estimated, for example, from the fading characteristics. On the other hand, a slow moving mobile should be handed over to microcells. Wrong decisions should be avoided; for example, a mobile station in a car that has stopped at traffic lights should be kept in macrocells since it will speed up in the next moment.

11.4.3.6 PN-Offset Planning

In IS-95 and cdma2000, pilot signals belonging to different base stations and sectors are distinguished by the use of different phase offsets of the same pilot pseudo-noise (PN) sequences. If a propagation delay between two base stations exceeds the pilot signal offset, the so-called PN confusion occurs. PN confusion might cause dropped calls due to high interference from another base station or a handover to a wrong target cell. Therefore, PN-offsets need to be planned [9,10].

The distance d a pilot signal travels can be converted to a distance in chips using the following formula [10]:

$$T_d \text{ (chips)} = \frac{B}{c} \times d \tag{11.5}$$

where B is the chip rate of the system and c is the speed of light (3×10^8 km/s). If the chip rate is 1.2288 Mcps as in IS-95 and the PN-offset index is 64 chips, one ON-offset index corresponds to a 15.6-km propagation distance. For cdma2000, one PN-offset index would correspond to 5.6 km. Figure 11.6 presents a scenario where the propagation distance difference of cell I and cell J is equal to PN-offset, therefore causing PN-confusion.

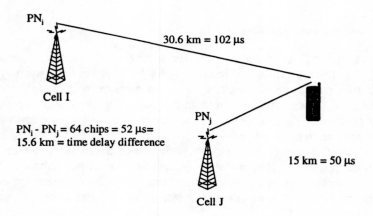

Figure 11.6 PN-confusion (*After*: [10]).

11.4.3.7 Iterative Network Coverage Analysis

Once the radio network environment has been characterized, control channel power allocation decided, and handover parameters planned, a detailed coverage analysis can be carried out. The ratio of in-cell to total interference is also referred to as the *frequency reuse coefficient* (see Chapter 7). The frequency reuse coefficient is unique for each cell. Starting from the initial configuration based on the preceding steps in the planning process (cell count, detailed characterization of the radio environment, control channel power planning, soft handover parameter planning, interfrequency handover planning), network quality is analyzed and new frequency reuse coefficients evaluated [5]. These are then used again to predict the coverage within different cells. This iterative process is repeated until convergence is achieved. A network planning tool can be used to automate the process. With this type of tool, quality plots can be created and gaps or holes in the coverage can be detected.

11.4.3.8   Nonuniform Traffic

In general, nonuniform traffic degrades the overall performance of the system. On one hand, the quality becomes poor due to increased interference in dense traffic zones; on the other hand, the quality becomes excessive. This dispersion of the communication quality restricts the system capacity [11]. The system efficiency can be improved by adaptive control of cell radius, antenna directivity, and uplink received power threshold. The cell radius is controlled by adapting the pilot transmission power. If the observed SIR is higher than required, the cell radius can be extended; and if it is lower, the cell radius is reduced. The desired received power threshold in the uplink is increased and reduced, respectively, to balance the cell radii of uplink and downlink. In a sector cell configuration, the central angle of each sector (i.e., the antenna directivity) is changed to equalize the communication quality in every sector belonging to the BS [11].

Of course, dynamic control of cell radius and antenna directivity requires careful planning and stable control system to prevent undesired effects. Also, interaction with handovers should be considered.

11.4.3.9   Radio Network Testing

After the network has been deployed, it is tested by drive tests to find out the network quality in practice. Furthermore, drive tests are used to collect data for optimization of the network performance. Before measuring, the optimization scope is determined so that correct measurements can be performed. Typical topics for measurements are coverage limitations, drop-call rate, quality, and pilot sets for soft handover.

Once the network has been set up, the work does not end. Testing continues as an essential part of the network operations. Therefore, proper operation and maintenance facilities are required to perform measurements and to collect data about network behavior. The measurement data are used to monitor the network quality and to locate congestion and quality gaps.

11.5   MICROCELL NETWORK PLANNING IN CDMA

Due to propagation environment and network topology, microcell deployment differs significantly from the macrocell deployment (see Chapter 4). Microcell base stations are installed at lamp post level and therefore signals tend to propagate along street canyons. Two critical aspects for microcell network planning are corner effect and soft handover design.

11.5.1   Corner Effect

Figure 11.7 illustrates the so-called *corner effect*. When a mobile moves around a corner, changes in received signal level at the mobile station happen very rapidly. In case there is a new base station behind the corner, the signal strength received by the mobile station rises very fast. If the mobile station could not acquire the new base station fast enough, the increased interference leads to a dropped call. On the other hand, since the new BTS cannot regulate the power of the mobile station, the high transmission power of the mobile station can block all users in the new cell. To reduce the impact of corner effect, a fast forward handover can be performed if the old base station is dropped. Corner effect can also be avoided by proper planning of cell locations and handover thresholds. Handover thresholds can be specified in such a way that the mobile station is in soft handover before and after a street corner (i.e., overlapping cells).

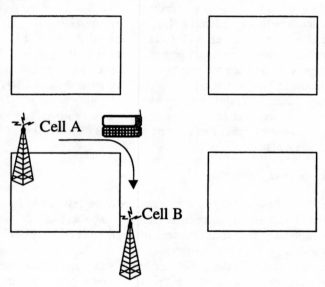

Figure 11.7 Corner effect.

11.5.1.1   Soft Handover Design in Microcell Environment

Since microcells are small, frequent handovers occur as the users move along the streets. The danger is that if traditional cell planning is used, then fast moving mobiles cause so much interference due to too slow handovers that capacity in microcells will drastically decrease. Furthermore, the signaling load increases considerably due to a large number of base stations. Thus, it is advantageous to deploy intelligence to the base stations.

The need for soft handovers in microcellular CDMA can be reduced by using distributed antenna and sectored cells. With distributed antennas, as shown in Figure 11.8, large coverage areas are achieved without use of many small micro- or picocells. The forward link capacity is not good, since the signal is transmitted to the whole coverage area even though the mobile is only in one location.

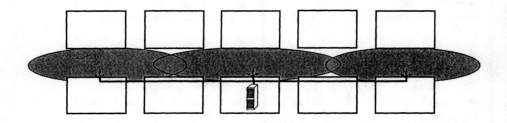

Figure 11.8   Street microcell with distributed antennas.

11.5.2   Micro/Macrocells in the Same Frequency

Placing micro- and macrocells in the same frequency is possible with low antenna installation and antenna tilting [12]. The danger is that the whole cell is in soft handover and no extra capacity is gained. In this solution, a fast soft handover from macrocell to microcell is required.

11.6   INDOOR PLANNING

The building types, sizes, construction, and materials exhibit a very large variation. Thus, no general propagation model that would be valid everywhere exists. Attenuation due to walls and floors depends on the construction of the building. Furthermore, propagation loss between floors depends also on structures like stairs and elevators. Therefore, field measurements or prediction of propagation by Ray-tracing are required to characterize individual buildings.

Due to short propagation distances, the values of indoor delay spreads are very small. Thus, the coherence bandwidth is large and consequently diversity order is low. Therefore, extra diversity may be added with delayed transmission or reception. One possibility is to use distributed antennas. By introducing time delay elements between the antennas in the distributed antenna system, deliberate multipaths are created that can be processed by the CDMA RAKE receiver [13]. However, creating too many multipaths actually degrades the performance due to uncaptured energy.

The deployment of indoor cells into the same frequency as outdoor cells presents some challenges for the system operation. Since mobiles have to be connected to a cell where they use minimum transmission power, outdoor users can occasionally get the lowest power with picocells, for example, through windows. This situation most likely will not last very long. How is the handover working in this situation? One way to avoid this kind of situation is to deploy indoor cells to a different frequency band. Another possibility is careful network planning to avoid this. This means low transmission powers indoors and antennas with their backs towards windows.

11.7   SECTORIZATION AND SMART ANTENNAS

Sectorization increases the capacity of a CDMA system. In an ideal case, perfect isolation between sectors is achieved and the capacity will increase in direct proportion to the number of sectors. Of course, in practice, isolation is not perfect and the capacity increase is smaller. Furthermore, the number of softer handovers increases with the number of sectors. Several factors impact the sectorization gain. These include interference distribution and antenna location. Obstacles and structures near the antenna tend to change the antennas impedance and radiation pattern (see also Section 7.3.2.8). This causes the coverage to deviate from the planned coverage. Another impact is that coupling between diversity antennas might increase, and consequently, correlation increases, degrading performance. Usually it is assumed that 3 sectors result to a capacity increase of 2.5.

Smart antennas (also called SDMA) are defined as multibeam or adaptive array antennas without handover between beams [14]. Multibeam antenna uses multiple fixed beams in a sector, while in an adaptive array the received signals by the multiple antennas are weighted and combined to maximize the SNR. The advantage of antenna arrays compared to fixed beam antennas that in addition to the M-fold antenna gain they provide M-fold diversity gain. However, they require a receiver for each antenna, and tracking the antenna weights at the rate of the fading.

The M-fold antenna gain will increase the range by a factor of $M^{1/\gamma}$ where γ is the pathloss exponent, and reduces the number of base stations to cover a given area by $M^{2/\gamma}$ [14]. A multibeam antenna with M beams can increase the capacity by a factor of M by reducing the number of interferers. Adaptive arrays can provide some additional gain by suppressing interferers further. However, since there are so many interferers the additional gain might not be worth the complexity.

11.8   NETWORK DIMENSIONING

11.8.1   BTS Channel Element Planning

The number of channel elements can be calculated when the offered traffic in Erlang, control channel requirements, and number of users in soft handover are known. Calculations are based on Erlang tables that give the number of required channels for a given blocking probability and offered traffic.

In addition to the traffic channels that can be determined from capacity calculations, common control channels in the downlink and access channels in the uplink need channel elements. Each wideband CDMA carrier requires one pilot and at least one common control channel, which consists of the paging channel, synchronization channel, and broadcast channel. In the uplink, the number of access channels depends on the expected access load and synchronization time requirements. In case an umbrella cell is used to overlay microcells, the microcells do not necessarily need that many common control channels and access channels if the initial access traffic is handled mainly through macrocell. However, for packet access, common control channels are required also in microcells.

The additional number of channel elements due to soft handover depends on the soft handover type. Softer handover does not need an additional channel element since the same channel element can handle signals from different sectors within the base station. Two-way soft handover needs two channel elements, and three-way soft handover needs three channel elements. As was discussed previously, the number of users in soft handover depends on the radio environment, handover thresholds, and antenna configuration and it can be found out from simulations or field measurements.

Due to the trunking gain effect, it is beneficial to pool channel elements between sectors and different carriers in the same cell site. This also applies to the channel elements due to soft handover. Therefore, the actual number of additional channel elements is less than the percentage of users in soft handover due to improved Erlang efficiency as will be illustrated by the following example.

Example: Table 11.2 shows a calculation for the number of required BTS channel elements. We consider a reference BTS implementation, which has one CDMA frequency and three sectors and is realized as one cluster with a common pool of "floating" channel elements. The total carried traffic per sector is 39 Erlang including soft handover overhead traffic of 9 Erlang. Thus, the total traffic for the cell site is 3×39 = 117 Erlang. For a blocking rate of 2%, the total amount of needed channel elements for the cluster is 130, or 43.3 per sector (using the Erlang-B formula applied for 117 Erlang). In addition, for the downlink one channel element for pilot channel and for common control channel is required. In the uplink, one channel element for access channel is assumed. Thus, the overall number of channel elements per sector is 46.

Without additional soft handover traffic, the required number of channel elements for traffic channels would be 34.3. The additional number of channel elements due to soft handover is 26.2%, which is less than the 30% overhead traffic and is due to the Erlang effect.

If there was no pooling between sectors, the required number of channel elements for traffic channels would be 49 (Erlang-B formula applied for 39 Erlang). Thus, the pooling reduces the required number of channel elements by 12%.

For high bit rate services, we have 8 Erlang traffic. Assuming 30% overhead traffic due to soft handover, we have 10.4 Erlang per sector and 31.2 Erlang per site, 41 channel elements = 41/3 = 13.7. The additional soft handover traffic 33/3=11 is 8 channel elements (i.e., 24.2%).

Table 11.2
Channel Element Calculation

	Speech 8 Kbps	Data 144 Kbps
Erlang/sector	30 Erl	8 Erl
Soft handover overhead	30%	30%
Total traffic (3 sectors)	117 Erl	31.2 Erl
Number of channel elements/sector from Erlang B table, pooling taken into account	44	14
Channel elements for pilot and common control channels	2	2
Total number of channel elements	46	16

11.8.2 Number of BSCs and Switches, HLR and VLR Signaling Traffic

The number of BSCs is determined based on the maximum configuration. Different vendors support different traffic loads through the BSC. Similar to BSC, switches from different vendors support different traffic loads, which also determines the number of required switches.

In order to determine signaling related to mobility management (i.e, between HLR and VLR), mobility models are required. The mobility models describe the mobility of wireless users (see Section 4.7).

11.8.3 Transmission Capacity

The required transmission capacity depends on the user and signaling data that needs to be transmitted between network elements. Connections include at least BTS-BSC, BSC-switch, and possibly BSC-BSC connections for handover purposes. Transmission capacity depends on the transmission technology. ATM, for example, provides good multiplexing gain. Soft handover traffic increases the required transmission capacity up to the network element, most likely to the BSC, where soft handover is terminated. In addition, soft handover between cells belonging to different BSCs over switches might require special arrangements. It also has to be considered whether additional blocking is allowed for transmission.

11.8.4 Transmission Network Optimization

When the rough network dimensioning has been obtained, an optimization phase is carried out. Depending on the costs of transmission, different network configurations can produce optimum overall cost. Sometimes it is beneficial to have more BSCs than required to tailor the transmission network for the most optimum configuration. If a BTS is situated in the middle of two BSCs, then routing of that BTS needs to be determined based on transmission costs that might be different for the two cells.

11.9 CO-EXISTENCE

When a wideband CDMA network is deployed over a given area, other networks will nevertheless continue to operate. Most likely the W-CDMA network will co-exist with GSM, IS-95, and PDC systems. Two main scenarios for the deployment of wideband CDMA are then predictable:

- Wideband CDMA introduced in third generation frequency band;
- Wideband CDMA introduced into the second generation frequency bands.

An existing cellular operator will most likely start UMTS with local coverage, relying on the second generation system for low bit rate wide area coverage. Therefore, it is wise to ease the dual-mode terminal design by properly selected air interface parameters. In the longer term, the replacement of a second generation system by wideband CDMA depends on the gained advantages, such as higher bit rates or increased spectrum efficiency. Furthermore, licensing and availability of spectrum also impact the decision. If the available bands of an operator are congested, the higher spectrum efficiency of wideband CDMA might allow the operator to pack existing users into less spectrum and to create room for higher rate services. However, the introduction of a wideband carrier into congested frequency bands is fairly difficult. First, the operator needs to preload the network with dual-mode terminals with new wideband capabilities. Next, he needs to release the spectrum from old carriers and switch the new carrier on almost simultaneously to avoid interruption in service.

A Greenfield third generation operator in a country with an existing cellular network also needs dual-mode terminals. Only if the data market emerges very fast and justifies the investment in a new nationwide network can an operator rely on single-mode terminals.

The transition to third generation systems depends on technical possibilities such as increased spectrum efficiency and data rates but most of all on market needs and regulatory conditions. In this section, we highlight some of technical aspects related to network planning facilitating smooth transition and co-existence. The transition from the first generation technology AMPS to the second generation digital technologies in the United States gives us some background for the transition to third generation as well. To avoid problems with co-existence of different air interfaces, the following aspects need to be considered from the network planning perspective:

- Intermodulation;
- Deployment scenarios;
- Guard zones and bands;
- Transition aspects;
- Handover between systems.

11.9.1 Intermodulation (IM)

A power amplifier is nonlinear near the saturation point. In this nonlinear area, the signal is distorted and produces harmonics. If two or more different frequencies are present at the input of an amplifier, intermodulation products are generated [15]. Figure 11.9 illustrates intermodulation power and its relation to the power of the original signal. As can be seen, the power of the third order intermodulation product increases three times faster than the power of the original signal. Equally, the power of the fifth order intermodulation product increases five times faster.

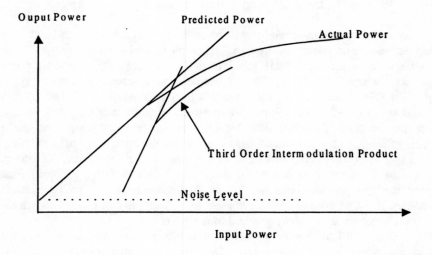

Figure 11.9 Intermodulation power (*After*: [15]).

In mobile radio systems, there are two types of intermodulation products: transmit IM and mobile generated IM [16]. The mobile station generated IM is more significant than transmit IM from the base station. Transmit IM is generated at the base station transmitter. Several carriers are passed through the same power amplifier creating intermodulation products that might fall in the wideband CDMA band. Mobile generated IM is caused by the nonlinear effects in the active stages of the receiver front-end; these stages typically consist of a RF low noise amplifier (LNA), mixer and IF amplifier. As was discussed previously, the power of the third order intermodulation products, considered as the primary source for IM interference, is increasing very fast as the RF gain increases. Thus, when the mobile is close to the interfering base station, the third order products become severe in relation to the desired CDMA signal causing possibly a dropped call. One proposed solution to avoid this is to switch out the LNA to eliminate strong IM signals from occurring at the mobile station front-end prior to mixing [17]. However, if the desired signal is already weak, this might lead to a dropped call as well.

If wideband CDMA is deployed within the same frequency band as an existing network, there are two possibilities to map CDMA cells and existing cells. One alternative is *one-to-one mapping* of the cells of the wideband CDMA and the other system. Another alternative is that one wideband CDMA cell can cover more than one cell of the existing system (1-to-N mapping), since, in the beginning, wideband CDMA deployment load is still low and the network can be built to be coverage limited. Furthermore, the existing system has already reached quite small cell sizes due to large capacity demand. Later, when wideband CDMA capacity demand increases, it is likely that it will have either the same cell size as the other systems or an even smaller cell size.

From the deployment point of view, one-to-one mapping is better with respect to intermodulation product avoidance. There should be no intermodulation interference problems since wideband CDMA has similar power than the interfering system. Of course, this will lead to increased network infrastructure costs in the beginning. In 1-to-N mapping, the CDMA mobiles with small signal power might receive high intermodulation products. When a wideband CDMA mobile at the CDMA cell edge is at the sensitivity limit, then high power intermodulation products might block the mobile. This kind of deployment requires more planning.

11.9.2   Guard Bands and Zones

A second issue is frequency co-existence. Due to spectrum regrowth proper guard bands need to be allocated in order to not block the system by adjacent spectrum or to avoid that the other system blocks the wideband CDMA system.

A guard zone around the CDMA coverage area is needed to avoid co-channel interference. An example of an uncoordinated co-channel interference situation is when different technologies in the same frequency band are not separated well enough by a geographical barrier. Such a situation has been experienced in the United States between PCS market areas. The Code of Federal Regulations (CFR) Title 47 Part 24 addresses issues related to interference caused by the system. It specifies, for example, attenuation rules outside the operators' own spectrum (Section 24.238) and allowed field strength at the border of an operator's licensed service area (Section 24.236). The

National Spectrum Managers Association (NSMA) has prepared a document titled "Inter PCS Co-block Coordination Procedures" addressing the intersystem coordination rules for PCS systems. It specifies coordination distance, allowed degradation due to external interference, interference calculation procedures, and other coordination practices to prevent harmful interference from one system to another.

The guard zone depends on the interference that the system can tolerate from other systems and vice versa.

Example: For GSM the required carrier-to-interference ratio is

$$\frac{C_{GSM}}{I} = 9\,dB \tag{11.6}$$

We allow 1 dB degradation in C/I due to interference from CDMA:

$$\frac{C_{GSM}}{I + I_{CDMA}} = 8\,dB \tag{11.7}$$

Thus

$$\frac{C_{GSM}}{I_{CDMA}} > 15\,dB \tag{11.8}$$

From this we can now calculate the minimum separation between GSM and CDMA base stations using the same frequency. The ratio between cell size R and distance D can be calculated from [18]

$$q = \frac{D}{R} = \left(6\frac{C}{I}\right)^{1/\lambda} \tag{11.9}$$

Deployment within existing footprint reduces deployment efficiency. If coverage- limited deployment is desired, this might not be feasible since existing sites and their antennas are used. Furthermore, guard zones decrease the capacity of existing systems and depend on the deployment scenario.

Guard zones between IS-95 and AMPS have been studied in [19]. For a coverage limited system, they are larger in order to maintain the quality of the IS-95 network. Within a guard zone, the CDMA carrier plus guard bands cannot be used. So, even if IS-95 had larger capacity and coverage and it could be deployed using a lesser number of sites than AMPS, it proves to be that due to guard zone reasons, one has to convert enough sites to IS-95 to increase the total traffic in the network [19].

11.10 FREQUENCY SHARING

Frequency sharing means sharing the allocated frequency spectrum in the same geographical area between two or more operators. It is one potential technology to increase the overall capacity of a set of mobile radio networks in the same area. By sharing the frequencies, the trunking loss due to the subdivided total frequency band between operators is avoided. The frequency sharing is possible if the co-channel interference power of the interfering systems is less than the co-channel interference power from the own system. However, since CDMA systems are due to have a DCA scheme, frequency sharing might be possible only if the operators use the same cell sites.

REFERENCES

[1]   Yang, S., *CDMA RF System Engineering*, Boston: Artech House, 1998.
[2]   Cheung, J. C. S., M. A. Beach, and J. P. McGeehan, "Network Planning for Third Generation Mobile Radio Systems," *IEEE Communications Magazine*, Vol. 32, No. 11, November 1994, pp. 54–59.
[3]   Jakes, W. C., *Microwave Mobile Communications*, New York: John Wiley & Sons, 1974.
[4]   Stellakis, H., and A. Girdano, "CDMA Radio Planning and Network Simulation," *IEEE Int. Symbosium on Personal, Indoor and Mobile Radio Communications PIMRC'96*, Taiwan, 1996, pp. 1160–1162.
[5]   Wallace, M., and R. Walton, "CDMA Radio Network Planning," *Proceedings of VTC'94*, Stockholm, Sweden, 1994, pp. 62–67.
[6]   Qualcomm, *CDMA Network Engineering Handbook*, 1993.
[7]   Chopra, M., K. Rohani, and J. D. Reed, "Analysis of CDMA Range Extension due to Soft Handoff," *Proceedings of VTC'95*, Chicago, IL, USA, May 1995, p. 917–921.
[8]   Rege, K., S. Nanda, C. F. Weaver, and W.-C. Peng, "Analysis of Fade Margins for Soft and Hard Handoffs," *Proceedings of IEEE 6th Personal, Indoor and Mobile Radio Communications Conference (PIMRC'95)*, Toronto, Canada, 1995, pp. 829–834.
[9]   Chang, C. R., J. Z. Van, and M. F. Yee, "PN Offset Planning Strategies for Non-Uniform CDMA Networks," *Proceedings of VTC'97*, Vol. 3, Phoenix, Arizona, USA, May 4-7, 1997 pp. 1543–1547.
[10] Yang, J., D. Bao, and M. Ali, "PN Offset Planning in IS-95 Based CDMA Systems," *Proceedings of VTC'97*, Vol.3, Phoenix, Arizona, USA, May 1997, pp. 1435–1439.
[11] Sato, S., and Y. Amezawa, "A Study on Dynamic Zone Control for CDMA Mobile Radio Communications," *Proceedings of ICC'97*, Vol. 2, Montreal, Quebec, Canada, June 8-12, 1997, pp. 984–988.

[12] Shapira, J., "Microcell Engineering in CDMA Cellular Networks," *IEEE Transactions on Vehicular Technology*, Vol. 43, No. 4, November 1994, pp. 817–825.

[13] Xia, H. H., A. B. Herrera, S. Kim, and F. S. Rico, "A CDMA-Distributed Antenna System for IN-building Personal Communications Services," *IEEE Journal on Selected Areas in Communications*, Vol. 14, No. 4, May 1996, pp. 644–650.

[14] Special Issue on Smart Antennas, *IEEE Personal Communications*, Vol. 5, No. 1, February 1998.

[15] Faruque, S., *Cellular Mobile Systems Engineering*, Boston: Artech House, 1996.

[16] Hamied, K., and G. Labedz, "AMPS Cell Transmitter Interference to CDMA Mobile Receiver," Proceedins of VTC'96, Vol. 3, Atlanta, Georgia, USA, April 1996, pp. 1467–1471.

[17] Gray, S. D., and T. Kenney, "A Technique for Detection AMPS Intermodulation Distortion in an IS-95 CDMA Mobile," *Proceedings of PIMRC'97*, Vol. 2, Helsinki, Finland, September 1-4, 1997, pp. 361–365.

[18] Lee, W. C. Y., *Mobile Cellular Telecommunications Analog and Digital Systems*, Second Edition, New York: McGraw-Hill, Inc., 1995.

[19] Ganesh, R., and V. O'Byrne, "Improving System Capacity of a Dual-Mode CDMA Network," *International Conference on Personal Wireless Communications*, Mumbai, India, December 1997, pp. 424–428.

Chapter 12

NETWORK ASPECTS

12.1 INTRODUCTION

In this chapter, we discuss the network aspects related to the development of third generation wideband CDMA systems. Since many of the network related aspects are independent of the radio interface, wideband CDMA itself is not discussed in depth in this section. The focus of this chapter is on the evolution from second generation networks towards third generation systems. It should be noted that most of the aspects described in this chapter are still under development in the standardization and thus are subject to change. This evolution proposal presents the author's view on the subject.

Section 12.2 describes a generic design methodology for mobile radio systems. This will guide the reader to understand the formal methodology used in the design of modern telecommunication systems. The different models used in the design methodology are defined. The open system interconnection (OSI) reference model and the Integrated Services Digital Network (ISDN) protocol reference model, as well as their application to the design of modern wireless systems, are introduced. The reference architecture defines the network elements and interfaces between them. The main network elements found in today's second generation systems are defined in the context of a network reference architecture. The already established functionality of second generation wireless systems is discussed with the help of functional planes. Finally, the preceding concepts are tied together by the stack and protocol architecture; the functions of different functional planes are mapped into the stack and protocol architecture, which is based on the OSI and ISDN protocol reference model principles.

The distinction of access and core network is one of the leading principles in third generation standardization work. Section 12.3 explains how this division impacts the network architecture of third generation systems. Furthermore, the Generic Radio Access Network (GRAN) and "family of systems" concepts defined in ETSI and ITU, respectively, are discussed.

In Section 12.4, several important technologies with respect to third generation networks are discussed, namely intelligent networks (IN), asynchronous transfer mode (ATM), SS7 signaling system, and mobile application part (MAP) protocol.

Section 12.5 describes the new service capabilities required for third generation networks. Based on this, the new requirements that these new capabilities put on the current second generation networks are identified.

In Sections 12.5 and 12.6, the network architecture and protocols of GSM- and IS-41/IS-95-based systems are described. The impact of third generation requirements on these networks is identified and evolution possibilities are suggested.

While this chapter covers the evolution of the second generation mobile radio systems to third generation systems, the basic principles of evolution apply to any system. We try to highlight the procedure of evolution, which consists of the identification of new needs, in the assessment of current capabilities, and in the mapping of the evolution needs into interfaces and protocols of the existing systems and standards.

12.2 DESIGN METHODOLOGY

Modern communications systems are very complex collections of different functions, and mobile radio systems are no exception. In order to manage the design of such systems it is essential to use some structured modeling and design methodology. There exist several different system design methodologies, for example, the IN conceptual model and the CCITT 3-stage methodology.

The CCITT 3-stage methodology was originally developed for the design of ISDN-based networks [1]. Since wireless networks contain many new elements compared to ISDN-based fixed networks, the 3-stage methodology is usually applied in an ad hoc manner in the design of wireless telecommunication networks. A simplified view of the 3-stage methodology is illustrated in Figure 12.1. First, the new services planned to be offered by third generation networks need to be characterized. Based on this, new functional capabilities to support these services are identified. Finally, the functions are mapped into a reference architecture, which defines network elements and interfaces. For protocol design, the identified functions are mapped into a combined stack and protocol architecture based on OSI and ISDN reference models.

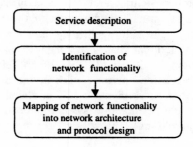

Figure 12.1   Design stages.

In the following five sections, the OSI and ISDN reference models, reference architecture, functional planes, and stack and protocol architectures are described.

12.2.1   OSI and ISDN Reference Model

The OSI model [2] shown in Figure 12.2 specifies seven different layers, each consisting of a set of functions that can be implemented independently from the functions in the other layers. The OSI model specifies communication principles between two network elements. The abstraction level of the model increases from bottom to top. This also means that in a time axis the span of events is shorter in lower layers and increases towards higher layers. In this way, communication between network elements can be defined in a modular manner. In addition to the peer entity, each layer communicates with the layers in the same network element immediately below and above it.

The physical layer of the OSI model is often referred as layer 1, link layer as layer 2, and network layer as layer 3. The rest of the layers are abstracted as higher layers. In the following, we use both naming conventions.

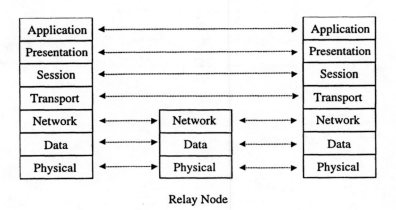

Figure 12.2   The OSI model.

The ISDN protocol reference model (PRM) is similar to the OSI reference model in the respect that it also organizes communication functions in layers and describes the relations of these layers with respect to each other [3]. In the modeling of wireless networks both models are used. The ISDN PRM introduces a division between the so-called *user* and *control planes*. The user plane consists of all functions being in charge of transferring user data, and the control plane consists of all functions being in charge of transferring information for the control of user plane data. Since usually at lower protocol layers a distinction of user and control data flows cannot be made, a third plane, *transport plane,* has been introduced. This is not explicitly described in the OSI reference model. It corresponds to the lower OSI layers (physical and data link and possibly the network layer). Figure 12.3 shows the scope of user, control, and transport planes.

In the modeling of wireless networks, both OSI and ISDN PRM models are used. GSM is the first system that has followed these principles. The OSI model forms the basis for vertical protocol division, while the ISDN PRM model forms the basis for defining different protocols for user and control planes. Therefore, usually two different protocol stacks, the user and control plane, are defined. Sometimes the control plane is referred to as the *signaling plane*.

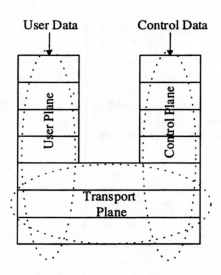

Figure 12.3 User, control, and transport planes in the OSI model.

12.2.2 Reference Architecture

The reference architecture specifies the network entities and the interfaces between these entities. Each network entity comprises a set of functions and is responsible for performing its allocated tasks. A sound architecture requires that related functions should be grouped in the same network entity. A typical mobile radio system contains at least the following network elements: mobile station, base station, and switch and data bases for roaming purposes such as home and visitor location registers (HLR and VLR). Figure 12.4 depicts an example of reference architecture. In the following, we depict the network entities found in digital mobile radio systems such as GSM, IS-95, IS-136, and PDC. These elements and interfaces (denoted by letters in Figure 12.4) will be then elaborated on in Sections 12.6.1 and 12.7.2 for GSM and IS-95/IS-41, respectively. The basis for the description is the GSM specification GSM 03.02 and the TR-45 reference model for IS-41.

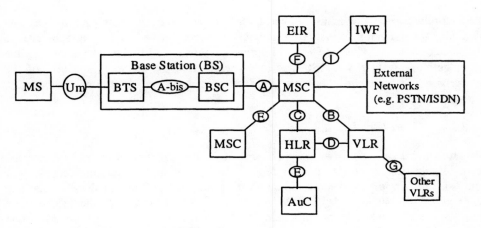

Figure 12.4   Example reference architecture.

12.2.3   Network Elements

In this section, different network elements in the reference model are described. The mobile cellular network itself is a public land mobile network (PLMN), which can connect either to external networks such as public switched telephone network (PSTN), ISDN, and public data network, or to another PLMN.

12.2.3.1   Mobile Station

The mobile station (MS) consists of the physical equipment used by a subscriber. The mobile station is usually divided into mobile termination (MT) and terminal equipment (TE) parts. MT terminates the radio path and contains functions such as modulation, error correction, and mobility management. TE is a terminal (e.g., data terminal) connected to a MT possibly via a terminal adaptation function.

12.2.3.2   Base Station System

The base station system (BSS) terminates the radio path in the network side. It communicates with the mobile station and connects to the mobile switching center through an interface termed the A-interface. Often, a so-called Abis-interface is implemented in such a way that the BSS consists of one base station controller (BSC) and one or more base transceiver stations (BTS). One BSC controls one or more BTSs. A BTS is a network component, which serves one cell.

12.2.3.3   Mobile-services Switching Center

The mobile-services switching center (MSC) is an exchange that performs all the switching and signaling functions for mobile stations located in a geographical area designated as the MSC area. The main difference between a MSC and an exchange in a fixed network is that the MSC has to take into account the mobile nature of the subscribers. Therefore, it has to perform procedures required for the location registration.

12.2.3.4   Home Location Register

This functional entity is a database in charge of the management of mobile subscribers. It maintains subscriber information (e.g., international mobile station identity, user profile). Furthermore, it contains some location information that enables the charging and routing of calls towards the MSC where the MS is located (e.g., the MS roaming number, the VLR address, the MSC address, the local MS identity). One home location register (HLR) can be associated with one or more MSCs.

12.2.3.5   Visitor Location Register

The visitor location register (VLR) is the location and management database for the mobile subscribers roaming in the area controlled by the associated MSC(s). Whenever the MSC needs data related to a given mobile station currently located in its area, it interrogates the VLR. When a roaming mobile station enters a new service area covered by an MSC, the MSC forwards a registration request (or a location update request) to the associated VLR, which informs/interrogates the HLR. Usually, one VLR is associated and co-located with one MSC.

12.2.3.6   Authentication Center

The authentication center (AC or AuC) is associated with an HLR, and it manages the authentication and encryption parameter for each individual subscriber.

12.2.3.7   Equipment Identity Register

The equipment identity register (EIR) is used for keeping track of mobile stations and their identities. Its purpose is to be able to record and prevent the use of possibly stolen or otherwise misused terminals.

12.2.3.8   Interworking Function

The interworking function (IWF) is a functional entity usually associated with the MSC. It allows interworking with external networks.

12.2.4   Functional Planes

Functional planes identify the functions performed in a cellular radio system to make the system work. The functions of each plane can then be mapped in the OSI model and the network architecture. This mapping serves as a basis for the detailed protocol and interface specification. In second generation systems, five generic functional planes are identified [4]:

- Communications management (CM);
- Mobility management (MM);
- Radio resource management (RRM);
- Transmission;
- Operations, administration, and maintenance (OAM).

The communication management layer includes functions related to setting up, maintaining, and releasing calls between users (i.e., call control (CC)). Furthermore, the CM layer has means to manage calls with so-called supplemental services. These include call forwarding, call waiting, and call hold. The third function of the CM layer is the management of short message services (SMS).

Mobility management covers the functions related to keep track of mobile subscribers within the home network and when they are roaming. Roaming is the use of services from other networks than the terminal's home network. Specific functions to manage these tasks are location update, paging, and cell and network selection. In order to obtain service from a cell, a user must register to the location area of that cell. This is called a location update. Before the network routes a call towards the subscriber, it pages the mobile station. The page is only sent to those cells belonging to the location area the mobile station currently belongs to. Obviously, the size of the location area is a trade-off between the signaling required for location updates and paging traffic. Roaming requires that there exist means to pass information between the home and visited networks. This is handled by the HLR and VLR, and signaling between them.

The task of radio resource management (RRM) layer is to maintain communication between terminals and base station. In order to do this, RRM performs the following tasks:

- Admission control;
- Channel assignment;
- Load control;
- Power control;
- Handover.

Admission control determines whether to accept the requested new communication, which can be based on several criteria. The channel assignment function assigns a physical channel for the connection. Load control manages the network load: it can reduce the traffic in the network, for example, by reducing the bit rate for certain users. The radio resource management functions, including power control and handover, were discussed in Chapter 5.

The transmission layer is concerned with the transfer of signaling and user data. For the user data, the system needs to provide a consistent end-to-end transmission path, therefore, translation functions are required between different parts of the network. For example, currently the speech over the GSM air interface is transmitted with a 13-Kbps data stream. On the other hand, in the infrastructure side 64-Kbps PCM links are used. Thus, we need adaptation between these two rates.

The OAM plane contains functions related to monitoring and controlling of the system. These functions include billing and accounting, and collecting performance data.

12.2.5 Stack and Protocol Architecture

Now we have defined the concepts required for understanding the definition of a modern telecommunications network: reference architecture, OSI and ISDN reference model, and functional planes. To put these concepts together we need to map the functions identified in different functional planes into the OSI model layers and into the network elements. This is performed with the allocation of the identified functions[1] to a network/physical architecture (i.e., a set of network elements) described by *stack and protocol architecture* illustrated in Figure 12.5. As can be seen from Figure 12.5, several protocols can cross one interface, and some of the protocols exchange messages between nonadjacent network elements. An interface is a point of contact between two adjacent entities. In the stack architecture, the physical architecture is presented by means of protocol stacks. The functions are allocated to network elements and to protocol layers. It represents a "vertical" view of protocol layers. The protocol architecture focuses on certain protocol layers (i.e., it shows where a given protocol layer or sublayer is terminated). It represents a "horizontal" view with respect to protocols. In Figure 12.5, the different protocols are denoted with a letter P and two numbers indicating the layer number and the protocol in question.

The CM, MM, and RRM planes concern the control of the system. They are mapped into layer 3 of the OSI model. In order to perform their tasks, CM, MM, and RRM protocols need to exchange messages among peer entities located in different network entities and among other protocol entities. Thus, their specification consists of a definition of *messages* and the use of these messages.

The transmission functions within the mobile network, comprising of a MS, BS, MSC, HLR, VLR, and some additional network entities, can be mapped into three different OSI layers: physical layer (layer 1), link layer (layer 2), and network layer (layer 3). The end-to-end connection between two terminals also uses higher OSI layers; for example, for data connections the transmission control protocol (TCP) is used in the transport layer.

---

[1] A function can be described by a state machine or an algorithm. Network protocol designers consider a function as a state transition description (i.e., a state machine), while radio interface designers identify a function as an algorithm [5].

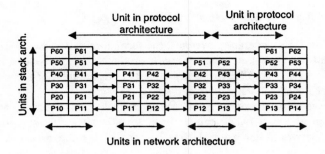

Units in network architecture

Figure 12.5 Stack and protocol architecture.

The link protocol structures the information to be transmitted in units bigger than a single bit. These units are called *frames*. When the maximum length of a message passed to the link layer exceeds the predefined maximum length, segmentation and reassembly have to be performed. The link layer also performs error detection and correction as well as quality monitoring and flow control.

While the link layer enables the exchange of information between directly interconnected network elements, the network layer provides additional transmission functions for information transmission with end-to-end connections through several network elements. The network layer covers aspects such as routing (i.e., how messages are passed from one point to another until they reach their final destination). Routing requires an addressing mechanism. For example, when a mobile station sends a message, it can go to the MSC, BSC, or HLR.

The above-described generic modeling based on application of the CCIT 3-stage model presents a simplified view of the actual system development. GSM is the first wireless system that has followed structured specification principles, and this is reflected in its sound architecture and protocols. However, even GSM has peculiarities where interfaces and protocols mix with each other and protocol boundaries are not clear. And sometimes, the conceptual distinction between protocols and interfaces is ill resolved in the actual GSM specifications [4].

The concepts of interface, protocol, and network elements will help us to understand how the second generation networks evolve towards the third generation. When new services such as multimedia calls are introduced, we need to first identify the new fuctionalities in the different planes required to support them. Then, these functions are mapped into protocols, interfaces, and network elements to see how they need to be developed. We might need a machine to fulfil a new function, for example the GPRS support node (GSN) in GSM (see Section 12.6.1), which was introduced in order to support packet switched services. On the other hand, the introduction of a new network element results in a new interface such as the Gb interface between BSS and GSN. An existing interface might also require to be changed to accommodate new services.

12.3   CORE AND ACCESS NETWORKS

When discussing third generation systems, the term core network and its evolution are frequently mentioned. A functional network architecture can be described with the concepts of access and core networks, as illustrated in Figure 12.6, in which the GSM network entities have been used as examples.

An access network comprises all functions that enable a user to access services. Furthermore, an access network can be used to hide all access-specific peculiarities from the core network. For example, in the case of a radio access network, all air interface–related functions should be kept within the access network part.

The core network comprises the switching network and service network. A switching network includes all the functions related to call and bearer control for fixed transmission. The MSC in GSM is an example of a switching network entity. A service network comprises all functionalities for the support services including location management.

Since most of the radio related parts are located in the access network, the core network should be little impacted by a new radio interface, and can evolve partly independently of the access network. Therefore, one of the drivers in third generation standardization, especially in Europe, has been a more clear separation between access and core network than that of the current systems. This is motivated by the isolation of radio-related functions from the switching functions. This separation has led to the concept of GRAN in ETSI, described in the next section.

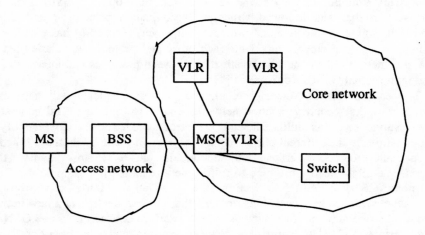

Figure 12.6   Functional network architecture with GSM entities used as an example.

12.3.1   Generic Radio Access Network Concept

In ETSI, the Generic Radio Access Network (GRAN) has been defined according to principles identified in the global mobile multimedia (GMM) report [6]. The basic idea of the GRAN is to separate the development of core and access networks. The benefit is that it leaves freedom in the future for new types of products, which utilize different core network solutions. Partly, the reasons to introduce the concept of GRAN have been political: to give freedom for standardization between rival camps. The interested reader can refer to [7] for different views about the GRAN concept.

The main focus of UMTS standardization will be in the radio access system, which will be connected to several core networks through interworking units (IWU), as depicted in Figure 12.7. The GSM BSS (Base Station Subsystem) is also part of GSM/UMTS system concept. The GRAN concept leads to an open, multivendor system that has proven to be attractive to operators. UMTS air interface will be an open and well-specified standard (U_u interface). Furthermore, the I_u interface specifies the interface between GRAN and core network, while the C_u interface specifies the interface between terminal equipment and UMTS subscriber identity module (USIM).

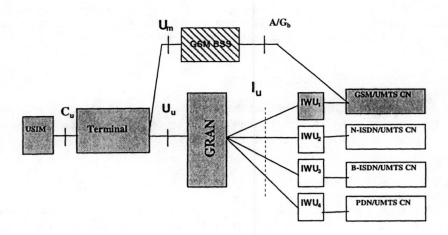

Figure 12.7   GRAN concept.

The internal structure of the GRAN comprises all the radio related functions. These include:

- All functions located in the MS terminating the lower air interface layers and controlling the channel management over the air interface from MS side;
- All functions located in the fixed part of the GRAN, called BSS, including all radio resource management functions.

In order to maintain the separation between radio dependent and independent parts, access and core networks, handover should be limited to BSS. This requires connections between BSSs.

12.3.2   ITU Family of Systems

The original idea of IMT-2000 was to create the specifications for a single global system. It has gradually been realized that changing overnight all networks to a new system would not work in practice due to commercial interests. This led to the development of the "family of systems" concept. The system belonging to ITU family of systems should be capable of supporting the IMT-2000 requirements and capabilities. Thus, the IMT-2000 target system has been replaced by a virtual IMT-2000 reference network that is primarily used for specifying the necessary interfaces for the IMT-2000 family members. The main candidates for IMT-2000 family members are GSM and IS-41 based core networks, which are evolving towards third generation.

The work to specify IMT-2000 family concept is currently underway. The ITU-T recommendation Q.FIN contains the definition of the family concept [8]. The main value of that work will most likely be the achieving of a consensus about the main third generation requirements, called capability sets, and, if desired, specification of roaming capabilities between IS-41 and GSM based IMT-2000 networks.

12.4   NETWORK TECHNOLOGIES

This section reviews some technologies that will be used in third generation networks, namely intelligent networks, asynchronous transfer mode, the signaling system 7, and mobile application part, and Mobile Internet protocol.

12.4.1   Intelligent Networks

The IN concept was developed to enable a fast deployment of new services in all telecommunication networks. The objective of IN is to allow the inclusion of additional capabilities to facilitate service provision, independent of the service/network implementation in a multivendor environment [9]. IN distinguishes between service switching, service control, and service data functions. The service control function (SCF) is the functional entity in the IN that contains the service logic for implementing a particular service. The service switching function (SWF) is the functional entity that interfaces the switch to the service control function. IN services are defined as sets of capabilities in ITU-T recommendations. Currently, Capability Set 1 (CS1) is stable.

An example of IN service is a simple number translation, like the number series "#61" would mean that your calls should be forwarded to a given number. Then, if the service provider would like to change what that number string means, the change could be easily programmed to the SCF without changing the basic telecommunication software in the switch.

12.4.2   Asynchronous Transfer Mode

Today's transmission protocols of many telecommunication networks are based on pulse code modulation (PCM), and mobile radio systems are no exception. The switching is also based on the switching of 64- or 56-Kbps PCM connections. ATM technology has received lot of attention during recent years as the next major transport technology. ATM has also been proposed for wireless applications (i.e., ATM cells are transmitted over the air interface). However, here we assume that it is only used as a transmission protocol in the infrastructure side.

ATM provides not only the multiplexing gains of packet switching, but also the guaranteed delay characteristics of circuit switching. The fundamental strategy behind ATM is to split the information into small fixed size units that are easy to handle. The fixed size of the cell allows efficient switching. ATM networks are high-speed switching systems offering large bit pipes, which allow *statistical multiplexing* (i.e., multiplexing of many connections with variable rate characteristics), which altogether reduces the overall bandwidth requirements. Since ATM is based on the transmission of fixed size cells, it can be easily evolved for future services.

The basic unit in an ATM is a cell of 53 bytes, illustrated in Figure 12.8. It consists of header (5 bytes) and information payload fields (48 bytes). The header consists of following fields:

- Generic flow control (GFC);
- Virtual path identified (VPI);
- Virtual channel identifier (VCI).

The VPI and VCI are used to identify the virtual path and virtual channels identified with that path to route the ATM cells from the source node to the destination node.

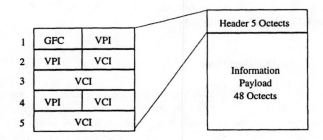

Figure 12.8   ATM cell format.

The ATM and the ATM adaptation layer (AAL) form the data link layer. AAL converts the arbitrarily formatted information supplied by the user into ATM cells. Various forms of AAL protocols are necessary to handle the different types of traffic.

- *AAL0* provides direct access to the ATM layer.
- *AAL1* assumes constant bit rate traffic, which is intolerant of missequenced information and variation in delay. It offers the following functions: segmentation and reassembly (SAR), handling of delay variation handling, handling of lost and misinserted cells, source clock recovery, monitoring for bit errors, and handling those errors.
- *AAL2* is used, for example, for voice and video. It assumes that traffic is bursty and intolerant of missequencing and that a time stamp is needed for packet reassembling. It offers the following functions: multiplexing, SAR, handing delay variation, handling cell lost/error, and source clock recovery. AAL2 will be used for compressing data in third generation mobile radio systems in the network infrastructure.
- *AAL3/4* and *AAL5* are geared to traffic that has bursty characteristics with variable frame length. Furthermore, delay is not critical and packets can be resequenced based on sequence numbers. AAL5 is expected to supersede AAL3/4 since it has lower overhead and TCP/IP acknowledgements fit into one cell in AAL5 instead of two cells in AAL3/4.

12.4.3   Signaling System 7

Both IS-41 and GSM make use of protocols specified for SS7, which is a signaling system designed for the transfer of control information between network elements. The layered structure of SS7 is shown in Figure 12.9. SS7 layers are called parts. The ISDN user part (ISUP) of SS7 contains messages carried from ISDN standard devices. For example, call-related signaling of GSM makes use of ISUP when connecting to external networks. The GSM and IS-41 MAP protocols use the transaction capabilities part (TCAP) for network control. Furthermore, the signaling connection control part (SCCP) and message transfer part (MTP) are used in the A-interface of GSM and in the IS-634 interface, which specifies the corresponding A-interface for the North American digital standards, especially for IS-95/IS-41. A tutorial of SS7 can be found in [10], and use of SS7 in IS-41-based systems is discussed in [11].

12.4.4   Mobile Application Part

A unique feature of mobile systems is the support of roaming (i.e., seamless provision of services for users who are subscribed to one operator's network while they are within the coverage of some other operator's network). The functions required for roaming are authentication of the subscriber, transferring subscriber data to the visited network, and mechanism for routing connections towards the subscribers. The functions to implement roaming are part of the mobility management. For this purpose, both GSM and IS-41 systems have defined mobile application part (MAP). MAP defines the application protocols between switches and databases (e.g., MSC, VLR, HLR) for supporting call management, supplementary service management, short message transfer, location management, security management, radio resource management, and mobile equipment management. In principle, the MAP of IS-41 and GSM perform similar functions. However, their implementations and the way that protocols are specified are different.

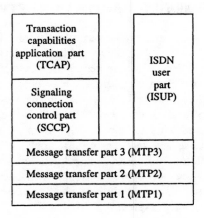

Transaction capabilities application part (TCAP)	ISDN user part (ISUP)
Signaling connection control part (SCCP)	
Message transfer part 3 (MTP3)	
Message transfer part 2 (MTP2)	
Message transfer part 1 (MTP1)	

Figure 12.9   SS7 protocols.

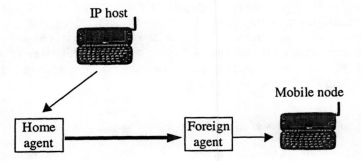

Figure 12.10. Mobile IP architecture.

12.4.5   Mobile Internet Protocol

In general, it can be stated that Mobile Internet protocol (IP) is for data networks what MAP is for telecom networks. Internet is built using software that relies on IP. Mobile IP is a modification to IP, which makes possible for a mobile node to visit other IP networks while still being reachable by its home IP address. Mobile IP architecture is based on the *home/foreign agent* concept depicted in Figure 12.10. The datagrams are routed to the home network using normal IP routing. If the mobile node is in a foreign network, the datagrams are encapsulated and tunneled towards the so-called *care-of address* assigned to the mobile node when away from the home network. For more details on Mobile IP, refer to [12,13].

12.5 EVOLUTION OF SECOND GENERATION NETWORKS

In the beginning of third generation system standardization, it was not clear that the evolution from second generation networks would be the right way to proceed nor even what was meant by the evolution of second generation networks. The original idea was to start from scratch and to define a completely new core network and protocols for mobility management and call control. However, during recent years it has become more evident that third generation core networks will be based on the evolution of second generation networks. This is due to the desire to utilize existing network investments, and also because many of the original goals defined for third generation networks have already been implemented in second generation networks. However, one still has to consider a balance between evolution, which is restricted by the existing implementations of the standard, and revolution, which can be done without limitations and thus results in a more efficient system.

In order to develop the technical solutions for third generation networks, we need to answer two questions: in what areas is evolution needed and how can this evolution be implemented. Therefore, we need to identify the capabilities that differentiate third generation from second generation networks. Once we have identified the new capabilities that need to be supported, we can analyze what changes and new functionalities are required in the existing systems and standards. Next, the distribution of these functionalities into protocols and network elements, along with possible new protocols, network elements, and interfaces, needs to be identified. We have to separate those aspects that impact the standard and those that impact the way networks are implemented. For example, manufacturers of wireless equipment usually design the detailed algorithms for radio resource management.

12.5.1 New Capabilities of Third Generation Systems

In this section, we discuss what the impact of new services is on the required system capabilities.

12.5.1.1 Multimedia

Third generation networks will support multimedia services, meaning sufficient bandwidth and bearer flexibility [14]. The first requirement, sufficient bandwidth, relates to transport technology. Since some third generation services are asymmetric, bursty, and demand high bandwidth, transmission capabilities need to be developed to match these requirements. Higher air interface rates together with variable bit rates will mean that new, more flexible transport and switching technology is required. The support of multimedia service calls for separation of the call control from the connection and bearer control. A call/session might use various connections at any one particular instant. Thus, it should be possible to add and remove bearers during such a call. Separation of call and connection/bearer control means that many connections such as speech, video, and data could be associated with one single call and these could be handed over separately [14]. Thus, handover algorithms need to support multiple simultaneous services and different handover schemes for them.

One of the main applications for third generation networks is assumed to be the transmission of Internet services. Currently, the IP makes no assumptions about the underlying protocol stacks and offers an unreliable, connectionless network-layer service that is subject to packet loss, reordering, and packet duplication [15]. Therefore, the IP-delivery model is referred to as a best-effort service. For non real-time services, this kind of transmission model is not a main problem. However, for multimedia services, it is not adequate, and enhanced QoS classes are required. The resource reservation protocol (RSVP) is specified to provide means to reserve network resources along the data path, and to ensure end-to-end QoS for the selected application. It could be a solution for this problem. Other protocols are also considered for this purpose [15].

12.5.1.2   Bearer Service Classification

In order to better understand what kind of new bearer services are required, we introduce the ATM service classification. The third generation service classes are not necessarily exactly the same, but, since ATM was designed to carry all kinds of traffic, it serves as a good starting point for wireless networks.

Constant Bit Rate Service (CBR): CBR may be used for any transparent data transfer. Resources are allocated on a peak bit rate basis.

Unspecified Bit Rate Service (UBR): UBR bearers use free bandwidth when available. If no resources are available, the information frames are queued.

Available Bit Rate Service (ABR): Resources for sending at the specified minimum bit rate are allocated to the user. Higher bit rates may be used on best effort basis, free bandwidth is used when available.

Variable Bit Rate Service (VBR): This service provides a variable rate characteristic based on statistical traffic management. Average throughput (sustainable bit rate) and frame loss ratio are two key parameters. A set of traffic parameters specify the source traffic characteristics. This service can be either real-time (with delay bound) or non real-time (without delay bound).

Currently, second generation networks support CBR and UBR (with and without priority) services. GSM HSCSD also supports ABR services. VBR service would require enhancements to second generation core networks.

12.5.1.3   Service Development

One of the major differences between second and third generation networks is that instead of standardized services, standardized bearer capabilities for supporting services are provided. This means that rather than standardizing teleservices, bearer services providing suitable "bit-pipe" for any kind of services will be specified. Nevertheless, the most common services like speech would still have to be standardized to ensure seamless roaming and spectrum efficiency.

A very important protocol for mobile systems is the mobility management, which facilitates seamless roaming. This requires certain agreements and also that networks support the same services. On the other hand, due to hard competition, operators would like to distinguish themselves from each other. This has led to the emergence of operator-specific services provided by INs. With IN the operator-specific services can be provided in a flexible way

A further extension of the IN concept is the virtual home environment (VHE) [16]. It is defined as a system concept for personalized service portability across network boundaries. The VHE concept will ensure uniform appearance and presentation of services and features. Thus, not only can the user access services in a similar way as in the home network, but, for example, response messages or error messages are also presented in a similar way.

12.6   GSM EVOLUTION TOWARDS UMTS

In this section, we analyze the existing architecture, interfaces, and protocols of the GSM and GPRS networks. Based on this, we introduce a possible evolution scenario and suggest some enhancements required to allow the evolution of GSM and GPRS towards the third generation.

12.6.1   Reference Architectures

For GSM, we need to consider two reference architectures, one for the basic GSM and one for GPRS to identify the interfaces subject to evolution. These are illustrated in Figures 12.11 and 12.12, respectively.

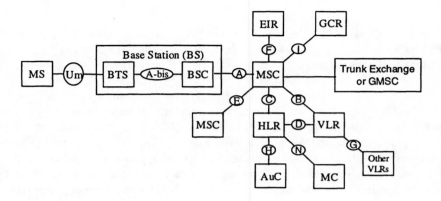

Figure 12.11   Basic GSM reference architecture.

In addition to the network elements specified in Section 12.2.3, we can identify the following GSM-specific network elements.

Subscriber Identity Module (SIM): This is part of the mobile station and is used to store all subscriber related information. It is either a smart card (the size of a credit card) or a so-called "plug-in" SIM, which was designed to facilitate an easier implementation of small terminals.

Group Call Register (GCR): The GCR is the management database for the voice group or broadcast calls in the area controlled by the associated MSC(s). Whenever the MSC needs data related to a requested voice group or broadcast call, it interrogates the GCR to obtain the respective voice group or broadcast call attributes. More information is provided in TS GSM 03.68 and 03.69.

Shared Interworking Function (SIWF): The SIWF is a network function that provides interworking for circuit data/fax calls. SIWF consists of a SIWF controller (SIWFC) functionality located in MSCs and SIWF server(s) (SIWFS) located in the PLMN. A SIWFS can be accessed by several other network nodes (e.g., any MSC in the same PLMN). More information is provided in TS GSM 03.54.

Gateway MSC (GMSC): If a network delivering a call to the PLMN cannot interrogate the HLR, the call is routed to an MSC. This MSC will interrogate the appropriate HLR and then route the call to the MSC where the mobile station is located. The MSC that performs the routing function to the actual location of the MS is called the gateway MSC (GMSC). For more information, refer to TS GSM 03.04.

SMS Gateway MSC (SMS-GMSC): The SMS-GMSC acts as an interface between a short message service center and the PLMN, to allow short messages to be delivered from the service center (SC) to mobile stations.

SMS Interworking MSC: The SMS interworking MSC acts as an interface between the PLMN and a short message service center to allow short messages to be submitted from mobile stations to the SC.

Interworking Function (IWF): The IWF is a functional entity associated with the MSC. The IWF provides the functions necessary to allow interworking between a PLMN and the fixed networks (ISDN, PSTN, and PDNs). The functions of the IWF depend on the services and the type of fixed network. The IWF is required to convert the protocols used in the PLMN to those used in the appropriate fixed network. The IWF may have no functionality where the service implementation in the PLMN is directly compatible with that at the fixed network. The interworking functions are described in GSM Technical Specifications GSM 09.04 – 09.

Interfaces: In the basic GSM configuration presented in Figure 12.10, all the functions are considered implemented in different equipments. Therefore, all interfaces within PLMN are external. Interfaces A and Abis are defined in the GSM 08-series of technical specifications. Interfaces B, C, D, E, F, and G need the support of the SS7 MAP to exchange the data necessary to provide the mobile service. No protocols for the H-interface and for the I-interface are standardized. We now describe the main interfaces defined between the different GSM network elements:

- The interface between BSC and the MSC (A-interface) is specified in the following GSM specifications: GSM 08.01, 08.02, 08.04, 08.06, 08.08, and 08.20.
- The interface between the BTS and the BSC (the Abis-interface) is specified in the following GSM specifications: GSM 08.51, 08.52, 08.54, 08.56, 08.58, 08.59, and 08.60.
- The interface between the MSC and the HLR (C-interface): signaling in this interface uses the MAP protocols specified in GSM 09.02.
- The interface between the MS and the BTS (air interface, Um-interface) is specified in the GSM 04 and 05 specification series. The GSM 04 series covers the general principles and higher protocol layers, and the GSM 05 series contains the physical layer specifications.
- The interface between the MSC and its associated VLR (B-interface) is internal to the MSC/VLR, and signaling on it is not standardized.
- The interface between the HLR and the MSC (C-interface): the GMSC must interrogate the HLR of the required subscriber to obtain routing information for a call or a short message directed to that subscriber. Signaling in this interface uses the MAP protocols specified in GSM 09.02.
- The interface between the HLR and the VLR (D-interface) is used to exchange data related to the location of the mobile station and to the management of the subscriber. Signaling in this interface uses the MAP protocols specified in GSM 09.02.

- The interface between MSCs (E-interface) is used for the exchange of data for handover between two MSCs. This interface is also used to forward short messages. Signaling on this interface uses the MAP specified in GSM 09.02.
- The interface between VLRs (G-interface): when an MS initiates a location updating using TMSI, the VLR can fetch the IMSI and authentication set from the previous VLR. Signaling on this interface uses the MAP specified in GSM 09.02.
- The interface between the HLR and the gsmSCF (J-interface) is used by the gsmSCF to request information from the HLR. The support of the gsmSCF-HLR interface is a network operator option. As a network operator option, the HLR may refuse to provide the information requested by the gsmSCF.

GPRS architecture: GPRS is logically implemented on the GSM structure through the addition of two network nodes, the serving GPRS support node (SGSN) and the gateway GPRS support node (GGSN) [17]. Therefore, several new interfaces have been defined. Figure 12.12 depicts these, together with the GSM interfaces explained previously.

- The interface between the SGSN and the BSC (Gb-interface).
- The interface between GGSN and HLR (Gc-interface) is optional.
- The interface between SMS-GMSC and SGSN, and between SMS-IWMSC and SGSN (Gd-interface) enables GPRS MSs to send and receive short messages over GPRS radio channels.
- The interface between two GSNs within the same PLMN (Gn-interface).
- The interface between two GSNs in different PLMNs (Gp-interface) has the same functionality as Gn-interface plus a security function required for inter-PLMN communication.
- The interface between SGSN and HLR (Gr-interface).
- The interface between SGSN and MSC (Gs-interface).
- The interface between GPRS and fixed network (Gi-interface) is actually called a reference point since it is not fully specified. This is described in GSM 09.61.

For more details on GPRS, refer to [18–20].

12.6.2   Protocol Stacks

Figure 12.13 illustrates the GSM protocol stack. The layer 3 protocols – radio resource management (RRM), mobility management (MM), and call control (CC) – are specified in GSM 04.08. LAPDm and LAPD are link layer protocols. SCCP' is the GSM specific implementation of the original SS7 SCCP protocol [4]. The A-interface is used for messages between BSC and MSC as well as for messages to and from the mobile station. Therefore, the BSS application part (BSSAP) specified in GSM08.08 is split into two subapplication parts, the BSS management application part (BSSMAP) and the

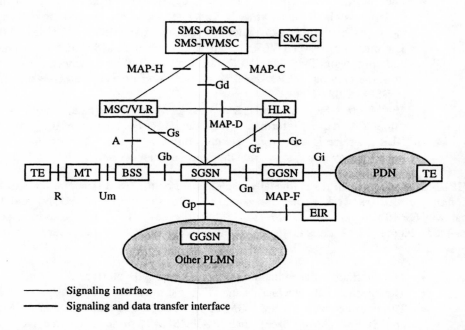

Figure 12.12   GSM/GPRS reference architecture. (*Source*: [17], reproduced with permission from ETSI.)

direct transfer application part (DTAP). DTAP is used to transfer the signaling for call and mobility management (CC and MM protocols) transparently to the mobile station. The BSSMAP supports all the procedures between the MSC and the BSS that require interpretation and processing of information related to single calls, and resource management. Both these protocols make use of the SCCP and MTP protocols, which are part of the SS7 protocols. The transaction capabilities application part (TCAP) is used by the MAP protocol for network control. Not shown in Figure 12.13 is the MAP protocol. GSM 09.02 specifies the application protocols between exchanges and data bases (MSCs, GMSCs, VLRs, HLRs, EIRs, and SGSNs) for supporting call management, supplementary service management, short message transfer, location management, security management, radio resource management, and mobile equipment management. It also specifies the applicability of SCCP and TCAP protocols to support these exchanges.

12.6.2.1   *GPRS Protocol Stacks*

The GPRS protocol architecture makes a distinction between the transmission and signaling planes. The transmission plane consists of a layered protocol structure providing user information transfer, along with associated information transfer control procedures (e.g., flow control, error detection, error correction, and error recovery).

Figure 12.14 shows the protocol stack for the GPRS transmission/user plane, while Figure 12.15 presents the GPRS signaling/control plane.

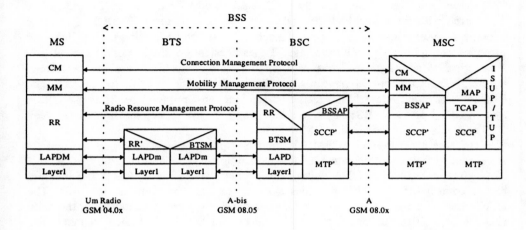

Figure 12.13 GSM protocol stack.

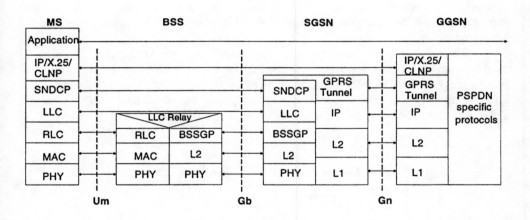

Figure 12.14 GPRS transmission/user plane. (*Source*: [17], reprinted with permission of ETSI.)

The GPRS tunneling protocol (GTP) transfers data between GPRS support nodes. It encapsulates the protocol data units (PDUs) of packet data protocols such as X.25 or IP. Furthermore, GTP provides a mechanism for flow control between GSNs, if required. The GTP protocol is defined in GSM 09.60. Below the GTP protocol are TCP and UDP protocols. TCP carries GTP PDUs for protocols that need a reliable data link (e.g., X.25) and UDP for protocols that do not need a reliable data link (e.g., IP). IP version 4 is used as a GPRS backbone network-layer protocol providing routing functions for user data and control signaling. The IP version 6 will replace version 4 in the future.

Between the SGSN and the MS, packet data protocol (PDP) PDUs are transferred with the subnetwork dependent convergence protocol (SNDCP) that maps network-level characteristics into the characteristics of the underlying logical link, and provides multiplexing of multiple layer 3 messages into a single virtual logical link connection. In addition, ciphering, segmentation, and compression are also performed by the SNDCP protocol. Logical link control (LLC) provides reliable logical link. The LLC relay function relays LLC PDUs between Um and Gp interfaces. The base station system GPRS protocol (BSSGP) conveys routing and QoS-related information between BSS and SGSN. It is specified in GSM 08.64.

In the radio interface GSM RF, the GSM physical layer specified in GSM 05 series performs modulation, demodulation, encoding, and decoding of data.

The data link layer has two sublayers: radio link control (RLC) and medium access control (MAC). RLC provides a radio solution-dependent reliable link. The MAC function controls the access signaling (request and grant) procedures for the radio channel, as well as the mapping of LLC frames into the GSM physical channel. The RLC and MAC functions are described in GSM03.64.

The signaling plane consists of protocols for control and support of the transmission plane functions such as attaching to and detaching from the GPRS network, activation of a PDP (e.g., IP or X.25) address, controlling the routing path to support user mobility, controlling the assignment of network resources, and providing supplementary services.

The previously described physical layer, RLC/MAC and LLC protocols are also used for signaling transmission between the MS, BSS, and SGSN. The layer 3 mobility management (L3MM) protocol supports mobility management functionality such as GPRS attach, GPRS detach, security, routing update, location update, PDP context activation, and PDP context deactivation between MS and SGSN.

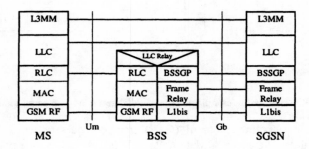

Figure 12.15   GPRS signaling/control plane between MS and SGSN. (*Source*: [17], reprinted with permission of ETSI.)

Signaling between the SGSN and HLR is performed using the same protocols, TCAP, SCCP and MTP, as for the non-GPRS GSM PLMNs. The MAP protocol supports signaling exchanges between SGSN and HLR with enhancements for GPRS.

12.6.3   Evolution of the GSM Architecture

In this section, we discuss the expected changes to GSM architecture and protocols due to the evolution towards third generation. These changes are currently discussed and developed in the standardization, and the reader is referred to the standardization documents in ETSI SMG3 AND SMG12 (see Chapter 14) for the latest information.

As shown in Figure 12.16, the first implementations of GRAN will be based on the integration of URAN and GSM/UMTS core network, which has been evolved from the GSM core network by integrating new third generation capabilities. The evolved GSM network elements are referred to as 3G MSC and 3G SGSN.

GRAN interfaces with GSM/UMTS core network via Iu-interface corresponding to GSM A-interface and GPRS Gb-interface [21]. As can be seen, radio access is isolated from the core network, and the goal is that the GSM/UMTS core network would have the flexibility to support any radio access scheme. Circuit switched services are routed via the GSM MSC, and the packet switched services via the GPRS part of the GSM/UMTS core network.

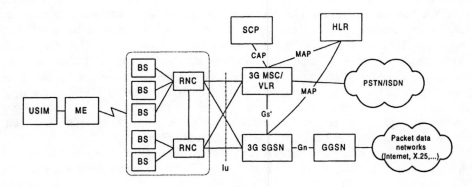

Figure 12.16   One possible scenario where GRAN is connected to an evolutionary GSM core network; 3G MSC/VLR and 3G SGSN provide the Iu interface.

It is not yet clear how tightly the Iu-interface will combine interfaces A and Gb. However, it seems desirable that they could be more integrated than today. This can be motivated from the service perspective. Today's clear separation of speech and circuit switched and packet data might not be true in the future. Developments such as RSVP make it possible to provide guaranteed QoS for packet networks, and thus, if UMTS/GSM is used to connect a user to packet access networks, similar capabilities need to be provided.

The question is also whether SGSN and MSC will be combined or will they remain separate entities. In this case, we would have within the same call connections to both the packet data side and the circuit data side. This would mean that MSC and SGSN would have to request abstract communication resources (i.e., in principle, QoS requirements only, not actual radio resources), and BSC would allocate radio resources for the communication according to current needs. As an example, a packet-based use of radio resources for IP traffic could be upgraded any time to circuit-based (without informing SGSN about it) if the MSC side wants to establish a call. The BSC would thus find the minimum radio resources that fulfil the sum of all requirements of MSC and SGSN services (the highest QoS requirement obviously influences the selection the most). Of course, if SGSN also supports other than best effort services, then it could also request higher priority services.

The clearer separation of the radio access and core networks in UMTS, when compared to GSM, results in some questions that need to be solved before paging procedures, and consequently the required signaling can be determined. For example, in what level does the core network need to know about the MS location inside GRAN? If the level is high, then this results in high traffic between the core network and GRAN when the mobile station is moving [5].

12.6.3.1   Handover

It is assumed that the handover decision is always made inside the GRAN. Thus, for inter-GRAN handover, functions to set up a path within the core-network, or some other arrangements, are required. Depending on the inter-GRAN handover type, different changes for GSM are required. The backward handover, where the handover signaling is performed through the old base station, is very similar to the current GSM handover. In the forward handover, the mobile station initiates the handover through the new base station. The need for this type of handover was discussed in Chapter 11: when trying to avoid corner effect, in which the connection to the old base station is lost, a very fast handover is required to prevent the blocking of the existing users in another cell. Forward handover requires a large number of changes in GSM.

To speed up handovers between two RNCs, direct interconnections similar to IS-634 (see Section 12.7.2) have been proposed. For this, a new interface between RNCs, as shown in Figure 12.16, is required. For services which have long packets or frequently transmit small packets, there is a physical connection established. Thus, a handover procedure is also required for packet service, except for short packets, which are transmitted in the random access message. If handover is performed between two BSC belonging to different GPRS support nodes, then there has to be a way to establish soft handover between these RNCs.

12.6.3.2   Transmission Infrastructure

The transmission infrastructure has to meet the new requirements imposed by wideband services. Since the data services are bursty and often asymmetric, the transmission solution has to be able to efficiently multiplex different types of information. ATM will provide efficient support for transmission of bursty wideband services. However, since

ATM was originally designed for very high-speed transmission in the fixed network, some modifications may be needed to accommodate cellular-specific infrastructure requirements.

12.6.4 Protocol Aspects

Figure 12.17 shows a possible protocol stack for UMTS, and Figure 12.18 shows the GRAN connected to a packet data network using the evolved GPRS protocols. Obviously, since there will be a new air interface, wideband CDMA, the radio-dependent protocols such as layers 1 and 2 will be completely replaced. The upper layer protocols, on the other hand, can be reused with some adaptation. For example, CC and MM protocols can be adapted, and thus, GSM 04.08 could form the basis for further development of UMTS CC and MM protocols denoted as MM' and CC' in Figure 12.17. The RR messages as such are not radio dependent and can be at least partly reused. The GPRS LLC and GSM RLP could be used as such for the wideband CDMA air interface. If ATM is used as the basic transport technique in the network side, the LAPD protocol will be replaced by AAL2/5, service specific convergence sublayer (SSCF), and service-specific connection oriented (SSCOP) protocols.

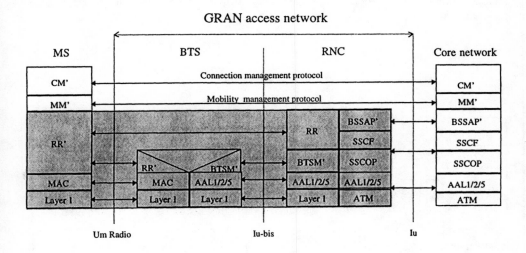

Figure 12.17 General protocol stacks of GRAN using GSM as a core network. The GRAN protocols are shaded.

12.6.4.1 Evolution of GSM BSSMAP and BSSGb Protocols

We can distinguish two different categories of functions over the GSM A-interface. The first category covers the functions to the BSS, while the second covers the functions to the MS. The functions related to the first category are used mainly to manage radio resources and are as follows:

- Bearer setup;
- Bearer release;
- Resource check;
- Paging;
- Ciphering;
- Handover.

BSSMAP (part of the BSSAP protocol) and BSSGP protocols specify these functions. The new BSSAP and BSSGP protocols are denoted by BSSAP' and BSSGP' in Figures 12.17 and 12.18. The information transmitted in messages supporting the above functions is specified by data elements. These data elements need to be updated. For example, bearer control messages need to be specified to include new channel types. There will then be new bit rates and new coding schemes. Furthermore, not all coding schemes are known today. The need for some generic bearer service classes exists, since wideband CDMA could support an almost infinite number of data rates due to the very high granularity. As there will be more mobile station classes due to the increased number of service possibilities, the GSM classmark information carrying the properties of the mobile stations should also be updated.

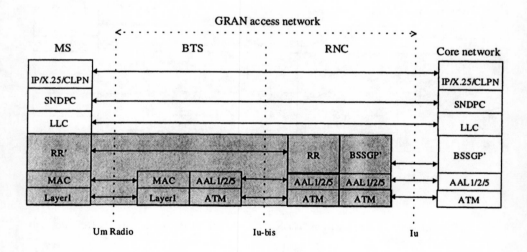

Figure 12.18   Transmission/user plane for the packet data interconnection.

Wideband CDMA GRAN will support QoS negotiation. Furthermore, as was discussed previously, it is expected that packet data networks will also evolve in this aspect, and thus QoS negotiation support from GPRS backbone network will be needed as well. Thus, the main change is that the GPRS BSSGB protocol should support some extra control messages for QoS negotiation purposes.

12.6.4.2 CAMEL

The purpose of Customized Applications for Mobile Network Enhanced Logic (CAMEL) is to provide a mechanism for supporting services consistently and independently of the subscribers' location [22]. It is based on IN principles and can be used to provide operator-specific services for roaming users. CAMEL relies on triggers within the supporting network infrastructure to suspend call processing and communicate with a remote computing platform before proceeding to handle the service functions. CAMEL uses the GSM MAP protocol to transfer information between appropriate network elements. The service switching function (SSF) in the visited network will have an interface directly to the service control function (SCF) in the home GSM network.

Non call-related events such as call independent supplementary service procedures, transfer of short messages, or mobility management procedures are within CAMEL's scope [22]. Thus, there is still need for further evolution towards the full implementation of the VHE concept, which would manage full transparency of offered services regardless of the network to which a user is currently attached.

12.7 EVOLUTION OF IS-41/ IS-95

In this section, we analyze the existing architecture, interfaces, and protocols of the IS-41/IS-95 networks. Based on this, we introduce a possible evolution scenario and suggest some of the required enhancements to allow the evolution of IS-95/IS-41 towards the third generation. An overview of the different standards required in the implementation of IS-41/IS-95-based systems is given in the next section. For a detailed treatment of the IS-41 standard, refer to [23].

12.7.1 Overview of IS-41/IS-95 Standards

When considering a complete IS-95 based network implementation, we need to consult several standards. The main standards are:

- EIA/TIA/IS-95B, Mobile Station-Base Station Compatibility Standard for Dual-Mode Wideband Spread Spectrum Cellular System;
- EIA/TIA/IS-41-C, Cellular Radiotelecommunications Intersystem Operations;
- TIA/EIA/IS-707, Data Service Options for Wideband Spread Spectrum Systems;
- TIA/EIA/IS-658, Data Services Interworking Function Interface for Wideband Spread Spectrum Systems;

- TIA/EIA/IS-634, MSC-BS Interface, Rev A.

12.7.2   Reference Architecture

Figure 12.19 shows a partial view of the reference architecture for the IS-41 core network. The message center (MC) is an entity that stores and forwards short messages. It may also provide supplementary services for SMS.

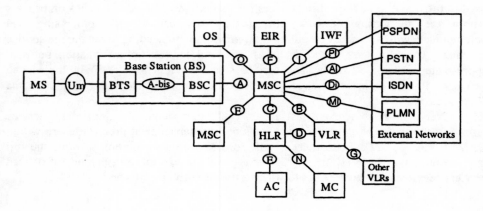

Figure 12.19   Partial view of IS-41 reference architecture.

The following main interfaces are defined between the network elements:

- The interface between the BSC and the MSC (A-interface) is specified in the IS-634 standard.
- The interface between the BTS and the BSC (the Abis-interface) is not specified.
- The interface between the MSC and the HLR (C-interface): signaling is specified in the IS-41 standard.
- The interfaces between the MS and the BTS (air interface, Um-interface) are specified in the IS-95, IS-97, and IS-98 standards.
- The interface between the MSC and its associated VLR (B-interface) is not specified.
- The interface between the HLR and the MSC (C-interface): signaling is specified in the IS-41 standard.
- The interface between the HLR and the VLR (D-interface) is used to exchange data related to the location of the mobile station and to the management of the subscriber. Signaling in this interface is specified in the IS-41 standard.
- The interface between the HLR and the VLR (D-interface): signaling is specified in the IS-41 standard.

- The interface between IWF and MSC (L-interface) is specified in the TIA/EIA/IS-658 standard.

12.7.2.1   Detailed A-interface Architecture

The first version (Rev 0) of the IS-634 standard defining the A-interface was published in 1995. The first implementations of IS-95-based systems were thus actually proprietary and, consequently, the standard was heavily influenced by the existing implementations. In the new revision of IS-634 Rev A, to be published in 1998, two distinct architectures have been defined: architecture A (with the selection/distribution unit (SDU) located inside of the base station) and architecture B (with the SDU located outside of the base station). Architecture A has a direct inter-BS connection option and an indirect (through MSC) BS-to-BS connection option. Architecture B has only a direct inter-BS connection.

There are six "parts" in the IS-634 Rev A. As compared to the earlier version of IS-634, a major restructuring has been performed:

- Part 0 (Base Part): high level function overview;
- Part 1 (Common Protocol): common protocol for architecture A and B;
- Part 2 (Architecture A): specific protocol for architecture A only;
- Part 3 (Architecture B): specific protocol for architecture B only;
- Part 4 (Interworking): interworking protocol between architecture A and B;
- Part 5 (Protocol Details): defines messages, information elements, and timers.

The detailed A-interfaces for architectures A and B are shown in Figures 12.20 and 12.21, respectively. The A-interface consists of A1- through A7-interfaces. A3- and A7-interfaces are both TCP/IP connections over ATM and are used for direct inter-BS connections for faster and more efficient soft handover. The A3-interface is divided into A3s (for signaling) and A3t (for traffic). Direct inter-BS connections (A3t, A3s, and A7) support faster soft handover and can add/drop multiple cells in same message.

The A6-interface is used in the indirect (through MSC) BS-to-BS connection option of architecture A (not shown). It is a 16-Kbps subrate circuit carrying user data/speech between the BS and the MSC during soft handover.

SDU is a new functional unit in IS-634. SDU contains functional entities such as traffic handler, signaling layer 2, multiplex sublayer, frame selection/distribution for soft handover, and power control. It also contains the transcoder functionality, which can be located either inside or outside of the base station.

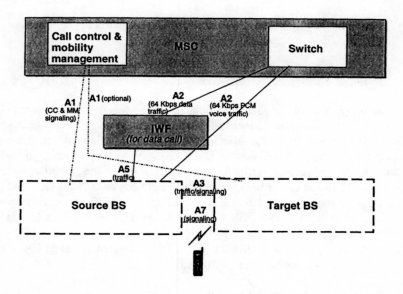

Figure 12.20   IS-634 architecture A.

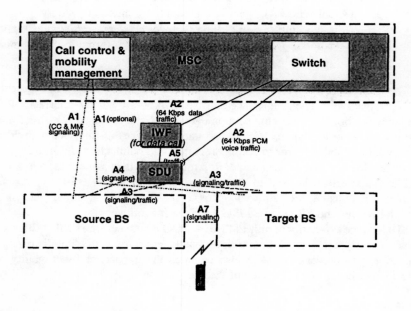

Figure 12.21   IS-634 architecture B.

12.7.3   Protocol Stacks

12.7.3.1   Air Interface

Figure 12.22 shows the IS-95 air interface protocol structure for the mobile station and the base station. Layer 1 is the physical layer of the digital radio channel, including those functions associated with the transmission of bits, such as modulation, coding, framing, and channelization via radio waves. Between layer 1 and layer 2 is a multiplex sublayer containing the multiplexing functions that allow sharing of the digital radio channel for user data and signaling processes. The multiplex sublayer provides several multiplex options. By *multiplex option* it is meant the ability of the multiplex sublayer and lower layers to be tailored to provide special capabilities. The multiplex option defines such characteristics as the frame format and the rate decision rules. The physical layer and multiplex sublayer are specified in Chapters 6 and 7 of the IS-95 specifications. 7. The IS-95 combines the operation of the network and data link layers and treats them as one layer.

Primary traffic is the main traffic stream carried between the mobile station and the base station on the traffic channel. Secondary traffic is an additional traffic stream that can be carried between the mobile station and the base station on the traffic channel. Signaling traffic means control messages that are carried between the mobile station and the base station on the traffic channel.

For user data, protocol layering above the multiplex sublayer is service option dependent and is described in standards for the service options. Service option means a service capability of the system. A *service option* may be an application such as voice, data, or facsimile.

For the signaling protocol, two higher layers are defined. Signaling protocol layer 2 is the protocol associated with the reliable delivery of layer 3 signaling messages between the base station and the mobile station, such as message retransmission and duplicate detection. Signaling layer 3 is the protocol associated with call processing, radio channel control, and mobile station control, including call setup, handover, power control, and mobile station lockout. Layer 3 signaling is mainly passed transparently to the MSC through the BS.

12.7.3.2   A-Interface

The IS-634 standard defines the MSC-BS messages, message sequencing, and mandatory timers at the base station and the mobile switching center. Figure 12.23 depicts IS-634 functions: call processing and supplementary services, radio resource management, mobility management, and transmission facilities management.

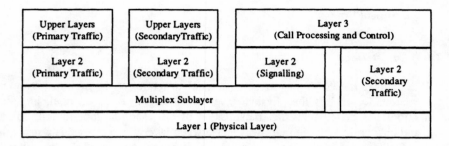

Figure 12.22 Mobile station and base station layers. (*Source*: [24], reproduced under written permission of the copyright holder (Telecommunications Industry Association). At the time of the publication, the standard which contains this figure was not finalized, please check with TIA for the correct version.)

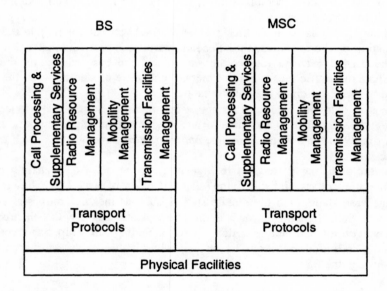

Figure 12.23 A-interface functions (*Source*: IS-634 Rev. A, reproduced under written permission of the copyright holder (Telecommunications Industry Association).)

Figure 12.24 shows the A1-interface signaling protocol reference model. Similar to GSM, IS-634 has a base station application part (BSAP), which contains BSMAP and DTAP. The BSAP corresponds to BSSAP in GSM and BSMAP to BSSMAP. Thus, DTAP is used to transfer the mobility management and call control related signaling between the MSC and the mobile station. BSMAP transfers, for example, radio resource management related signaling between BS and MSC. The transport protocols for user and signaling data are shown in Figures 12.24 and 12.25, respectively. The physical layer is based on the use of T1 digital transmission system interface (1.544 Mbps providing 24×56 or 64-Kbps channels), E1 digital transmission interface (consisting of 30×56 or 64Kbps channels), OC1 digital transmission interface (51.84 Mbps), and OC3 digital transmission interface (155.52 Mbps).

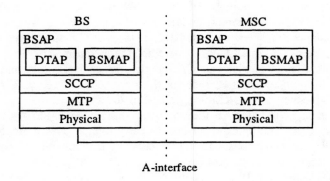

Figure 12.24 A1 interface signaling protocol reference model.

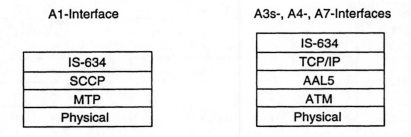

Figure 12.25 Transport protocol options for signaling connections.

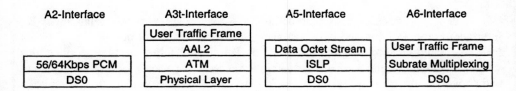

Figure 12.26 Transport protocol options for user traffic connection.

12.7.3.3   Data Service Protocols

The data service protocols for IS-95 are specified in TIA/EIA/IS-707 "Data Service Options for Wideband Spread Spectrum Systems." The IS-707.4 specifies asynchronous fax and data services, and IS-707.5 specifies packet data services. The L-interface is specified in TIA/EIA/IS-658.

Figure 12.27 shows an overview of data service protocols based on IS-658. The physical layer in the network side is based on B-channels or H-channels with unrestricted digital information as defined in ANSI T1.607-1990, or an ANSI/IEEE 802.3 LAN. FR SVC is frame relay switched virtual circuit, which conforms to ANSI T1.618. The adaptation layer accepts data octets from the link layer and the relay function, assembles them into blocks of data octets for transmission in the information field of FR SVC frames, or vice versa, accepts blocks of data octets, and disassembles them. Radio link protocol (RLP) reduces the error rate by applying retransmission protocol for erroneous packets. RLP is specified in 707.2.

The link layer is based on the point-to-point protocol (PPP), defined in RFC 1661. PPP provides a multiplexed method to carry higher layer protocols over serial links. Above PPP, asynchronous data and fax data applications use link control protocol (LCP) and internet protocol control protocols (IPCP). LCP provides a mechanism for the mobile station and the IWF to negotiate various options provided by PPP. IPCP allows the mobile station to request a temporary IP address from the IWF.

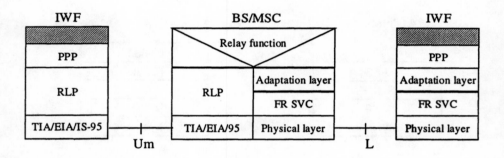

Figure 12.27   Mobile data protocol overview. (*Source*: TIA/EIA/IS-658, reproduced under written permission of the copyright holder (Telecommunications Industry Association).)

IS-707.5 specifies two protocol options for packet services: relay layer R_m interface protocol option and network layer R_m interface protocol option. The network layer R_m interface protocol option supports Terminal Equipment (TE) applications where the mobile termination is responsible for aspects of packet mobility management and network address management. The relay layer R_m interface protocol option supports TE applications where the TE is responsible for aspects of packet mobility management and network address management.

12.7.4   Evolution of IS-41/IS-95 Towards Third Generation

The standardization work of IS-41/IS-95-based networks towards third generation has recently started. In this section a few aspects relevant to the evolution are discussed.

12.7.4.1 Architecture Evolution

The ATM transmission infrastructure will provide efficient support for transmission of bursty wideband services. IS-634 is already using ATM in BS-to-BS communication. However, this needs to be extended into the A-interface towards MSC and possibly to the L-interface.

12.7.4.2 Protocol Evolution

Since the IS-41 or IS-95 were not specified using some formal specification language, the partitioning between protocols is sometimes difficult to identify. Interactions between different protocols make their independent evolution difficult. Thus, a clearer separation between different protocol layers is required before introducing third generation enhancements.

The BSSAP protocol in IS-634 needs to be evolved in a similar manner as in GSM. In addition, IS-41 call processing and mobility management protocols can be reused with improvements. If global roaming is desired, modifications are necessary for IS-41. For example, interworking with GSM MAP should be developed. One improvement for IS-41 could be the specification of soft handover between different MSCs. To simultaneously support multiple traffic channels changes are required in IS-41.

Most likely the PPP layer will be common to all data services, as in the current IS-707 standard. However, below the PPP protocol, a MAC layer will be specified to support packet and possibly circuit switched data services. Furthermore, the RLP protocol needs revisions to accommodate new frame sizes.

REFERENCES

[1] ITU-T Recommendation I.130, "Method for the Characterization of Telecommunication Services Supported by an ISDN Network Capabilities of an ISDN."

[2] ITU-T Recommendation X.200, "Information Technology – Open Systems Interconnection – Basic Reference Model: The Basic Model."

[3] ITU-T Recommendation I.320, "ISDN Protocol Reference Model."

[4] Mouly, M., and M.-B. Pautet, *The GSM System for Mobile Communications*, published by the authors, 1992.

[5] Haferbeck, R., L. Laitinen, and S. Huusko, "BSS Aspects on the Radio Dependent Functionality," Issue 1, *CEC Deliverable, Future Radio Wideband Mobile Access System – FRAMES*, November 30, 1997.

[6] ETSI, "Global Multimedia Mobility (GMM), A Standardization Framework," Part A and B, 1996.

[7] E. Berruto, G. Colombo, P. Monogioudis, A. Napolitano, and K. Sabatakakis, "Architectural Aspects for the Evolution of Mobile Communications Towards UMTS," *IEEE Journal on Selected Areas in Communications*, Vol. 15, No. 8, October 1997, pp. 1477–1487.

[8]   ITU-T Recommendation Q.FIN, "Framework of IMT-2000 Networks."

[9]   ITU-T Recommendation I.312/Q.1201, "Principles of Intelligent Network Architecture."

[10]   Modarressi, A. R., and R. A. Skoog, "Signaling System No. 7: A Tutorial," *IEEE Communications Magazine*, July 1990, pp. 19–35.

[11]   Lin, Y.-B., and S. K. DeVries, "PCS Network Signaling Using SS7," *IEEE Personal Communications*, Vol. 2, No. 3, June 1995, pp. 44–55.

[12]   Perkins, C. E., *Mobile IP: Design Principles and Practises*, Reading MA: Addison-Wesley Publishing Company, 1998

[13]   Perkins, C. E., "Mobile IP", *IEEE Communications Magazine*, May 1997, pp. 84–99.

[14]   ETSI, UMTS 23.20, "Universal Mobile Telecommunications System (UMTS), Evolution of the GSM Platform towards UMTS."

[15]   White, P. P., "RSVP and Integrated Services in the Internet: A Tutorial," *IEEE Communications Magazine*, Vol. 35, No.5, May 1997, pp. 100–106.

[16]   ETSI, UMTS 23.70, "Universal Mobile Telecommunications System (UMTS); Service Aspects; Virtual Home Environment."

[17]   GSM 03.60, "General Packet Radio Service (GPRS); Service Description; Stage 2 ETSI," GSM Technical Specification 03.60, Version 5.0.0, June 1997.[2]

[18]   Cai, J., and D. J. Goodman, "General Packet Radio Service in GSM," *IEEE Communications Magazine*, October 1997, pp. 122–131.

[19]   Hämäläinen, J., *Design of GSM High Speed Data Services*, Ph.D. Thesis, Tampere University of Technology, Finland, 1996.

[20]   Jung, P., Z. Zvonar, and K. Kammerlander (eds.), *GSM: Evolution Towards 3rd Generation,* Boston: Kluwer Academic Publishers, 1998.

[21]   ETSI, UMTS 23.30, "Universal Mobile Telecommunications System (UMTS) Iu Principles."

[22]   Verkama, M., "CAMEL and Optimal Routing, Evolution of GSM Mobility Management towards UMTS," in *GSM: Evolution towards 3rd generation*, P. Jung, Z. Zonar, and K. Kammerlander (eds.), Boston: Kluwer Academic Publishers, 1998.

[23]   Gallagher, M. D., and R. A Snyder, *Mobile Telecommunications Networking with IS41*, New York: McGraw-Hill, 1997.

[24]   TIA/EIA/SP-3693, "Mobile Station-Base Station Compatibility Standard for Dual-Mode Wideband Spread Spectrum Cellular Systems," to be published as ANSI/TIA/EIA-95B.[3]

---

[2] The present technical specifications are the property of ETSI. No reproduction shall be made without ETSI authorization. The original version of these ETSI technical specifications can be obtained from the Publication Office of ETSI.

[3] To purchase complete version of any TIA document, call Global Engineering Documents at 1800-854-7179 or send a facsimile to 303-397-2740.

Chapter 13

SYSTEM COMPARISON

13.1 INTRODUCTION

This chapter presents comparisons between wideband CDMA and other third generation air interface candidates. The comparison criteria used can be divided into objective and subjective criteria. A comparison based on a single objective criterion is free of interpretations if the results have been obtained using the same assumptions for the different considered technologies. Examples of objective criteria are capacity and coverage. An evaluation based on subjective criteria is not free of interpretations. Examples of subjective criteria are the impact of the air interface in the fixed network and flexibility of the air interface. Very often criteria are also related to each other, for example, more sophisticated receiver techniques, such as interference cancellation, result in better performance but also in larger complexity. The overall judgment depends on the weight given to each criterion. Furthermore, a comparison is always a snapshot of the available technologies at a specific moment, and new information might change the judgment. It is also important to understand which technologies form an essential part of an air interface and which technologies are air interface independent. For example, speech coding is not specific to the air interface.

Due to the complex nature of the third generation systems, we do not try to present a comprehensive comparison of the proposed technologies. Rather, we explain and interpret the results of earlier comparisons. Based on this, we assess the feasibility of wideband CDMA for third generation air interface.

The chapter begins with a discussion of the comparisons performed for the second generation systems. Second, the SIG5 comparison results for ATDMA and CODIT air interfaces are presented. Next, a performance comparison of wideband TDMA and wideband CDMA for 2-Mbps transmission is described. The performance of wideband CDMA and GSM air interfaces is discussed. The FRAMES project compared different multiple access schemes, the results of which are presented here. A short review of ARIB air interface comparison follows. Furthermore, we discuss the ETSI UMTS radio access (UTRA) scheme evaluation. Based on the review of the different air interface comparisons, we conclude with a discussion on the feasibility of wideband CDMA as a third generation air interface.

13.2   SECOND GENERATION AIR INTERFACE COMPARISONS

So far, the probably most well-known comparison of air interfaces, or rather the most well-known debate, has been between GSM and IS-95. Without going into the details, we can note that the published results from these comparisons are very controversial [1–5]. The main deficiency of these comparisons is that no common assumptions were specified, and thus, the obtained results were not directly comparable. Furthermore, since source codecs, not part of the air interface as such, were included, these comparisons do not actually reflect the performance differences between TDMA- and CDMA-based air interfaces. A more realistic comparison can be found in [5].

For the US PCS system selection, an extensive list of criteria was developed. The advantages and disadvantages of different air interfaces were compared based on these criteria. Originally, 16 proposals were submitted [6]. The merging of the system porposals was based on a voluntary consolidation, which resulted in reducing the number of air interfaces to seven. These are GSM1900, IS-95, IS-136, Digital European Cordless Telephone (DECT), Personal Access Communication System (PACS), W-CDMA, and a composite CDMA/TDMA.

13.3   RACE II - SIG5 COMPARISON

SIG5, a special interest group in the partly-EU funded RACE II program in Europe, evaluated the ATDMA and CODIT air interfaces [7]. CODIT is a wideband CDMA system described in Chapter 6. In the comparison, a 5-MHz system bandwidth was used. ATDMA is an advanced TDMA system with a bandwidth of 270.92307 kHz and 1107.692308 kHz in macro- and micro/picocell environments, respectively [8].

In order to achieve compatible results, SIG 5 specified a set of common assumptions [9]. Services used in the comparison were 12-Kbps speech service and 64-Kbps data services. However, since ATDMA and CODIT did not agree on these common assumptions in the beginning of the projects, it was not possible to unify all assumptions and models in SIG5. The specification of QoS for CODIT and ATDMA, as well as the system simulation tools, were different. Signaling overheads were calculated in different ways. In addition, the speech codecs of the systems were different, and thus, no real comparison could be performed for speech services.

Since the assumptions to generate spectrum efficiency results were not exactly the same, it was difficult to draw any definitive conclusions from the SIG5 performance comparison. However, it was noted that CODIT performed better for data services in a macrocell environment since it had stronger channel coding [7]. Furthermore, CODIT performed better for mixed services (speech and data) than ATDMA, mainly due to better interference averaging. In micro- and picocells, the system deployment had a large impact on the results. ATDMA performed better, but CODIT would have gained more from sectorized antennas. In a microcell environment, the isolation between cells is high and the intracell interference limits the CDMA performance. On the other hand, TDMA is orthogonal within a cell and, since the intercell interference is low, it achieves good spectrum efficiency. The gain from sectorized antennas in the CDMA case is translated into direct capacity gains due to the reduction of intracell interference, while in TDMA the reuse factor cannot be reduced in direct proportion to the number of sectors. The same conclusions apply for a picocell environment.

Later, an ATDMA and CODIT comparison was continued and reported in [10]. For this study, more detailed common assumptions were agreed upon. The results were similar to the SIG5 comparison (i.e., CODIT performed better for data services in macrocell). In addition, CODIT performed also better for 64-Kbps data service in a microcell environment. This difference compared to the SIG5 results was most likely due to slightly different deployment scenarios and assumptions used in the two studies.

In addition to the quantitative evaluation of performance, SIG5 evaluated the CODIT and ATDMA concept qualitatively using subjective criteria. A summary of this comparison is presented in Table 13.1. As can be seen, drawing any hard conclusions from this kind of subjective comparison is difficult. However, service flexibility seems to be the main advantage of CODIT.

13.4   TDMA VERSUS CDMA FOR 2 MBPS SERVICE

The performance of four wideband TDMA and one wideband CDMA air interfaces for 2-Mbps transmission has been evaluated in [11]. The details of the wideband TDMA schemes have been presented in [12] and the wideband CDMA schemes in [13]. Two of the TDMA schemes (2-MHz bandwidth with B-O-QAM and Q-O-QAM modulation) were the basis for the FRAMES FMA1 scheme, which was later adopted as the IS-136 indoor carrier (with a bandwidth of 1.6 MHz) [14]. The wideband CDMA scheme was later submitted to the FRAMES project, where it was adopted with some modifications as the FMA2 wideband CDMA scheme [14]. The FMA2 uplink specification was then adopted for the ETSI WCDMA uplink scheme. In this study, the wideband CDMA system had bandwidths of 5, 10, and 20 MHz.

The link level performance was simulated in macro-, micro-, and picocell environments. The cell capacity was simulated in micro- and macrocell environments. In addition, link budget (range) was analyzed for macro-, micro-, and picocells.

In general, the capacity and range of the TDMA and CDMA schemes were comparable, with just some minor differences. In the uplink microcell channel with 3 km/h mobile speed, the wideband CDMA performance with 5- and 10-MHz bandwidths was somewhat better than TDMA due to fast power control. In this channel, frequency diversity alone did not give full benefit. The downlink capacity of TDMA with reuse factor of one was slightly better than the CDMA capacity. The reuse factor of one was possible due to interference cancellation [15]. Without it, the reuse factor would have been larger and thus the capacity comparable to the wideband CDMA capacity.

Table 13.1
Summary of the Qualitative Comparison Between CODIT and ATDMA Concepts

	CODIT	ATDMA
Linear transmitter and receiver requirements	Linear modulation QPSK and OOPSK used	Linear modulation B-O-QAM and Q-O-QAM used
Handportable requirements – need for duplexer	Required	Required for higher bit rates
Power control issues	Open and closed loop Dynamic range in UL 80 dB for open loop and 12 dB for closed loop Command rate 2 Kbps (0.5 ms) Step size 1 dB	Open loop, optional fast loop in uplink Dynamic range 30 dB Command rate 80 – 320 ms or no power control Step size 1 dB
Dynamic channel allocation	N/A	Used to achieve higher capacity
Frequency hopping issues	N/A	Used, gain depends on correlation bandwidth and available bandwidth
Sharing frequency band capability	CDMA less sensitive to narrowband interference; Sharing very difficult due to near-far effect, might be possible if same cell sites or interference cancellation used	Sharing might be possible With FDMA or CDMA if CDMA power density low enough
Synchronization requirements for base station	Asynchronous; DL synchronization required during handover on call-by call basis	No synchronization required; Might be beneficial to speed up scanning and handover especially in micro- and picocells
Frequency and code planning	No frequency planning except for fitting CODIT frequency channels (1, 5, and 20 MHz) into the allocated bandwidth (e.g., 30 MHz) Pilot code channels need to be planned	Either fixed frequency plan or DCA; With DCA common control channels need to be planned.
Service flexibility	Universal transport layer Variable rate services frame-by-frame basis Simultaneous speech and data Connectionless data	Data rates for low delay and long delay constrained services: 9.6, 19.2, 64 Kbps, up to 2 Mbps Unconstrained delay data
Handover performance	Soft handover Interfrequency handover, using compressed mode results to 1.5 dB degradation in C/I ratio	Backward handover Forward handover PRMA++ allows reservation of resources in a new cell before releasing the old traffic channel Macro diversity specified, but not as beneficial as in CDMA
Distributed and adaptive antennas	Not studied	Not studied
EMC issues	CDMA has continuous transmission in reverse link, power control modifies very rapidly the total transmission power	Pulsed transmission

13.5 COMPARISON OF WIDEBAND CDMA AND GSM

The impact of different radio channel and service conditions on wideband CDMA and GSM air interfaces has been studied in [16]. The wideband CDMA air interface is the same as in [16], which was later used as the basis for the FMA2 scheme in FRAMES [17]. Evaluation of spectrum efficiency was carried out for speech, 12-Kbps data service, and 96-Kbps data service. Furthermore, a mixed service case with 80% speech and 20% 12-Kbps data users was evaluated. The required BER is 10^{-3} for all services. The evaluated services are summarized in Table 13.2.

Table 13.2
Evaluated Services

	Data rate	BER	Maximum delay (ms)
Speech (GSM)	6.5, DTX = 0.5	10^{-3}	60
WB-CDMA	8, DTX = 0.4	10^{-3}	30
Low rate data	12	10^{-3}	100
High rate data	96	10^{-3}	100

For both air interfaces, the spectrum efficiency of the downlink was less than in the uplink. To remove the imbalance between up and downlink, introduction of antenna diversity at the terminal should be considered. For speech and data services, the downlink of the GSM air interface provided equal performance compared to wideband CDMA. However, the uplink performance of wideband CDMA was better that the performance of GSM. This was due to multiuser detection. Furthermore, the reduction of E_b/N_0 due to antenna diversity used in the uplink turns into higher capacity gain for CDMA. In general, all radio access technologies can provide equal spectrum efficiency, as was shown by this study. However, by analyzing the performance of different services, critical design factors can be identified and thus further improvements developed. For GSM/DCS, the following learned lessons are summarized below:

- Interference cancellation can substantially improve the performance for both up and downlink.
- Burst-by-burst hopping improves performance of high data rate services.
- DTX gain does not always turn into capacity improvement due to nonlinear increase in interference as the reuse factor is reduced.
- The GSM half-rate codec introduces unnecessary delay and better interleaving scheme would improve performance.

The following lessons for wideband CDMA were learned:

- All energy from the channel needs to be collected to obtain good performance.
- Power control is essential to provide good performance.
- Uplink power control signaling reduces the spectrum efficiency of downlink especially for low bit rate speech service.

- In a channel with a large number of equal gain taps, soft handover does not provide additional diversity due to limited number of RAKE fingers in the mobile station.

13.6   FRAMES MULTIPLE ACCESS COMPARISON

The FRAMES project investigated hybrid multiple access technologies in order to select the best combination as a basis for further detailed development of UMTS radio access system. The multiple access evaluation results have been reported in [17]-[19]. The FRAMES evaluation campaign consisted of two stages. At the first stage, several candidate schemes were compared and schemes with similar characteristics were combined. Based on the results of the first stage, the multiple access schemes were divided into two groups:

- Multicarrier TDMA (multiples of 200 kHz), single wideband TDMA (WB-TDMA, bandwidth 1-2 MHz), and hybrid CDMA/TDMA (bandwidth 1.6 MHz);
- Asynchronous CDMA (WB-CDMA, bandwidth 6 MHz), OFDM/CDMA, and synchronous CDMA.

At the second stage, these schemes were evaluated against criteria derived from the UMTS requirements. Based on this evaluation, a harmonized multiple access platform was defined. The FRAMES multiple access (FMA) platform consists of two modes, FMA1 with and without spreading (bandwidth 1.6 MHz) and a wideband CDMA mode (bandwidth 6.4 MHz). In the following, we summarize the FRAMES evaluation of the FMA platform based on [17].

Provision of Various Data Rates in Different Environments and Bearer Service Flexibility. Both FMA1 and FMA2 can support the UMTS bit rates from low bit rates up to 2 Mbps. FMA1 supports variable bit rates with low granularity by DTX or resource re-assignment together with adaptive coding. The slotted structure of FMA1 suits well bursty packet type services. In FMA2, variable bit rates are supported by adaptive coding and power assignments. Due to this power sharing and long initial synchronization, FMA2 is more suitable for moderately varying circuit switched services. For both modes, different operating points for the services are obtained by different combinations of coding and link adaptation. In FMA1, mixed bearer services can be provided to one user by packing them into different slots/codes; while in FMA2, mixed bearer services for one user can be multiplexed and have different operating points. For both modes bearers/services can be added and dropped during a call.

Spectrum Efficiency (capacity). The spectrum efficiency for FMA1 is presented in Table 13.3 and for FMA2 in Table 13.4. The prevailing conditions for the spectrum efficiency simulations are summarized below:

- Hexagonal cell layout;
- Uniformly distributed mobile stations;
- 5% outage probability;
- Pathloss law with a decay factor of 3.6;
- 10 dB shadowing parameter;
- Power control used (except for FMA1 mixed services);
- Errors due to power control and handover taken into account;
- FMA1 with spreading evaluated (orthogonal codes);
- For FMA1 frequency reuse is optimized for each case;
- FMA2 had frequency reuse factor of 1.

Table 13.3
Spectrum Efficiency of FMA1

Service	Spectrum Efficiency (Kbps/MHz/cell), load/reuse factor	
	Downlink	Uplink
12 Kbps, BER 10^{-3}, 40 ms	124 , 77%/3	250, 52%/1
144 Kbps, BER 10^{-3}, 40 ms	126, 70%/4	240, 33%/7
Mixed services 90% / 10% 12 Kbps / 144 Kbps	57 (no power control)	95 (no power control)
2 Mbps, BER 10^{-3}, 100 ms	Downlink is limiting direction; Spectrum efficiency 100 - 150 Kbps/MHz/s.	

Table 13.4
Spectrum Efficiency of FMA2

Service	Spectrum Efficiency (Kbps/MHz/cell)	
	Downlink	Uplink (with multiuser detection)
Speech/low rate data 12 Kbps, BER 10^{-3}, 40 ms	108	192
Medium data 144 Kbps, BER 10^{-3}, 100 ms	108	389
Mixed services (12 Kbps/144 Kbps)	115	322
2 Mbps, BER 10^{-3}, 100 ms	Downlink is limiting direction Spectrum efficiency 100 - 150 Kbps/MHz/s.	

Spectrum efficiency figures in the downlink are very close to each other. Poor performance of mixed service in FMA1 is due to missing power control, which was not used due to modeling difficulties. For both modes, downlink is the limiting direction since uplink receiver antenna diversity was used. The better performance of FMA2 in the uplink is due to multiuser detection. Multiuser detection efficiency (i.e., the amount of own-cell interference that can be removed) varies in different environments and here it was assumed to be 60%.

Coverage. Coverage evaluation was carried out for the same services and conditions as for spectrum efficiency. The access scheme -dependent parameters determining range are:

- Link level performance E_b/N_0;
- The amount of overhead transmission.

E_b/N_0 figures taking into account the overhead transmission due to power control and training bits are presented in Table 13.5. The differences are very small, and the uplink is the limiting direction due to lower transmission power. For the low bit rate services, FMA1 has a slight advantage; while for the 144-Kbps service, FMA2 has an advantage.

Table 13.5
E_b/N_0 results

FMA1		
E_b/N_0	Downlink	Uplink
Speech/low rate data 12 Kbps, 10^{-3}, 40 ms	7.3 dB	3.4 dB
Medium rate data 144 Kbps, 10^{-3}, 40 ms	5.9 dB	4.4 dB
FMA2		
E_b/N_0	Downlink	Uplink
Speech/low rate data 12 Kbps, 10^{-3}, 40 ms	6.9 dB	7.2 dB
Medium data 144 Kbps, 10^{-3}, 100 ms	6.5 dB	3.1 dB

Support for Adaptive Antennas. Both modes support adaptive antenna techniques. In FMA1, capacity gains are realized through smaller cluster sizes and in FMA2 by allocating more codes. SDMA can be used for C/I and capacity improvements. In FMA2, a user dedicated reference signal instead of a common pilot signal is needed in the downlink.

Hierarchical Cell Structures (HCS). Hierarchical cells are supported by interfrequency handover, which in FMA1 is an inherent part of the system design and in FMA2 is supported through discontinuous uplink transmission and dual receivers in the downlink. In both modes, the mobile terminal assists with measurements for handover (MAHO). Handover between FMA-based UMTS and second generation (GSM) is supported through dual-mode terminals together with the capability of measuring GSM BCCH frequencies from FMA, owing to the choice made above for the clock rate and frame structure.

Duplex Method, Support for Public and Private Environments. FMA1 is better suited for operation in private environments than FMA2, since it requires less coordination. FMA1 supports asymmetric data services in TDD mode since the number of slots between uplink and downlink can be varied. FMA2 could, in principle, operate in TDD as well, but the basic transmission scheme should be modified in that case, knowing also that flexible allocation of resources between up and downlink is difficult.

Terminal Impacts (Power Consumption and Complexity, GSM/UMTS Dual-Mode Terminals). In the evaluation of terminal impacts, it can be noted that cost, size, and power consumption for baseband complexity decrease drastically with time, enabling more complex baseband algorithms like multiuser detection and interference cancellation. RF power consumption is mainly determined by the power level and linearity requirements of the output amplifier. In macrocells with low bit rates, RF power consumption surpasses baseband power consumption. In FMA1, wide bandwidth and high data rates give stringent A/D converter requirements. In addition, higher order modulation and multicode transmission increase the power amplifier linearity requirements. FMA2 power amplifier requirements are less stringent due to variable spreading gain in the uplink. In FMA1, a dual receiver is needed for inter-frequency handovers at higher data rates, while in FMA2 a dual receiver is always needed to support inter-frequency handover. A dual-mode UMTS and GSM/DCS terminal needs additional GSM/DCS duplexer and RF filters, but it can reuse the same reference oscillator for both modes with a careful parameter design.

BSS Impacts (Evolution from Existing Systems, BSS Complexity and Cost). Regarding BSS impacts, FMA1 operates with hard handover. In FMA2, soft handover is required, necessitating additional costs due to 1.5 times more transmission capacity plus diversity combining. In FMA1, integration of new cells can be handled by slow dynamic channel allocation (DCA) to simplify network planning, and in FMA2 by automatic power planning.

13.7   ARIB COMPARISON

ARIB has reported a performance comparison between TDMA (MTDMA system), OFDM (BDMA system), and CDMA technologies [21]. The comparison was based on simulations and experimental systems. The conclusions were as follows [21]:

- Experiment data suggest that the CDMA system has a good feasibility although some details require further study.
- The major technical contention of MTDMA system is high bit rate transmission. MTDMA could be an alternative to the CDMA system in office and pedestrian environments.
- BDMA offers potentiality for capacity expansion. However, it is difficult to judge at this stage, due to the lack of data based on theoretical studies and experimental verification.

Later, it was decided not to adopt BDMA and MTDMA for further standardization. The advantages of wideband CDMA for IMT-2000 were stated to be flexible high data rate services along with high quality, improved multipath resolution and increased interference averaging. Time-to-market aspects could also be counted as advantages for the CDMA solution. No detailed technical results are publicly available from the ARIB comparison.

13.8   ETSI AIR INTERFACE COMPARISON

During 1997, ETSI SMG2 performed an air interface concept evaluation in order to select the UMTS terrestrial radio access (UTRA) scheme. Four concepts were compared in this process, namely, wideband CDMA (concept α), OFDM (concept β), wideband TDMA (concept γ), and TD-CDMA (concept δ). The spectrum efficiency results are summarized in Table 13.6, and the qualitative comparison results in Table 13.7. LCD means low constrained delay data and UDD unconstrained delay data (i.e., packet data). The evaluation criteria, common assumptions, and models are described in [22]. The evaluation results have been presented in [23].

The level of evaluation details for different concepts is highly variable. However, it can be concluded that all schemes seem to satisfy the criteria. Wideband CDMA seems to be the most flexible for variable bit rate and mixed service applications. However, wideband CDMA requires the largest minimum spectrum allocation. The spectrum efficiency of wideband CDMA was best for speech and 384-Kbps services in the downlink. In an indoor office, wideband TDMA had comparable spectrum efficiency to wideband CDMA. Furthermore, downlink antenna diversity improved the spectrum efficiency of all schemes considerably.

In general, all schemes evaluated in ETSI could have formed the basis for UMTS air interface. However, wideband CDMA was selected for FDD due to its technical merits, such as flexibility and high spectrum efficiency, as well as its international support and the large amount of research that had resulted in a detailed specification. This was judged to reduce the risk for commercial systems. TD-CDMA was selected for TDD, as a result of a compromise, since it had the largest support after wideband CDMA in the final voting. Furthermore, as discussed in Chapter 9, wideband CDMA TDD mode naturally contains time division principle, and thus, the political compromise was also viable from a technical point of view.

13.9   CONCLUSIONS

What is the best or the most suitable air interface for third generation wireless communication systems? From a purely technical point of view, there is no definitive answer. Each scheme has some strong and weak points. In addition, if there are differences, it is a very subjective matter to determine how significant they are.

In general, we can state that any of the proposed technologies, TDMA, CDMA, OFDM, or hybrid CDMA/TDMA could be used as an IMT-2000 air interface. The technical differences are not that large. Furthermore, we can state that many of the

Table 13.6
Spectrum Efficiency Results in Kbps/MHz/cell for UTRA Concepts

Environment	Service	WCDMA (concept α) Uplink/downlink	OFDMA (concept β) Uplink/downlink	WB-TDMA (concept γ) Dl results only since limiting direction	WB-TD-CDMA (concept δ) Dl results only since limiting direction
Vehicular	Speech	98 / 78	33/31	55	72 (FH) 68 (no FH)
	LCD384	138/85 204/123 (30 dBm MS, 8 Mcps) 138/211 175/211 (30 dBm MS) 204/250 (30 dBm MS, 8 Mcps)	152/208 (with optimized transmitter power)	113	129/176 176 (DL diversity)
Outdoor-to-indoor pedestrian	UDD384	470 / 565	440/465	811	812 (DL ant div, reuse 1) 387 (DL ant div, reuse 3)
	Speech	127 / 163 189 / - (C/I based HO)	30.75/32.25	190	75 (FH) 73 (no FH)
Indoor office	service mix - speech - UDD384	315 / 207 315 / 460 (DL ant div)	TBD	60	110 (FH) 104 (no FH)
	UDD2048	300 / 230 300 / 500 (DL ant div)	240/240	332 743 (wall attenuation 5 dB)	170 (reuse 1) 195 (reuse 3) 405 (DL ant div, reuse 1) 132 (DL ant div, reuse 1)

Source: [23].

noticed weaknesses of the previous schemes can be overcome by further development. For example, in the FRAMES comparison, CDMA was noted to be better suited for moderately varying bit rates rather than for packet data. Since then, however, a new packet access mode for WCDMA in ETSI and ARIB has been developed (see Section 6.3.6) and this drawback has been circumvented. In FRAMES, no antenna diversity was used in the downlink, and thus, it was concluded that the downlink limits capacity. However, different solutions such as antenna diversity in the mobile station, base station transmit diversity, and interference cancellation in the mobile station are being discussed as a solution for this problem.

Table 13.7

Summary of the Comparison Results for the UTRA Concepts

	WCDMA (concept α)	OFDMA (concept β)	WB-TDMA (concept γ)	WB-TD-CDMA (concept δ)
Bearer capabilities • Rural outdoor: at least 144 Kbps (goal to achieve 384 Kbps), maximum speed: 500 km/h • Suburban outdoor: at least 384 Kbps (goal to achieve 512 Kbps), maximum speed: 120 km/h • Indoor/low range outdoor: at least 2Mbps, maximum speed: 10 km/h • UTRA should allow evolution to higher bit rates.	Supported with 4.096 Mcps chip rate. With 8.196 and 16.392 Mcps bit rates up to 4 and 8 Mbps supported	Supported. In indoor, over 2 Mbps supported	Supported. Up to 4 Mbps for short range	Supported. Higher bit rates can be supported using higher order modulation or higher RF bandwidth
Flexibility • Negotiation of bearer service attributes • Parallel bearer services, real-time/non real-time communication modes • Adaptation of bearer service bit rate • Circuit and packet oriented bearers • Supports scheduling of bearers according to priority • Adaptation of link to quality, traffic, and network load, and radio conditions • Wide range of bit rates should be supported with sufficient granularity • Variable bit rate real time capabilities should be provided • Bearer services appropriate for speech shall be provided	Supported. Bit rates from 100 bps up to 2 Mbps with a granularity of 100 bps, change of bit rate on 10-ms basis. TDD mode can be used for asymmetric services	Supported. Granularity of 13 Kbps, finer granularity by varying channel coding. TDD mode can be used for asymmetric services	Supported. Two different size time-slots, granularity by varying channel coding. TDD mode can be used for asymmetric services	All supported. Variable bit rates provided by pooling of time slots and CDMA codes, granularity by varying channel coding. TDD mode can be used for asymmetric services
Handover • Provide seamless (to user) handover between cells of one operator • The UTRA should not prevent seamless HO between different operators or access networks • Efficient handover between UMTS and second generation systems (e.g., GSM) should be possible	Supported. Soft handover and inter-frequency handover with slotted mode for single receiver mobile stations. Either by dual receiver or by slotted mode for GSM carrier measurements. WCDMA and GSM have same multi-frame structure	Supported. Hard handover. Same frame structure allows synchronization of UMTS and GSM cell sites for easier handover	Supported. Hard handover. Same frame structure allows synchronization of UMTS and GSM cell sites for easier handover	Supported. Hard handover. Same frame structure allows synchronization of UMTS and GSM cell sites for easier handover

Table 13.7 (continued)

	WCDMA (concept α)	OFDMA (concept β)	WB-TDMA (concept γ)	WB-TD-CDMA (concept δ)
Compatibility with services provided by present core transport networks • ATM bearer services, GSM services, IP based services, and ISDN services	Supported by all schemes			
Radio access planning • If radio resource planning is required automatic planning shall be supported	In general, all schemes need coverage planning and planning of adequate capacity for different services (see Chapter 11). The overall planning effort is expected to be the same for all schemes			
Public network operators • It shall be possible to guarantee predetermined levels of QoS to public UMTS network operators in the presence of other authorized UMTS users.	To guarantee predetermined QoS levels, UMTS public operators require dedicated frequency bands with appropriate guardbands. Therefore, this requirement was not supported			
Private and residential operators • The radio access scheme should be suitable for low-cost applications where range, mobility, and user speed may be limited • Multiple unsynchronized systems should be able to successfully co-exist in the same environment • It should be possible to install base stations without coordination • Frequency planning should not be needed	Frequency avoidance techniques (e.g., not make an access on a frequency that is too disturbed) Power control is used to minimize interference but is still able react on increased received interference Multi-user detection and interference cancellation techniques Spectrum sharing between TDD WCDMA and FDD WCDMA systems possible	DCA, frequency hopping for interference averaging	DCA, interference averaging by frequency and time hopping	DCA, TX power limitations for private system
Variable asymmetry of total band usage • Variable division of radio resource between uplink and downlink resources from a common pool (NB: this division could be in either frequency, time, or code domains)	This is primarily supported by the TDD mode in all schemes. TDD modes can have variable switching point to provide asymmetry (see Chapter 9). For FDD, it is possible to pair different uplink and downlink portions by using a variable duplex distance CDMA based scheme can have a slightly easier trade-off between coverage and bit rate between up and downlink			

Table 13.7 (continued)

	WCDMA (concept α)	OFDMA (concept β)	WB-TDMA (concept γ)	WB-TD-CDMA (concept δ)
Spectrum utilization • Allows multiple operators to use the band allocated to UMTS without coordination[1] • It should be possible to operate the UTRA in any suitable frequency band that becomes available such as first and second generation system's bands.	Sharing supported in limited scenarios Carrier spacing between UMTS operators 4.4 to 5 MHz Reframing requires 5.2 MHz (GSM on one side and UMTS on the other band-edge) or 5.6 MHz if GSM is on both sides	DCA	DCA	Minimum spectrum for reframing is 3×1.6 MHz + guardbands, hot spot reframing with 1.6 MHz + guardbands
Coverage, capacity • System should be flexible to support a variety of initial coverage/capacity configurations and facilitate coverage/capacity evolution • Flexible use of various cell types and relations between cells (e.g., indoor cells, hierarchical cells) within a geographical area without undue waste of radio resources • Ability to support cost effective coverage in rural areas	Supported HCS supported within 15 MHz	Supported Flexible resource allocation	Supported DCA	Supported HCS supported with at least 3 layers, DCA can be used, also between layers
Mobile terminal viability • Handportable and PCMCIA card size UMTS terminals should be viable in terms of size, weight, operating time, range, effective radiated power, and cost	No major difference between technologies. This depends also on future developments			
Network complexity and cost • The development and equipment cost should be kept at a reasonable level, taking into account the cost of cell sites, the associated network connections, signaling load, and traffic overhead (e.g., due to handovers)	All supported			

---

[1] Spectrum sharing, without any coordination, in the same geographical area and still guarantee a level of quality of service to the users is impossible in any system. This requirement is for further study.

Table 13.7 (continued)

	WCDMA (concept α)	OFDMA (concept β)	WB-TDMA (concept γ)	WB-TD-CDMA (concept δ)
Mobile station types • It should be possible to provide a variety of Mobile Station types of varying complexity, cost, and capabilities in order to satisfy the needs of different types of users	No real differences			
Alignment with IMT-2000 • UTRA shall meet at least the technical requirements for submission as a candidate technology for IMT 2000 (FPLMTS)	All meet.			
Minimum bandwidth allocation • It should be possible to deploy and operate a network in a limited bandwidth	Minimum bandwidth 5 MHz including guardbands. For cosited UMTS operators 4.4 MHz carrier spacing can be used	Minimum bandwidth 5 MHz (no detailed guardband analysis) 0.8 – 1.6 MHz in single isolated cell, small network with 3.2 MHz	Minimum bandwidth 3x1.6 MHz.	Minimum bandwidth 3×1.6 MHz + guardbands
Electromagnetic compatibility • The peak and average power and envelope variations have to be such that the degree of interference caused to other equipment is not higher than in today's systems	Continuous transmission improves the peak-to-average power ratio and envelope variations compared with GSM and similar TDMA-based systems	Similar to GSM	Similar to GSM	Similar to GSM. Multicode and 16QAM modulation cause additional envelope variations
RF radiation effects • UMTS shall be operative at RF emission power levels which are in line with the recommendations related to electromagnetic radiation	In principle the average power levels are independent of RTT			
Security • The UMTS radio interface should be able to accommodate at least the same level of protection as the GSM radio interface does	From a ciphering point of view, all radio interface technologies offer the same level of protection as good as of GSM			

Table 13.7 (continued)

	WCDMA (concept α)	OFDMA (concept β)	WB-TDMA (concept γ)	WB-TD-CDMA (concept δ)
Coexistence with other systems • The UMTS terrestrial radio access should be capable to co-exist with other systems within the same or neighboring band, depending on systems and regulations	See spectrum utilization	No detailed guardband analysis done	No detailed guardband analysis done	No detailed guardband analysis done
Dualmode terminals • It should be possible to implement dual mode UMTS/GSM terminals cost effectively	To measure the GSM carriers in WCDMA mode, a slotted mode has been defined GSM RX RF and IF filters required	Harmonized clocks GSM RX RF and IF filters required	Harmonized clocks GSM RX RF and IF filters required	Harmonized clocks GSM RX RF and IF filters required

REFERENCES

[1] Qualcomm, "A Comparison of the Six C's of Wireless Communications: Coverage, Capacity, Cost, Clarity, Choice, Customer Satisfaction," *Technical Report*, Qualcomm, 80-12589, Revision x1, 1994.

[2] Mohr, W., and J. Farjh, "GSM and CDMA, an Objective Comparison," *GSM MoU/ECTEL MRSG, Latin American Seminar*, Buenos Aires, Argentina, December 4-5, 1995.

[3] Ericsson, "PCS 1900 The New Personal Communications System for North America," *Technical Report*, Ericsson, December 12, 1994.

[4] Ericsson, "IS-95 and PCS-1900 Capacity Comparison," *Technical Report*, Ericsson, July 6, 1994.

[5] Fagen, D., A. Aksu, and A. Giordano, "A Case Study of CDMA and PCS-1900 Using the GRANET Radio Planning Tool," *Proceedings of ICPWC*, Mumbai, India, December 1997, pp. 505–509.

[6] Cook, C. I., "Development of Air Interface Standards for PCS," *IEEE Personal Communications*, Vol. 1. No. 1, 1994, pp. 30–34.

[7] Pizarroso, M., and J. Jiménez (eds.), "Preliminary Evaluation of ATDMA and CODIT System Concepts," *SIG5 Deliverable MPLA/TDE/SIG5/DS/P/002/b1*, September 1995.

[8] Urie, A., M. Streeton, and C. Mourot, "An Advanced TDMA Mobile Access System for UMTS," *IEEE Personal Communications*, Vol. 2, No. 1, February 1995, pp. 38–47.

[9] Pizarroso, M., and J. Jiménez (eds.),"Common Basis for Evaluation of ATDMA and CODIT System Concepts," *SIG5 deliverable MPLA/TDE/SIG5/DS/P/002/b1*, September 1995.

[10] Barberis, S., P. Blanc, L. Bonzano, E. Gaiani, V. Palestini, and G. Romano, "Radio Transmission and Capacity Comparison between ATDMA and CODIT Systems for UMTS," *Proceedings of VTC'97*, Vol. 2, Phoenix, USA, May 1997, pp. 815–819.

[11] Pehkonen, K., H. Holma, I. Keskitalo, E. Nikula, and T. Westman "A Performance Analysis of TDMA and CDMA Based Air Interface Solutions for UMTS High Bit Rate Services," *Proceedings of PIMRC'97*, Helsinki, Finland, September 1997, pp. 22–26.

[12] Nikula, E., and E. Malkamäki, "High Bit Rate Services for UMTS Using Wideband TDMA Carriers," *Proceedings of ICUPC'96*, Vol. 2, Cambridge, Massachusetts, USA, September/October 1996, pp. 562–566.

[13] Westman, T., and H. Holma, "CDMA System for UMTS High Bit Rate Services," *Proceedings of VTC'97*, Phoenix, Arizona, USA., May 1997, pp. 824–829.

[14] Ojanperä, T., and R. Prasad, "Overview of Air Interface Multiple Access for IMT-2000/UMTS," *IEEE Communications Magazine*, 1998.

[15] Ranta, P., A. Lappeteläinen, and Z.-C. Honkasalo, "Interference cancellation by Joint Detection in Random Frequency Hopping TDMA Networks," *Proceedings of*

ICUPC'96, Vol. 1, Cambridge, Massachusetts, September/October 1996, pp. 428–432.

[16] Ojanperä, T., P. Ranta, S. Hämäläinen, and A. Lappeteläinen, "Analysis of CDMA and TDMA for 3rd Generation Mobile Radio Systems," *Proceedings of VTC'97*, Vol. 2, Phoenix, Arizona, USA., May 1997, pp. 840–844.

[17] Ojanperä, T., J. Sköld, J. Castro, L. Girard, and A. Klein, "Comparison of Multiple Access Schemes for UMTS," *Proceedings of VTC'97*, Vol. 2, Phoenix, Arizona, USA, May 1997, pp. 490–494.

[18] Ojanperä, T., K. Rikkinen, H. Häkkinen, K. Pehkonen, A. Hottinen, and J. Lilleberg, "Design of a 3rd Generation Multirate CDMA System with Multiuser Detection, MUD-CDMA," *Proceedings of ISSSTA'96*, Vol. 1, Mainz, Germany, 1996, pp. 334–338.

[19] Ojanperä, T., M. Gudmundson, P. Jung, J. Sköld, R. Pirhonen, G. Kramer, and A. Toskala, "FRAMES - Hybrid Multiple Access Technology," *Proceedings of ISSSTA96*, Vol. 1, Mainz, Germany, 1996, pp. 320–324.

[20] Ojanperä, T., P.-O. Anderson, J. Castro, L. Girard, A. Klein, and R. Prasad, "A Comparative Study of Hybrid Multiple Access Schemes for UMTS," *Proceedings of ACTS Mobile Summit Conference*, Vol. 1, Granada, Spain, 1996, pp. 124–130.

[21] Ishida, Y., "Recent Study on Candidate Radio Transmission Technology for IMT-2000," *First Annual CDMA European Congress*, London, UK, October 1997.

[22] ETSI UMTS 30.03, "Selection procedures for the choice of radio transmission technologies of the Universal Mobile Telecommunications System (UMTS)," *ETSI Technical Report*, v. 3.0.0, May 1997.

[23] ETSI UMTS 30.06, "UMTS Terrestrial Radio Access Concept evaluation," *ETSI Technical Report*, to be published by ETSI, 1998.

Chapter 14

STANDARDIZATION WORK AND FUTURE DIRECTIONS

14.1 INTRODUCTION

In Chapter 1, we divided the chronology of CDMA development into three eras: the CDMA pioneer era, the narrowband CDMA era, and the wideband CDMA era. The wideband CDMA era began with research activities in the early 1990s; it is now continuing with the adoption of wideband CDMA by several standards bodies. So far, ETSI, ARIB, and TR45.5 have defined wideband CDMA frameworks. This is only a starting point for detailed standardization. In order to move from the standardization phase into the implementation and commercial operations phases, detailed standards need to be ready. Furthermore, after a standard has been specified, a number of other elements need to be investigated and finalized in order to deploy commercial systems: spectrum needs to be acquired, applications developed, markets analyzed. Otherwise, a standard just gathers dust on a bookshelf among numerous other unused standards.

This chapter addresses the development of wideband CDMA proposals into commercial standards. We first describe the structure of standards bodies and industrial interest groups related to the third generation standardization in different regions. This establishes the basis for the discussion on third generation radio and network standardization. We address the role of regional standard bodies and the ITU in this process. Special attention is given to the ITU radio transmission technology (RTT) evaluation process. Regulatory developments such as spectrum licensing are described.

The third generation market, and the role of second generation systems in establishing it, are discussed. The roles of other technologies such as wireless LANs, wireless ATM, and satellites, which can either complement or compete with IMT-2000, are presented. We conclude this chapter by discussing enabling technologies and by describing a third generation service/application scenario.

14.2   STANDARDIZATION BODIES AND INDUSTRY GROUPS

This section introduces the different standardization bodies for the third generation systems, their responsibilities, and their organization. We start by describing the IMT-2000 standard organization in the ITU. Next, the standards bodies and the third generation industry interest groups in Europe, the United States, Japan, Korea, and China are presented. Furthermore, each subsection starts with a description of the general approach each region is taking to the development of mobile radio systems.

14.2.1   ITU

Figure 14.1 shows the organization of ITU for the IMT-2000 standardization. The telecommunications standardization in the ITU is divided into two sectors: Radiocommunications Sector (ITU-R) and Telecommunication Standardization Sector (ITU-T). The Intersector Co-ordination Group (ICG) is coordinating the IMT-2000 radio and network standards. The IMT-2000 radio aspects are being standardized in the Task Group (TG) 8/1 in ITU-R. TG8/1 is also responsible for the overall system architecture for IMT-2000 within ITU. In ITU-T, SG8 has been identified as the lead study group for coordination of standardization activities on IMT-2000 network standards. For more information on the ITU standardization see [1].

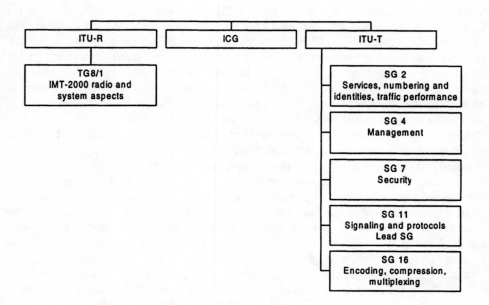

Figure 14.1   IMT-2000 standardization structure in ITU.

The role of the ITU in the standardization of third generation systems has been under discussion several times. The work in TG8/1 started in 1986. The telecommunications environment has changed dramatically since then. Therefore, it has been necessary to consider ITU's role with respect to regional standards bodies. The creation of the ITU Family of Systems concept has marked a clear shift in the ITU role (see Section 12.5.2). It has been recognized that second generation standards will evolve further and most likely form the basis for the third generation networks. Given the diverse need and the pace of development in different regions, a single global approach does not seem to be adequate. Furthermore, commercial interests due to the current investments into the second generation networks are further obstacles towards a single third generation standard. Based on these considerations, the role of the ITU seems to move towards setting global frameworks for requirements, spectrum allocations, interworking, and so on, rather than to the detailed drafting of standards.

14.2.2 Europe

In Europe, telecommunication standards are developed by ETSI. The general approach in Europe has been building consensus with emphasis on long-term solutions. The European Union is involved in the standards setting process by developing enabling regulatory policies. The global success of GSM is partly based on this type of standardization approach, and the Europeans are trying to carry this over to the third generation.

The ETSI structure related to UMTS developments is depicted in Figure 14.2. The most relevant technical committee (TC) from the UMTS point of view is TC SMG (Special Mobile Group). TC SMG is responsible for GSM and UMTS development. The actual standardization work is carried out in subtechnical committees. SMG2 has been responsible for the UMTS air interface evaluation and will standardize UMTS radio access network. SMG3 is responsible for GSM/UMTS core network standardization, and SMG12 is responsible of the overall UMTS architecture. SMG5, the original subtechnical committee responsible for UMTS, has been discontinued in 1997. TC NA (Network Aspects) has some network activities related to UMTS. TC SES (Satellite) is partly responsible for standardization of the UMTS satellite component. ETSI Project (EP) Digital Enhanced Cordless Telecommunication (DECT) is also considering some UMTS aspects. The work in TCs and STCs is based on the voluntary effort from ETSI members. In addition to STCs, a permanent nucleus, the Special Task Force, is supporting the TC SMG. For more information on the ETSI standardization see [2].

The UMTS Forum is a nonprofit association under the Swiss law. The UMTS Forum was established in 1996 on the recommendation of the UMTS Task Force [3]. It provides advice and recommendations to the European Commission, European Radiocommunications Office (ERC), ETSI, and national administrations. The UMTS Forum is divided into working groups (e.g., WG1 Regulatory, WG2 spectrum, and WG3 Market Aspects). For more information on the UMTS Forum see [4].

The GSM MoU Association is the GSM operators' organization for promoting and evolving the GSM cellular platform world-wide. Regional interest groups have been established to take requirements and special needs from the various regions into account: Asia Pacific Interest Group (APIG), Arab Interest Group (AIG), European Interest Group (EIG), North American Interest Group (NAIG), Central/Southern African Interest Group (CSAIG), East Central Asia Interest Group (ECAIG), and India Interest Group (IIG). In addition to GSM operators, government regulators/administrations can be members of the GSM MoU Association. Within GSM MoU, the Third Generation Interest Group (3GIG) was formed in 1994. The 3GIG is responsible for conveying GSM MoU vision and requirements into the standardization of UMTS. Furthermore, it addresses third generation market aspects and licensing issues. For more information on the GSM MoU see [5].

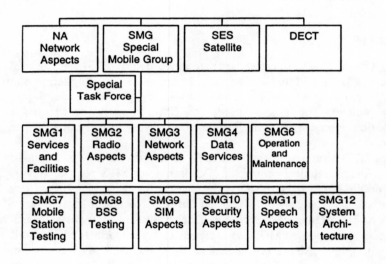

Figure 14.2   Partial view of the ETSI organization relevant to UMTS.

14.2.3 The United States

In the United States, the standards setting process is more driven by market inputs and industry interests than in Europe. On one hand, this leads to faster reaction to changes, but on the other, it leads to a larger number of standards. The selection between standards is then left to the market forces. The market-driven approach is reflected also in the standardization bodies. There are two main standardization bodies for mobile radio systems. The Telecom Industry Association (TIA) is a trade organization that provides its members with numerous services including standardization. The committee T1 on telecommunications was established in 1984 as a consequence of the breakup of the Bell System [6]. Both the TIA and T1 can set American National Standards Institute (ANSI) accredited standards. Within TIA, the Wireless Communications Division is responsible for standardization of wireless technologies. The main committees from a third generation point of view are TR45 and TR46 (Public Mobile and Personal Communications Standards). Within T1, subcommittee T1P1 is responsible for management activities for personal communications systems (PCS). For more information on the TIA standardization activities see [7], and on T1P1 activities see [8].

14.2.3.1 TR45

Figure 14.3 shows the structure of TR45. In consists of six permanent subcommittees and six ad-hoc groups not shown in the figure. The responsibilities of the subcommittees are as follows:

- TR45.1 is responsible for the AMPS standardization.
- TR45.2 is responsible of the development of the IS-41 standard.
- TR45.3 is responsible for IS-136, the US TDMA system, and its evolution to UWC-136.
- TR45.4 is responsible for the IS-634 standard (i.e., the A-interface standard between BS and MSC). For the third generation, TR45.4 will most likely develop the interface between the access and core networks.
- TR45.5 is responsible for IS-95 technology and its evolution to cdma2000.
- TR45.6 work items include support for packet data and Internet access. TR45.6 coordinates its activities with TR45.2 to define network architecture for IMT-2000 that complies with packet data requirements.

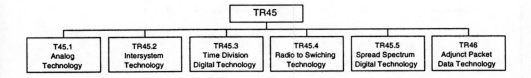

Figure 14.3   Structure of TIA TR45.

TR45.5 consists of four subcommittees as shown in Figure 14.4. TR45.5.1 is responsible for user needs and services such as multimedia services, speech, feature requirements, and data services. TR45.5.2 covers signaling and protocols including authentication/privacy, intersystem issues, call processing, and protocol tests. TR45.5.3 is responsible for the physical layer and has five working groups: RF parameters, supervision and malfunction detection, handoff and system timing, MS minimum performance, and BS minimum performance. TR45.5.4 is developing alternative

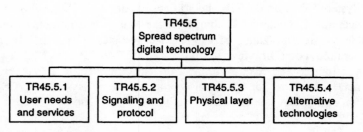

Figure 14.4   Structure of TIA TR45.5.

technologies and has been responsible for development of the framework for Wideband cdmaOne.

14.2.3.2   TR46 and T1P1

TR46 was created to address the TIA standards development for the US PCS. The main subcommittees within TR46 are listed below and represented in Figure 14.5.

- TR46.1 WIMS (Wireless ISDN and Multimedia Service);
- TR46.2 Cross technology;
- TR46.5 PCS1900;
- TR46.6 Composite CDMA/TDMA.

Standards committee T1P1 – Wireless/Mobile Services and Systems belonging to T1 organization – is closely related to TR46. They have an almost similar structure and jointly manage the above-mentioned TR46 standards. T1P1 has a cooperative arrangement with ETSI SMG regarding PCS1900/GSM technology.

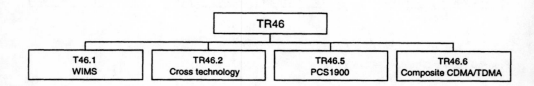

Figure 14.5   Structure of TIA TR46.

14.2.3.3 Industry Interest Groups

Each of the main PCS standards in the United States has its own industry interest group: the CDMA Development Group (CDG) for IS-95, Universal Wireless Communications Consortia (UWCC) for IS-136, GSM North America (GSM MoU regional interest group), and GSM Alliance. The last two are consortia for North American PCS1900 operators using GSM-technology.

CDG is an industry consortium of companies developing IS-95. The CDG is comprised of IS-95 service providers and manufacturers of subscriber and network infrastructure equipment. For more information on the CDG activities see [9]. The CDG organizational structure consists of the following entities [9]:

- Executive Board: a carrier-only board that oversees the CDG process. Participation on the board is limited to chief technical officers, chief operating officers, and chief executive officers.
- Steering Committee: a committee consisting of executives from member organizations. The objective of this committee is to review and approve the work of the supporting teams and address resource issues.
- CDG Working Groups (Teams): teams consisting of carrier and manufacturer business and technical resources. Advanced systems group is focusing on third generation technologies.

The UWCC is a limited liability corporation in Washington state, established to support an association of carriers and vendors developing, building, and deploying products and services based on IS-136 TDMA and IS-41 wireless intelligent network (WIN) standards. The UWCC structure consists of a board of governors and three forums: Global TFMA Forum (GTF), Global Win Forum, and Global Operators Forum. For more information on the UWCC activities see [10].

14.2.4 Japan

In Japan, the telecommunication operators and manufacturers have traditionally moved as one single entity. NTT has been in the leading role, dominating the wireless standardization scene, and the others have followed NTT in unison. The selection of IS-95 by DDI and IDO, two Japanese wireless operators, for replacing the analog system was an unusual break from the tradition of following NTT [11]. It has been very difficult for international companies to participate in the closed standards setting. One further obstacle has been the language. For third generation, Japan has opened standardization in order to create a global standard. Instead of Japanese, English has been selected as the working language for the documentation of the third generation standards.

In Japan, the IMT-2000 standardization is divided between two standardization organizations: Association of Radio Industries and Businesses (ARIB) and Telecommunication Technology Committee (TTC). ARIB is responsible for radio

414

standardization and TTC for network standardization. Contributions to ITU-R and ITU-T are done through the Telecommunication Technology Council of the Ministry of Posts and Telecommunications (MPT). MPT has established a Study Group on Next-Generation Mobile Communication, which gives recommendations to TTC and ARIB.

The organization of IMT-2000 Study Committee within ARIB is described in Figure 14.6. The Coordination Group (CG) in the Standards Subcommittee is responsible for international coordination with other standards bodies such as TTA, ETSI, and TIA. The harmonization discussions concerning the WCDMA and cdma2000 have been carried out in this group. SWG1 is responsible for system description; SWG2 is responsible for the specification of the air interface layer 1; and SWG3 is responsible for the air interface specification for layers 2 and 3. The requirements and objectives are established in the Application Group. For more information on the ARIB activities see [12].

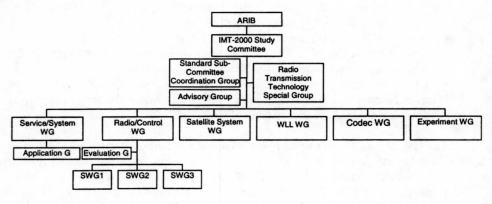

Figure 14.6   ARIB organization for IMT-2000.

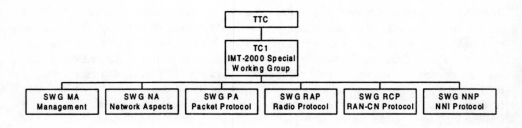

Figure 14.7   TTC organization for IMT-2000.

TTC, represented in Figure 14.7, is responsible for the development of system architecture, information flows, requirements for network control, layer 2 in cooperation with ARIB, layer 3, RAN-CN interface, network-to-network interface (NNI), call associated signaling, and interworking with other networks. Since Japan has chosen to base third generation network on GSM core network, close cooperation with ETSI is under way. For more information on the TTC standardization see [13].

14.2.5 Korea

In Korea, the government sets the policies for the mobile radio system standardization. For example, the adoption of IS-95 over other wireless technologies for Korean PCS was decided by the Korean government. The Korean goals for IMT-2000 are a global standard, global roaming, low cost terminal, to meet the ITU requirements, and backwards compatibility, if possible. Korea has a parallel approach to standards: international cooperation and development of domestic standards for implementation after 2002. The international cooperation includes relations with ARIB, ETSI, and TIA, as well as participation in the ITU RTT process.

The Telecommunication Technology Association (TTA) was established by the Korean Ministry of Communications (MIC) in 1988 to develop Korean Information Communication Standard (KICS) related to telecommunications. Within TTA, the IMT-2000 subcommittee has been established for the development of third generation standards. The structure of the subcommittee is depicted in Figure 14.8. For more information on the TTA standardization see [14].

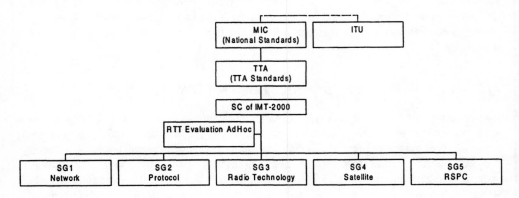

Figure 14.8   Structure of TTA.

The Electronics and Telecommunications Research Institute (ETRI) is an R&D organization developing technologies considered important for Korea. The Mobile Telecommunications Technology Division of ETRI is developing technology for IMT-2000. For more information on ETRI see [15].

14.2.6 China

In China, the Ministry of Post and Telecommunications (MPT) is driving IMT-2000 standardization. The Research Institute for Telecom & Transmission (RITT) is responsible for domestic standardization. It has actively cooperated with outside standardization bodies such as ETSI and ARIB. A MPT delegation has participated, for example, in the ETSI/SMG meetings as an associate member. MPT participates also actively the ITU process.

14.2.7 Other Groups

The Radio Standardization Meeting (RAST) is a group where different regional radio standardization organizations exchange views and plans regarding different areas of radio standardization, including third generation systems. The RAST activities are reported in [16].

The Future Advanced Mobile Wireless Universal System (FAMOUS) is a yearly meeting of administrations from Japan, the United States and the European Union. Its objectives are to provide a forum for discussion of matters related to the third generation mobile communications systems and to foster international cooperation regarding the interoperability of third generation systems on a global basis.

Asian-Pacific Wireless Forum (APWF) is a discussion forum for the purpose of Asian-Pacific cooperation in the R&D of wireless personal communications systems [17]. Based on the initiative of APWF, it has been proposed to set up a new standardization organization: Asian-Pacific Telecommunications Standards Institute (ATSI). Whether there is a need for an Asian-Pacific standard, and what the relation of the new organization in respect to the existing standards bodies would be, are still under discussion [17].

14.3 RADIO ACCESS NETWORK STANDARDIZATION

Both regional standards bodies (ETSI, ARIB, TIA, and TTA) and ITU-R TG8/1 have work programs to develop detailed specifications for the IMT-2000 radio access network. This raises a question: Which standards bodies are actually going to specify the IMT-2000 air interface and which are merely going to adopt already-made specifications? Since the ITU RTT evaluation process is aimed to produce a global standard, we will describe it and discuss its role with respect to regional standards activities.

Figures 14.9 and 14.10 illustrate the time schedule and the different steps of IMT-2000 air interface development, respectively. As can be seen, the consensus building within ITU starts only after evaluation documents have been submitted. However, the consensus building has already started based on initiatives from the regional standardization bodies and within the framework of FAMOUS meetings. In particular, Japan has established the CG group for harmonization purposes. In the beginning of 1998, TIA TR45.5 established an ad-hoc committee for IMT-2000 coordination and harmonization process of CDMA proposals. The role of the ITU process has thereby been to accelerate the regional developments towards closer cooperation.

Detailed standardization of the third generation systems, which are of an order of magnitude more complex than today's systems, will place a heavy burden on standards organizations. Not only do the new features need to be standardized, but also backwards compatibility aspects need to be taken into account. Although the parameters of some wideband CDMA proposals have been harmonized, maintenance of this harmonization during the standards process presents a major challenge. Small details need to be agreed on, and change requests will pour into standards bodies continuously. This high level harmonization can take place in ad-hoc groups, and to some extent, it has already happened. However, detailed standardization requires much more strict rules and framework. Having two or more standard bodies to do their detailed work separately and trying to achieve joint specification through ad-hoc meetings might prove to be very complex, if not impossible. Therefore, a joint forum for detailed standard development would be preferable.

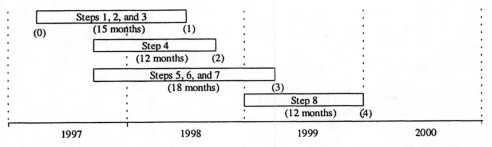

Steps in the Radio Interface Development Process:

Step 1: ITU BR issues request for submission of candidate radio transmission technologies (RTTs)
Step 2: Manufacturers or other parties develop candidate RTTs
Step 3: Proponents submit candidate SRTT proposals
Step 4: Evaluation groups evaluate candidate RTTs
Step 5: TG 8/1 reviews evaluation activities
Step 6: TG 8/1 reviews candidate RTTs to ensure minimum performance capabilities are met
Step 7: TG 8/1 considers evaluation results, builds concensus, and determines the key technical characteristics of the IMT-2000 radio interface
Step 8: TG 8/1 develops radio interface specification Recommendations

Critical milestones in the Radio Interface Development Process:

(1) Issue request for candidate RTTs March 1997 ITU proposed cut off for submission of candidate RTT proposals, to be confirmed by evaluation groups June 1998
(2) Cut off for submission of evaluation reports to ITU September 1997
(3) TG 8/1 determines the key characteristics for the IMT-2000 radio interface March 1999
(4) TG 8/1 completes development of radio interface specification Recommendations December 1999 (approx.)

Figure 14.9   Time schedule of the radio interface development process (*Source*: [18], reproduced with permission of ITU).

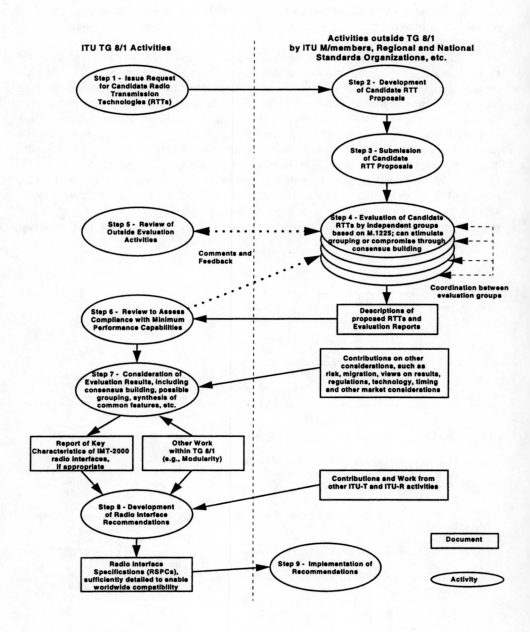

Figure 14.10   IMT-2000/FPLMTS radio interface development process (*Source*: [18], reproduced with permission of ITU)

One model for standards development is as follows. The core-standard would be developed by a joint standards body ensuring global conformance and minimum compatibility. Regional standard bodies would focus to specific issues related to their region's specific needs.

ITU activities encompass all third generation proposals. Given the current situation, where we have several proposals not all based on wideband CDMA, the ITU may not be flexible enough for this type of standards development. This seems to be the view of regional standards bodies, which are seeking cooperation outside ITU. Thus, a more likely development is that a global standard would be specified in a joint industry forum, for example.

14.3.1 ITU Evaluation Groups

The following evaluation groups have been formed [1]:

- IMT-2000 Evaluation Groups;
- Australia: Cooperative Research Centre for Broadband Telecommunications and Networking (CRC-BTN);
- Austria: ETSI Project DECT;
- Brazil: ANATEL;
- Canada: Canada Evaluation Group (CEG);
- China: China Evaluation Group (ChEG);
- Europe: ETSI SMG2, and INMARSAT;
- Japan: Association of Radio Industries & Businesses (ARIB);
- Korea: TTA Evaluation Ad-hoc;
- Malaysia: ITU;
- New Zealand: Radio Spectrum Management;
- USA:TIA/TR-45 Ad-hoc International Standards Development Group, and T1P1/TR46 International Standards Ad-Hoc;

14.3.2 IMT-2000 Radio Transmission Technology Proposals

The following proposals were submitted for terrestrial radio access by June 1998:

- ARIB/Japan: W-CDMA;
- CATT/China: TD-SCDMA;
- TTA Korea: Global CDMA I and II;
- ETSI project DECT: DECT;
- ETSI: UTRA;
- TIA/USA: TR45.3 (UWC-136) and TR45.5 (cdma2000), and TR46 (WIMS W-CDMA);
- T1P1-ATIS/USA: WCDMA/NA.

For satellite radio access the following proposals were submitted:

- European Space Agency (ESA): SW-CDMA and SW-CTDMA;
- ICO: ICO RTT;
- INMARSAT: Horizons;
- TTA Korea: satellite RTT.

14.4   CORE NETWORK STANDARDIZATION

The focus of third generation standards has been on the air interface. However, for a complete system the core network is equally important. There are two main views for the core network development. The first builds on a clean table approach where the core network would be based on B-ISDN. The second approach builds on the evolution from the second generation systems. Recently, the second approach has been gaining more support. For example, the adoption of the Family of Systems concept in ITU was a recognition of the evolution of second generation core networks.

The standardization of the evolving second generation core networks is driven by regional standardization bodies. ETSI SMG is responsible for GSM/UMTS core network standardization, and TIA TR45.2 for the IS-41 core network standardization. Japan has also decided to adopt the GSM/UMTS core network. Similar considerations as for the radio standardization apply for the core network standardization. A joint forum may need to be established for detailed specification development or one group must take a clear lead. The role of ITU-T is seen by many parties to be in global issues, such as common numbering and identifying interfaces required for interworking, rather than in developing detailed standards.

14.5   REGULATION

Fundamentally, regulatory matters are national issues. However, in some cases regional and global harmonization of regulations and policy issues is desirable [19]. Areas where common regulations may be required are licensing, services, provisioning, interconnection, infrastructure, frequencies, numbering, security, and policies [19]. With regard to a radio access system, the most important regulatory areas are spectrum coordination and licensing.

The process of obtaining frequencies can be divided into three phases: identification of spectrum, allocating the spectrum for specific purposes, and licensing the spectrum. Since the identified spectrum is usually already used for some other purposes, it needs to be cleaned from existing users before it can be used for the new purpose. Therefore, proactive long-term planning is required to ensure the availability of spectrum when required.

ITU plays an important role in spectrum regulation. The Radio Regulations (RR) of ITU are updated in World Radio Conference (WRC). The next WRC is scheduled for 1999. The national regulators are not bounded to follow the ITU guidelines for spectrum allocation. However, the ITU RRs form a tool to encourage them to do that in order to achieve global harmonization of spectrum [19]. The IMT-2000 spectrum was identified in the year 1992 by WARC92 as a result of ITU studies on IMT-2000. These studies indicated that the minimum spectrum for IMT-2000 should be 230 MHz.

Spectrum is allocated for a specific purpose. It has to be decided what services can be provided, what technology is allowed to be used, and how the spectrum is allocated between operators. In addition to the IMT-2000 services, it might be desirable to provide some other services such as broadband data for indoor service within the IMT-2000 spectrum [20]. Whether this should be allowed or not must be decided. The spectrum use can be bounded to a specific standard, or the operator may be allowed to use any technology to deliver their services. The latter approach was taken in the US PCS spectrum assignments. The spectrum can be allocated on an exclusive basis or by using spectrum sharing. For an exclusive allocation, what is the minimum bandwidth for a single operator and what is the maximum number of operators that can be accommodated within the given frequency allocation?

The last phase in spectrum allocation is licensing. It has to be decided who can operate the networks. This can happen by "beauty contests" (i.e., who is evaluated to be most suitable to run a network), by lotteries, or by auctions. It has been argued whether the spectrum auction is a fair and effective way of assigning licenses for third generation mobile operators [20]. Furthermore, since spectrum is so scarce, it is a valuable asset. Thus, governments are eager to cash in that asset, as was seen in the United States for the PCS licenses. If it is decided to auction the spectrum, it has to be decided how auctions will be held and who can bid. The are three options for auctions: multiple round open auctions (used in the United States for the PCS frequencies), sealed bid auction, or royalty auction, where the payment takes form of a percentage of qualifying revenue. An important question is whether an existing operator can bid for the third generation license. This has to be weighted against the competitive advantage they might gain since they already have a nationwide coverage while the new operators have to build the coverage from scratch. The regulation could require national roaming between operators to alleviate the newcomers' position [20]. The regulation can also rule what the conditions are for a license. These include coverage requirements and the length of a license.

In general, Europe follows ITU recommendations for spectrum issues. The Europe-wide harmonization is carried out by the Conference of European Post and Telecommunications (CEPT). The European Commission can issue a directive to create harmonized frequency allocations for specific technologies. Examples of such directives are GSM in 1987 and DECT in 1991. Recently, the commission has published its proposed UMTS decision, which sets in place timescales and actions for national licences and spectrum harmonization by the year 2000 [21]. The European Radiocommunications Committee (ERC) of CEPT makes decisions which usually form the basis for the harmonized spectrum designations. CEPT has designated most of the IMT-2000 spectrum for UMTS with the adoption of the ERC Decision on UMTS, which identifies a total of 155 MHz of spectrum for terrestrial UMTS services, with an additional 60 MHz set aside for UMTS satellite services [22]. For the third generation frequencies, the United Kingdom intends to hold spectrum auctions in summer 1999. UMTS Auction Consultative Group (UACG) has been established to facilitate the UK government's ongoing consultation with industry and other interested parties on matters relating to the proposed auction of spectrum licences for terrestrial UMTS (for more details, refer to website http://www.open.gov.uk/radiocom/rahome.htm).

In the United States, the Federal Communications Commission (FCC) is managing the spectrum issues. The Inter-American Telecommunication Commission (CITEL) is issuing recommendations on spectrum issues for North and South America. The recommendations of CITEL are nonbinding but carry significant weight. The third generation spectrum is for the most part already used by PCS systems. However, the United States is currently developing spectrum positions for the third generation including the identification of new frequency bands.

In Japan, the Ministry of Post and Telecommunication (MPT) is responsible for spectrum regulation. The Japanese spectrum plan follows the ITU recommendation for IMT-2000 frequencies.

14.6 MARKET CONSIDERATIONS

In this section, we discuss the wireless market development. Prediction of the future is always a difficult task, and the wireless communications field is not an exception to this rule. Long-term market projections for mobile radio systems are difficult when companies are struggling even with short-term projections. In this section we present some subscriber predictions for third generation systems in order to assess potential users at the time of the deployment of third generation systems.

There are considerable differences between the several market estimates for mobile users. The UMTS Forum has predicted that there would be 1.7 billion mobile users by 2010. This is 20% of the world's user population. The UMTS Forum also estimates that 45% of the mobile users will use high speed data services by 2010. It is clear, however, that there is no single right figure for the IMT-2000 market. The development of the third generation market depends on several factors including coverage, tariffing, lifestyle changes, and terminal offerings. Table 14.1 presents world subscriber growth for different compounded annual growth percentages for 1998-2001. According to [23], growth of 40 to 45 percent is considered most plausible. Between 1996 and 1997, the annual growth worldwide averaged 51.6 percent and since 1990, compounded annual growth has averaged 50.9 percent.

Table 14.1
Forecast of World Subscriber Growth (Millions)

| Year | Assumed Compounded Annual Growth | | | | |
	25%	30%	35%	40%	45%
1998	258	268	279	289	299
1990	322	349	376	404	434
2000	403	453	508	566	629
2001	504	589	685	793	912

Source: [23].

The market for mobile multimedia services will depend on several factors. The market analysis group in the UMTS Forum has conducted a study that analyzed four different market scenarios for the mobile multimedia market [24]. These scenarios give an indication of the factors that will determine market development. A short summary of these scenarios follows.

Scenario 1, Slow Evolution: Mobile multimedia development is slow due to limited applications and high service and terminal prices. Unsuccessful liberalization has led to low competition. Fragmented standards have resulted in no economies of scale. In addition, fragmented standards hinder global application development. No global standard has been achieved, which has led to high equipment prices. Applications are difficult to reconfigure and to adapt to personal needs, and there have been no breakthrough developments for critical terminal technologies such as displays.

Scenario 2, Business Centric: Mobile multimedia takes off in the business sector but not in the consumer sector, as there is a lack of innovation in consumer applications. Lack of competition due to unsuccessful liberalization contributes to high premiums to access the service. However, high access costs can be afforded by business sector users. Terminals are intelligent and configurable to user needs but still difficult to use.

Scenario 3, Sophisticated Mass Market: Mass market for mobile multimedia has emerged. Terminals and applications have a large set of features and can be customized for personal needs. However, user interfaces are still rather complicated, being suitable for IT-literate users. Liberalization has resulted in high competition and thus low access premiums. Global standards for both systems and application platforms have facilitated low equipment costs and rich application developments. Spectrum was auctioned and the resulting costs are reflected in initially high prices that quickly drop.

Scenario 4, Commoditized Mass Market: A real mobile multimedia mass market has emerged and comprises both business and consumer users. Liberalization and adoption of global standards have resulted in economies of scale. Spectrum is cheap; simple and cheap terminals are available from the beginning.

Table 14.2 shows the number of mobile users and multimedia users by 2005 for these different scenarios. The essential differences between the first two and the last two scenarios concern the worldwide platform in both radio and traffic delivery and in application. The common platform facilitates economies of scale and sets a standardized basis for application development. The main difference between scenarios 1 and 2 is that in the second scenario users can configure their terminals according to their needs since they understand IT. In the first scenario, the user interface is more limited and thus not adequate even for most business users. The main difference between scenarios 3 and 4 is the higher initial cost in scenario 3 due to spectrum pricing and the price and ease of use for terminals. In scenario 3, spectrum is auctioned and terminals are bulky and expensive in the beginning. Furthermore, in scenario 3, tariffing is initially high due to high investment into spectrum and equipment. These factors contribute to the slower development of mass market as with scenario 4.

Table 14.2
Number of Mobile and Multimedia Users in Europe by 2005

Scenario	Mobile users by 2005 (penetration)	Multimedia users by 2005
Slow evolution	82 M (22%)	7.5 M
Business centric	82 M (22%)	9 M
Sophisticated mass market	123 M (35%)	19 M
Commoditized mass market	140 M (40%)	27 M

Source:[24].

The potential new users for voice for data services can also be viewed from the perspective of the penetration rate by the year 2000. The higher the penetration rate of a service, the less potential there is for new business. On the other hand, a very high penetration rate might drive a deployment of a new system to support the required capacity.

Second generation systems are at the same time a threat and an opportunity for IMT-2000 systems. If data rates are enough for current applications and there is no lack of spectrum, the deployment of third generation systems will be slower. We should remember the predictions for analog systems in the beginning of the 1990s: they were supposed to be phased out by the middle of the decade. However, in 1998 AMPS has still a significant part of the cellular subscribers. We should always remember that we can learn from history, but we cannot predict the future based on history. In a chaotic process, even the smallest change in starting conditions will lead to a totally different end result.

Second generation systems are also tools to educate operators and customers for the data market. For example, operators can test the impact of tariffing on service demand, without the heavy investments required for third generation services. When users are already used to new types of services, adoption of data services will be easier. For many services, third generation will just mean better quality of service (i.e., faster service and higher reliability) when compared to services provided by second generation systems. Thus, second generation systems can be viewed as a logical stepping stone towards third generation systems, and certainly not something we need to discard as quickly as possible.

14.7 WIRELESS BROADBAND NETWORKS

Wireless broadband networks will provide user bit rates up to 20 Mbps, with a maximum range of some tens of meters. Wireless ATM is the most promising technology to implement wireless broadband networks. The basic idea of wireless ATM (WATM) is to extend the communications capabilities of wired ATM such as provision of different QoS classes to mobile users [25].

In Europe and the US, there is common allocation for mobile broadband systems in the 5-GHz frequency band. Similar allocation is also considered in Japan. The very large amount of unlicensed spectrum facilitates simple air interface design because limited attention needs to be paid into the spectrum efficiency as in the cellular systems using licensed frequency bands. Furthermore, the tariffing structure for high speed data services could be different from the cellular since there is no price for the spectrum and allocating spectrum for data services does not decrease revenues from other services such as speech.

The 20-Mbps data rate and large allocations of unlicensed spectrum make the integration of WATM radio access with IMT-2000 an attractive opportunity to extend the current cellular business to a new market. The IMT-2000 would offer wide area mobility and coverage, while the WATM would offer hotspot broadband data services. This would require the integration of mobility management of IMT-2000 and WATM networks. This issue is considered at least in the ETSI Broadband Radio Access Network (BRAN) project [25].

14.8 SATELLITES

Satellites can provide coverage to remote, rural, and maritime areas where it would be too expensive or impossible to set up terrestrial infrastructure. Satellite-based systems providing mobile services are termed mobile satellite services (MSS) or satellite personal communications networks (S-PCN).

The planned S-PCN systems use nongeostationary orbits (Non-GEO). Non-GEOs can be subdivided into low earth orbit (LEO) and medium earth orbit (MEO), also called intermediate circular orbit (ICO), systems. LEO systems have altitudes from 800 to 1600 km and MEO systems around 10,000 km. Examples of non-GEO systems are Globalstar, Iridium, Loral, and Odyssey [26]. It should be noted that these systems are being deployed within the time frame of 1998–2002 (i.e., close to IMT-2000 deployment). The first services are speech and data up to 9.6 Kbps. The European third generation satellite activities are addressed in [27].

14.9 ENABLING TECHNOLOGIES

The applications and terminals need to be developed to a new level in order to utilize the extensive capabilities of third generation networks. The improvement in the processing power of DSPs and ASICs makes the radio part smaller and smaller, thus creating room for additional devices such as better displays, digital cameras, and many others. However, the increased software, higher number of application devices, and higher data call for improvements in battery technologies. Solar power technology might be used as a complementing energy source.

Digital cameras facilitate digitization of images. Although IMT-2000 will bring significant improvements to data rates, multimedia applications such as image transmission will require even higher bandwidths. Thus, efficient compression technologies are vital. Fractal coding is a new technology that could facilitate a breakthrough in this area [24].

Users viewing multimedia information will want to do it through a high-quality, high-resolution display. Mobile users use their terminals in very variable lighting conditions. Consequently, the display has to be capable of presenting the picture clearly regardless of external lighting. Of course, size is of key importance for portable devices.

In order to capitalize on mass market, user interfaces of wireless devices must be developed far beyond today's standards. Applications have to be easy to use, non-technical and understandable to a lay person. Voice recognition is one possible technique that can help with building user-friendly applications. Virtual reality is used to create a virtual environment for one user: a mobile user could imitate office conditions, for example, in a hotel room and could see the others in a realistic meting environment. Interactive virtual reality opens new possibilities for developing more attractive games that can be played against other users over the wireless link.

Context-aware applications are applications that change their behavior according to the user's present context—their location, who they are with, what is the time of the day, and so on [28]. In wireless systems, the context information can be used to either transfer information from the network to the user, or transfer information from the user to the network. One challenge to implement such applications is how to sense the user's context (location, surrounding people, time of the day, etc.). The Global Positioning System (GPS) is one way of locating users. However, it may not be accurate enough. Another possibility is to use the location information from the mobile radio system. The environment can be sensed by active badges, attached to people, identifying their carrier [28]. A standard way of creating context-aware applications (i.e., creating an actual program triggered by the context) has been presented in [28]. The methodology is based on a language similar to HTML used to create World Wide Web pages. The language is used to create *notes* that can be then executed based on the context. The context and notes can be passed from the user terminal to the backbone network and vice versa. An example would be when a user enters a certain area, notes for that area are passed from the network and then triggered by the terminal when arriving at specific locations within the given area. GSM's Short Message Service (SMS) has been used to pass context and notes between a GSM terminal and backbone network [28]. Some of the features for context-aware applications can be realized by the multicast feature of wireless networks (i.e., information is transmitted only within certain areas, as for price information in a supermarket). However, a more advanced environment for context-aware applications can take other factors than location into account, such as surrounding people, and thus, the applications can be tailored for individual users.

Operating systems form a platform for application development. EPOCH is an operating system for wireless information devices such as communicators and smart phones. Windows CE is an operating system developed for palmtop computers. The new version 2.0 adds wireless connectivity to this product. Sun Personal Java 1.0 is a similar type of product for designing application programming interfaces (API). The real value of these products are is that they create platforms for application development with which many are already familiar. These developments indicate that wireless Web services are going to be taken seriously. The key to mass market development is to create open interfaces for application developers.

A very significant development for creating wireless connectivity is the wireless application protocol (WAP), which aims for a global wireless network language and protocol stack. This would facilitate a common platform for development of applications for wireless Internet devices. The applications developed on top of WAP can run on future bearers and devices. The WAP Forum was created with the goal to make WAP an industry standard [29].

The new type of content places requirements for new types of charging mechanisms. Minute-based charging is not appropriate for data calls, but, for example, volume-based charging could be used. The importance of security will increase along with the number of users in wireless networks, as computer and wireless fraud are serious problems. Not only over-the-air security but also end-to-end security is important.

14.10 A WIRELESS DAY

Mike wakes up in the morning in Dallas and makes a video call to his wife Maureen, who is traveling and is on the other side of the globe in Tokyo. She answers from her hotel room and high-quality video connects them in a nice discussion. After the call, while getting ready for work, Mike wakes up his children, Amy and Matt, who want to first connect their Tamagotchi virtual pets with their friends to exchange vital information about their pets' dreams. This happens using improved real-time short message service, consuming hardly any bandwidth, and thus being cheap.

During his lunch break, Mike contacts his friend to play an interactive virtual reality video game through the wireless network. Bursts of images are transmitted through the wireless link. On the way back home Mike remembers that he needs to transmit his latest technical report to his boss. He stops on the road, connects to the corporate intranet, and a 10-Mbytes engineering document is transferred instantly through the IMT-2000 wireless connection. After driving a while, his car stops. What now, he wonders. There seems to be something wrong. He connects through video to his car service to get instructions. He focuses the digital camera in his wireless device into the car alarm lights. This view is transmitted to the car service, and based on it, they give him instructions on how to fix the problem.

Finally at home, Mike reads his email. Something from mom, he notices. A photograph taken in Hawaii, where she is on holiday. She took the photo with a digital camera and said, "send this to my son." Her IMT-2000 wireless terminal compressed the image, connected to the backbone network, and emailed the image to Mike over the wireless link.

14.11 CONCLUDING REMARKS

The industry has taken a big step forward to implement third generation systems. However, before commercial systems can emerge, an enormous effort is still required in standardization, regulation, and implementation. IMT-2000 has been the big business hype and will remain so for the next five years. Within that time span, we will see how successful third generation will be in creating new business opportunities. However, human nature is such that part of the pioneers who were involved in setting up third generation standard frameworks will want to focus on new technologies and leave the details for other people. Thus, the next hype is already waiting for us beyond the horizon.

REFERENCES

[1] ITU website, http://www.itu.int/imt
[2] ETSI website, http://www.etsi.fr
[3] UMTS Task Force Report, February 21, 1996.
[4] The UMTS Forum website, http://www.umts-forum.org/
[5] GSM MoU website, http://www.gsmworld.com/
[6] Dimolitsas, S., "Standards Setting Bodies," in *The Mobile Communications Handbook*, J. D. Gibson (ed.), Boca Ratton: Florida: CRC Press, 1996.
[7] TIA website, http://www.tiaonline.org/
[8] T1 website, http://www.t1.org/
[9] CDG website, http://www.cdg.org
[10] UWCC website, htpp://www.uwcc.org
[11] Donegan, P., "Domestic Rivalries Unsettle Japan's Third Generation Push," *Mobile Communications International*, No. 40, April 1997, pp. 51–53.
[12] ARIB website, http://www.arib.or.jp/arib/english/index.html
[13] TTC website, http://www.ttc.or.jp/
[14] TTA website, http://www.tta.or.kr/
[15] ETRI website, http://www.etri.re.kr/
[16] RAST Website, http://www.rast.etsi.fi/
[17] Kohno, R., and M. Nagakawa, "International Cooperative Research and Development of Wireless Personal Communications in Asian-Pacific Countries", *IEEE Personal Communications*, Vol. 4, No. 2, April 1997, pp. 6–12.
[18] ITU Circular Letter 4 April 1997 (8/LCCE/47), "Request for Submission of Candidate Radio/Transmission Technologies (RTTs) for IMT 2000/FPLMTS Radio Interface".[1]
[19] Leite, F., R. Engelman, S. Kodama, H. Mennenga, and S. Towaij, "Regulatory Considerations Relating to IMT-2000," *IEEE Personal Communications*, August 1997, pp. 14–19.
[20] UK Department of Trade and Industry, "Mobile Communications on the Move," *A Consultation Document from the Department of Trade and Industry*, July 1997, can be found from website http://www.open.gov.uk/radiocom/rahome.htm.

---

[1] The complete volume of [18] can be obtained from ITU, General Secretariat – Sales and Marketing Service, Place des Nations - CH – 1211 GENEVA 20 (Switzerland), Tel. +41 22 730 61 41 (English)/ +41 22 730 61 42 (French), Fax. +41 22 730 51 94, Telex: 421 0000 uit ch, e-mail: sales@itu.int). It should be noted that the sole responsibility for selecting extracts from [18] for reproduction lies with the authors and can in no way attributed to ITU.

[21] European Commission, "Proposal by the Commission for a Decision of the European Parliament and Council on the co-ordinated introduction of mobile and wireless communications (UMTS) in the Community," 1997.

[22] "European Radiocommunications Committee ERC Decision of 30 June 1997 on the frequency bands for the introduction of the Universal Mobile Telecommunications System (UMTS)," ERC/DEC/(97)07 can be found from ERO website http://www.ero.dk/

[23] Shosteck, H., "The Explosion in World Cellular Growth", *The 2nd Annual UWC Global Summit*, Vancouver, BC, Canada, April 15-17, 1998.

[24] The UMTS Market Aspects Group, *UMTS Market Forecast Study*, 1997.

[25] Mikkonen, J., C. Corrado, C. Evci, and M. Prögler, "Emerging Wireless Broadband Networks," *IEEE Personal Communications*, Vol. 36, No. 2, February 1998, pp. 112–117.

[26] Comparetto, G., and R. Ramirez, "Trends in Mobile Satellite Technology," *IEEE Computer*, February 1997, pp. 44–52.

[27] Guntsch, A., M. Ibnkahla, G. Losquadro, M. Mazzella, D. Rovidas, and A. Timm, "EU's R&D Activities on Third-Generation Mobile Satellite Systems (S-UMTS)," *IEEE Personal Communications*, Vol. 36, No. 2, February 1998.

[28] Brown, P. J., J. D. Bovey, and X. Chen, "Context-Aware Applications: From the Laboratory to the Marketplace," *IEEE Personal Communications*, October 1997, pp. 58–63.

[29] WAP Forum website, http://www.wapforum.org

ABOUT THE AUTHORS

Tero Ojanperä was born in Korsnäs, Finland, on November 12, 1966. He received his M.Sc. degree from the University of Oulu, Finland, in 1991. He joined Nokia Mobile Phones, Oulu, Finland, in 1990 and worked as a research engineer from 1991 to 1992. From 1992 to 1995, he led a radio systems research group concentrating on wideband CDMA, GSM WLL, and US TDMA. From 1994 to 1995, he was also a project manager of the wideband CDMA concept development within Nokia. Later, this concept formed the basis for the FRAMES Wideband CDMA.

From 1995 to 1997, he was a research manager in Nokia Research Center, Helsinki, Finland, heading Nokia's third generation air interface research program, consisting of several air interface projects such as wideband CDMA, packet data, radio resource management algorithms, OFDM, and wideband TDMA. He was actively involved setting up the FRAMES project and during 1995-1996, he was also leader of the technical area air interface and the multiple access work package in the project, responsible for the selection of the FRAMES Multiple Access (FMA) scheme. The FRAMES Wideband CDMA was the basis for the UMTS WCDMA concept in ETSI. From 1994 to 1997, he was a Nokia representative for the UMTS radio interface issues in the ETSI SMG5 and SMG2 standardization committees.

From August 1997 to August 1998, he worked as a principal engineer in Nokia Research Center, Irving, TX. His was involved in the US third generation standards activities for the cdma2000. In addition, he was involved in technical/strategic work for Nokia's proposal for the UWC-136.

Since September 1998, he has been with Nokia Telecommunications, Finland, as a Head of Research, Radio Access Systems.

He has authored several conference and magazine papers as well as a chapter in two books *Wireless Communications TDMA vs. CDMA* (Kluwer Academic Publishers, 1997) and *GSM: Evolution towards 3rd generation* (Kluwer Academic Publishers, 1998). He has three patents. During 1996-1997, he was a member of the Nokia patent committee and during 1997-1998, he was member of the patent committee of Nokia Mobile Phones. His research current interest lies in third generation wireless networks and multiuser detection.

Mr. Ojanperä is a member of the IEEE and during 1996 he was secretary of the IEEE Finland Section. He is a member of the International Advisory Committee of the Vehicular Technology Conference 1999 Fall.

Ramjee Prasad was born in Babhnaur (Gaya), Bihar, India on July 1, 1946. He is now a Dutch citizen. He received the B.Sc. (Eng.) from Bihar Institute of Technology, Sindri, India, and the M.Sc. (Eng.) and Ph.D. degrees from Birla Institute of Technology (BIT), Ranchi, India in 1968, 1970 and 1979, respectively.

He joined BIT as Senior Research Fellow in 1970 and became Associate Professor in 1980. While he was with BIT, he supervised many research projects in the area of Microwave and Plasma Engineering. During 1983-1988 he was with the University of Dar es Salaam (UDSM), Tanzania, where he became Professor in Telecommunications at the Department of Electrical Engineering in 1986. At UDSM, he was responsible for the collaborative project "Satellite Communications for Rural Zones" with Eindhoven University of Technology, The Netherlands. Since February 1988, he has been with the Telecommunications and Traffic Control Systems Group, Delft University of Technology (DUT), The Netherlands, where he is actively involved in the area of wireless personal and multimedia communications (WPMC). He is Head and Program Director of the Centre for Wireless and Personal Communications (CEWPC) of International Research Centre for Telecommunications-Transmission and Radar (IRCTR). He is currently involved in the European ACTS project FRAMES (Future Radio Wideband Multiple Access Systems) as a Project Leader of DUT. He has published over 200 technical papers as well as two books, *CDMA for Wireless Personal Communications* (Artech House, 1996) and *Universal Wireless Personal Communicatiions* (Artech House, 1998). His current research interest lies in wireless networks, packet communications, multiple access protocols, adaptive equalizers, spread-spectrum CDMA systems and multimedia communications.

He has served as a member of advisory and program committees of several IEEE international conferences. He has also presented keynote speeches, invited papers and tutorials on PIMRC at various universities, technical institutions and IEEE conferences. He was also a member of a working group of European cooperation in the field of scientific and technical research (COST-231) project dealing with "Evolution of Land Mobile Radio (including personal) Communications" as an expert for the Netherlands. He is now a member of the COST-259 project.

He is listed in the *US Who's Who in the World*. He was Organizer and Interim Chairman of IEEE Vehicular Technology/Communications Society Joint Chapter, Benelux Section. Presently, he is the elected chairman of the joint chapter. He is also the founder of the

IEEE Symposium on Communications and Vehicular Technology (SCVT) in the Benelux and he was the Symposium Chairman of SCVT'93. He is the Co-ordinating Editor and Editor-in-Chief of a Kluwer international journal on "Wireless Personal Communications" and also a member of the editorial board of other international journals including *IEEE Communications Magazine* and *IEEE Electronics Communication Engineering Journal*. He was the Technical Program Chairman of PIMRC'94 International Symposium held in The Hague, The Netherlands during September 19-23, 1994 and also of the Third Communication Theory Mini-Conference in conjunction with GLOBECOM'94 held in San Francisco, California during November 27-30, 1994. He is the Conference Chairman of the 50th IEEE Vehicular Technology Conference to be held in Amsterdam, The Netherlands, during September 19-22, 1999. He is a Fellow of IEE, a Fellow of the Institution of Electronics & Telecommunication Engineers, a Senior Member IEEE and a Member of NERG (The Netherlands Electronics and Radio Society).

INDEX

Recent Titles in the Artech House
Mobile Communications Series

John Walker, Series Editor

Mobile Data Communications Systems, Peter Wong and David Britland

Mobile Telecommunications: Standards, Regulation, and Applications, Rudi Bekkers and Jan Smits

Personal Wireless Communication With DECT and PWT, John Phillips and Gerard Mac Namee

Practical Wireless Data Modem Design, Jonathan Y.C. Cheah

Radio Propagation in Cellular Networks, Nathan Blaunstein

RDS: The Radio Data System, Dietmar Kopitz and Bev Marks

Resource Allocation in Hierarchical Cellular Systems, Lauro Ortigoza-Guerrero and A. Hamid Aghvami

RF and Microwave Circuit Design for Wireless Communications, Lawrence E. Larson, editor

Spread Spectrum CDMA Systems for Wireless Communications, Savo G. Glisic and Branka Vucetic

Understanding Cellular Radio, William Webb

Understanding Digital PCS: The TDMA Standard, Cameron Kelly Coursey

Understanding GPS: Principles and Applications, Elliott D. Kaplan, editor

Wireless Communications in Developing Countries: Cellular and Satellite Systems, Rachael E. Schwartz

Wireless Technicians's Handbook, Andrew Miceli

For further information on these and other Artech House titles, including previously considered out-of-print books now available through our In-Print-Forever® (IPF®) program, contact:

Artech House	Artech House
685 Canton Street	46 Gillingham Street
Norwood, MA 02062	London SW1V 1AH UK
Phone: 781-769-9750	Phone: +44 (0)20 7596-8750
Fax: 781-769-6334	Fax: +44 (0)20 7630-0166
e-mail: artech@artechhouse.com	e-mail: artech-uk@artechhouse.com

Find us on the World Wide Web at:
www.artechhouse.com

Printed in the United States
103883LV00004B/16/A

9 780890 067352